FOUNDATIONS OF REAL ANALYSIS

FOUNDATIONS OF REAL ANALYSIS

Expanding Horizons Beyond the Click

WILLIAM R. BRIAN

ELSEVIER

ACADEMIC PRESS

An imprint of Elsevier

Academic Press is an imprint of Elsevier
125 London Wall, London EC2Y 5AS, United Kingdom
50 Hampshire Street, 5th Floor, Cambridge, MA 02139, United States

ISBN: 978-0-443-48527-5

For information on all Academic Press publications
visit our website at https://www.elsevier.com/books-and-journals

Publisher: Peter Linsley
Acquisitions Editor: Stephanie Xu
Editorial Project Manager: Prashant Mahajan
Publishing Services Manager: Deepthi Unni
Production Project Manager: Ashok Ravi
Cover Designer: Bakyalakshmi S.

Typeset by VTeX

Contents

Introduction *ix*

1. Beginnings **1**

The axioms 1
A noncomplete ordered field 5
The Archimedean property and the density of the rationals 8
The uniqueness of the real line 12
The cardinality of a set 15
Countable and uncountable sets 16
The uncountability of the reals 20
Exercises 22

2. Metric spaces **25**

Fundamentals of metric spaces 25
Continuous functions 30
When are two metric spaces the same? 32
Connectedness and the Intermediate Value Theorem 34
Compactness 38
The Heine–Borel Theorem and the Extreme Value Theorem 43
The Cantor space and the Cantor set 45
Exercises 50

3. Completeness **53**

Sequences and subsequences 53
Completeness, compactness, and the real line 57
Limits of sequences and the extended real line 59
Infinite series 63
The Baire Category Theorem 72
Meager and comeager sets 75
Banach–Mazur games 77
Exercises 86

4. Continuity **89**

Continuity via limits 89

Uniform continuity 92
Some functions that are not continuous 94
Oscillation 98
Space-filling curves 103
Extending continuous functions 104
Cantor's function 108
A space-filling curve 110
Exercises 113

5. Differentiation **115**

Definition and preliminaries 116
Differentiation rules 117
The Mean Value Theorem 123
Must a derivative be continuous? 126
A continuous, nowhere differentiable function 129
Exercises 133

6. Integration **137**

Definition and preliminaries 137
The Cauchy criterion 140
Some integrable functions 144
Combining integrable functions 147
The fundamental theorems of calculus 155
Lebesgue null sets 163
Lebesgue outer measure 165
The Lebesgue–Vitali theorem 167
Volterra's function 170
Exercises 174

7. Subsets of $\mathbb{R}$ **177**

σ-algebras and Borel sets 177
The property of Baire 180
Lebesgue measurable sets 186
The Axiom of Choice 195
Vitali sets 197
Exercises 202

8. Function spaces **205**

Pointwise and uniform convergence 205

A topological perspective on uniform convergence 210
The Weierstrass Approximation Theorem 214
The Arzelà–Ascoli Theorem 219
How discontinuous can a derivative be? 226
Exercises 229

Index *233*

Companion site URL:
https://www.elsevier.com/books-and-journals/book-companion/9780443485275

Instructor site URL:
https://educate.elsevier.com/9780443485275

Introduction

Real analysis is, for many students, their first close encounter with rigorous mathematical reasoning. The goal of this book is to make that encounter as clear and inviting as possible.

Of course, some kind of mathematical reasoning is needed to work out complicated problems in algebra and calculus, and anyone wishing to study real analysis has usually done their fair share of this sort of thing. But real analysis is not about doing ever more complicated calculations. It is about understanding why those techniques work the way they do, and where they fail—the strange situations at the edges of intuition where things break down—and why. It is the difference between learning to drive a car and learning to build a car.

Rigorous reasoning begins with a few basic assumptions, and moves on from there with precise definitions and deductive proofs. This way of thinking is not meant to replace intuition—just the opposite. Those who have studied a bit of calculus often form mathematical intuitions that are incomplete or even wrong. This book challenges those intuitions, helping readers build a fuller, more robust understanding of ideas they may have thought they already understood. The goal is not to move away from thinking intuitively and creatively, but to do it better.

Who and what is this book for?

Foundations of Real Analysis is written for students at the advanced undergraduate or beginning graduate level. The text has taken shape slowly over the last eight years, during which evolving versions of this book have been used to teach my university's 5000-level, two-semester sequence in analysis. All the proofs, arguments, and pictures in the book have been tested over these years on real students, often going through many different versions before settling down into a form that the students understand best.

This book is designed to be read. I have an unusual habit of reading for fun—even mathematics texts. Consequently, a large amount of importance was placed on writing a book that might be fun to read. Care has been taken to make the text engaging and clear, suitable both for classroom use and for self-study. Like most books, this one has the best effect when read actively. Students are encouraged to work through each proof with a pencil in hand, to attempt every exercise, and (if possible) to work with other students.

This book is an introductory text, and does not assume any particular prior knowledge of real analysis. However, we must start somewhere, and some basic mathematical ideas and vocabulary are assumed. In order to help the student who may be missing part of this background, there is a glossary at the end of the book. This glossary contains definitions and short explanations of the mathematical background ideas necessary for reading this book (definitions of *injective* and *surjective* functions, a few notes on *equivalence relations*, set operators like *union* and *intersection*, etc.). Although most students will already know most of the material in the glossary, my experience is that few know all of it completely, so nearly everyone benefits from a review.

What is different about it?

The aim of a first course in real analysis should be more than proving familiar theorems with added rigor. It should reveal the surprising and subtle aspects of analysis—the places where intuition bends and where rigor shows itself to be necessary because an informal understanding of things is insufficient. This is what gives the topic real life.

To this end, this text is unusually rich with examples and counterexamples. These range from relatively simple constructions that illustrate core ideas to more elaborate and surprising examples: non-Archimedean fields, space-filling curves, the Weierstrass function, and Volterra's example of an everywhere-differentiable function whose derivative fails to be integrable, to name a few. The intention is not to let the theory drift too far from the concrete, to ground everything in visual, real-feeling examples. I also hope to capture some of the magic that comes from seeing an intuition-bending thing and wondering what it's all about.

Another distinguishing feature of this book is the selection of topics. The book places a strong emphasis on topology, particularly the topological underpinnings of the real line and a bit more set theory than usual. This includes some standard topics in an analysis course, such as a proof of the uncountability of the reals or an introduction to metric spaces and metric properties like completeness, compactness, and connectedness. Some less standard topics are also included, like an in-depth look at the Baire category theorem, the property of Baire, and the infinite games that arise naturally in this context (the Banach–Mazur games). These topics illustrate foundational aspects of real analysis, with the added benefit of being completely new to most students at the advanced undergraduate or early graduate level.

About 20 problems appear at the end of each chapter. These range from straightforward exercises that check understanding to very challenging problems that may take several days to solve. The purpose of the more difficult exercises is, of course, to challenge students, but also to give them a feeling for real mathematical research, where a single problem may defy one's best efforts for years at a time. I have not included any exercises that will take years to solve, but I have included some that are far from straightforward, that even my best students will not see how to solve right away. In my own courses, students often work collaboratively on these problems, and by working together they usually solve all of them.

This book is deliberately slim. It does not attempt to cover every topic that might appear in a course on real analysis. By design, it is possible—though not easy—to read every page and work every exercise in the span of two semesters. Some readers may feel some of their favorite topics have been left out: some important things have been, as there simply is not room for everything. Some topics, such as differentiation and integration, are included because they are indispensable. Others, such as Banach–Mazur games, are included to introduce ideas and proof schemes that are exciting but rarely seen at this level.

The study of real analysis rewards patience, curiosity, and precision. It asks the reader to move slowly, to look closely at things we thought we already understood and to see them anew, and also to find stranger things not imagined beforehand. My hope is that this book helps its readers experience that combination of clarity and surprise that makes mathematics so extraordinary.

Acknowledgments

I have many people to thank for helping to bring this book into being. I am especially grateful to the publication team at Elsevier, who have been enormously helpful; to those who proofread some of the early drafts of certain chapters, in particular Jocelyn Bell, who provided feedback after teaching through several chapters at Hobart and William Smith Colleges; to Lora Abernathy, a tireless hunter of typos; to my wife Caitlin, who, as always, supported me throughout the writing process; and to my kids, who asked lots of genuine and curious questions about what is in this book—it was both fun and illuminating trying to explain. And I am perhaps most indebted to the more than 100 students in my Analysis I and II classes over the last nine years, whose insightful questions and varied learning styles shaped this text into something far better than I could have produced on my own.

CHAPTER 1

Beginnings

In this first chapter, we take a look at two important properties of the real number line $\mathbb{R}$: It is *complete*, and it is *uncountable*. The first of these properties is order-theoretic, and it means roughly that there are no "gaps" in $\mathbb{R}$, places where numbers seem to be missing. The uncountability of $\mathbb{R}$ is a set-theoretic property, and means that, in some sense that we will soon make precise, there are very many real numbers.

These two things highlight some major differences between $\mathbb{R}$ and $\mathbb{Q}$. While the real number line $\mathbb{R}$ is both complete and uncountable, the rational number line $\mathbb{Q}$ is neither. In fact, one of the most important theorems in this chapter is that the completeness property of $\mathbb{R}$, together with its basic algebraic structure, gives a perfect description of it, uniquely identifying it among all possible number systems.

The axioms

If we wish to prove anything about the real line, we must begin with some assumptions. What axioms do we take as the starting point of our investigation? It turns out that we need only two, two big ideas from which flow all the other big ideas of analysis we are setting out to study. Here they are:

(A1) The real number line $\mathbb{R}$ is an ordered field.

(A2) $\mathbb{R}$ satisfies the *Axiom of Completeness.*

Each of these two axioms requires some explanation. Let us begin with (A1). What is an ordered field? Roughly speaking, an ordered field is a possible number system. It is a set of objects (which we think of as numbers) that we can add, subtract, multiply, divide, and compare as bigger or smaller, and these operations have to behave in a sensible way, the way we expect numbers to behave. More precisely, here is the full definition.

Definition 1.1. An *ordered field* is a set F together with two different functions $F \times F \to F$, denoted $\cdot$ and $+$, and a relation $<$ on F, such that the following rules are all satisfied:

(1) $(a+b)+c = a+(b+c)$ and $(a \cdot b) \cdot c = a \cdot (b \cdot c)$ for every $a, b, c \in F$.

(2) $a+b = b+a$ and $a \cdot b = b \cdot a$.

(3) $a \cdot (b+c) = (a \cdot b) + (a \cdot c)$.

(4) There is a unique member of F, denoted 0_F or just 0, such that $a+0 = a$ for every $a \in F$.

(5) For every $a \in F$, there is a unique member of F, denoted $-a$, such that $-a+a = 0$.

(6) There is a unique member of F, denoted 1_F or just 1, such that $a \cdot 1 = a$ for every $a \in F$. Furthermore, $0 \neq 1$.

Foundations of Real Analysis
https://doi.org/10.1016/B978-0-44-348527-5.00014-7

(7) For every $a \in F$ except for 0, there is a unique member of F, denoted a^{-1}, such that $a^{-1} \cdot a = 1$.

(8) For every $a, b \in F$, exactly one of the following three things is true: either $a < b$, or $a = b$, or $b < a$.

(9) If $a < b < c$, then $a < c$.

(10) If $a < b$, then $a + c < b + c$.

(11) If $0 < a$ and $0 < b$, then $0 < a \cdot b$.

It should not be hard to convince yourself that the real line $\mathbb{R}$, equipped with the familiar operations $+$ (addition) and $\cdot$ (multiplication), and the familiar order $<$, satisfies all 11 of these rules. At any rate, this is what we are assuming with our Axiom (A1). Another familiar example of an ordered field is the rational number line.

Example 1.2. Let $F = \mathbb{Q}$, and consider the usual operations $+$ and $\cdot$ and the usual order $<$ on $\mathbb{Q}$. Then, all 11 of the rules in Definition 1.1 are satisfied. Thus, the rational number line $\mathbb{Q}$ is an ordered field.

The study of ordered fields belongs to the area of mathematics known as *abstract algebra* (often called just "algebra" among mathematicians). We do not intend to dig too deeply into this subject. Our main reasons for defining ordered fields are (1) to make our assumptions about $\mathbb{R}$ precise and explicit and (2) to highlight the special role of $\mathbb{R}$ among ordered fields (which, again, we think of as "possible number systems").

Because our goal in this course is not to do lots of algebra and study ordered fields as such, we will not be proving much about $\mathbb{R}$ directly from Definition 1.1. Instead, we plan to explore the theory of $\mathbb{R}$ using the following somewhat vague pseudo-axiom.

Proposition 1.3. *An ordered field obeys all the rules of basic arithmetic that one learns before reaching college.*

This proposition captures the real meaning of Axiom (A1), i.e., basic arithmetic is valid in $\mathbb{R}$. This is how we intend to use (A1) moving forward. Proposition 1.3 also captures what we mean by calling ordered fields "possible number systems": It means that one can do arithmetic in an ordered field, and everything will behave as expected.

Example 1.4. For example, consider the following statement:

$$\text{for any } a \in F, \text{ if } 0 < a, \text{ then } -a < 0.$$

This statement can be proved from the rules in Definition 1.1 as follows: Rule 5 tells us there is some $-a \in F$ such that $-a + a = 0$. Starting with the statement $0 < a$, rules 4 and 10 tell us that $-a = -a + 0 < -a + a = 0$. QED!

The displayed statement in this example is the kind of thing we mean by "basic arithmetic." The point of Proposition 1.3 is that we plan simply to assume facts like this rather than proving them from Definition 1.1. The reason for this is that proofs like this one aren't really part of analysis. A course in abstract algebra might include such things (although it would be a mistake to think this more than scratches the surface of what that subject is really about). But for an analysis course, we wish to take Proposition 1.3 for granted and to have basic arithmetic available to use from the start.

Additionally, we plan to take for granted the language and abbreviations that come along with learning basic arithmetic, like writing "$b \geq a$" to mean that either $a < b$ or $a = b$, writing "$5x^2$" to abbreviate $5 \cdot x \cdot x$, etc.

Our second Axiom (A2) requires some explanation as well. Roughly, this axiom says that there are no "gaps" in the real number line, places where it seems a number should be but isn't. Before stating (A2) more precisely, we need a few definitions.

Definition 1.5. Suppose F is an ordered field and $A \subseteq F$. We say that $b \in F$ is an *upper bound* for A if $a \leq b$ for every $a \in A$. If A has an upper bound, we say that A is *bounded above*. Similarly, we say that $b \in F$ is a *lower bound* for A if $a \geq b$ for every $a \in A$. If A has a lower bound, we say that A is *bounded below*.

Example 1.6. Consider the ordered field $F = \mathbb{R}$. The set $\mathbb{Q}$ is neither bounded above nor bounded below. The set $\mathbb{N}$ of natural numbers is bounded below but not bounded above. The closed unit interval $[0, 1]$ is bounded both above and below.

Definition 1.7. Suppose F is an ordered field and $A \subseteq F$. We say that $s \in F$ is the *supremum* of A if

(i) s is an upper bound for A, and

(ii) if b is any upper bound for A, then $s \leq b$.

If this is the case, we write $\sup A = s$. Not every set has a supremum, for example, the set $\mathbb{N}$ has no supremum because it is not bounded above. When a set A does have a supremum, $\sup A$ is sometimes also called the *least upper bound* of A.

Similarly, $t \in F$ is called the *infimum* of A if

(i) t is a lower bound for A, and

(ii) if b is any lower bound for A, then $t \geq b$.

If this is the case, we write $\inf A = t$. When a set A has an infimum, $\inf A$ is sometimes also called the *greatest lower bound* of A.

Observe that a set A can have at most one supremum: If s and s' were both to satisfy the definition of being a "supremum" of A, then we could deduce that $s \leq s'$ (because s' satisfies condition (i) and s satisfies condition (ii)), and $s' \leq s$ (by the same logic, but interchanging the roles of s and s'); hence, $s = s'$. Similarly, if t and t' were both to satisfy the definition of an infimum for A, then $t = t'$. This justifies the language in our definition, referring to *the* supremum and *the* infimum of a set.

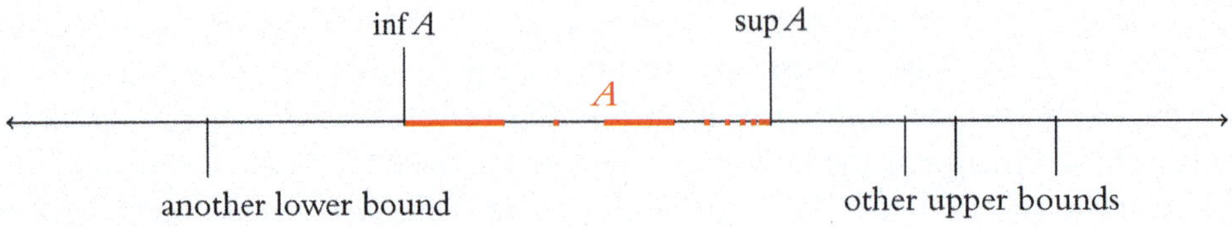

Example 1.8. In the ordered field $\mathbb{R}$, the set $\mathbb{Q}$ is bounded neither above nor below, so it has neither a supremum nor an infimum. The set $\mathbb{N}$ of natural numbers has no supremum, but $\inf(\mathbb{N}) = 1$. The intervals $[0, 1]$ and $(0, 1)$ both have supremum 1 and infimum 0. Notice that a set may or may not contain its infimum and its supremum.

The supremum of the set $\{1, 1.4, 1.41, 1.414, 1.4142, \ldots\}$ is $\sqrt{2}$. This is also the supremum of the sets $\{r \in \mathbb{Q}: r^2 < 2\}$ and $\{r \in \mathbb{R}: r^2 \leq 2\}$. ☕

Finally, we are ready to state our Axiom (A2) in full.

Definition 1.9. The *Axiom of Completeness* states:

If A is non-empty and bounded above, then A has a supremum. ☕

If $A \subseteq \mathbb{R}$ is not bounded above, it cannot have a supremum. Also, observe that while every number is an upper bound for the empty set $\emptyset$, no number is a *least* upper bound, so $\emptyset$ does not have a supremum either. This explains the restriction in the Axiom of Completeness to non-empty sets that are bounded above.

One might think at first that this axiom looks like only half of what we really want. Why should we assert that non-empty sets with upper bounds have supremums, and not also assert that non-empty sets with lower bounds have infimums? There is an apparent lopsidedness here. The reason for this lopsidedness is simply that we have no need to include the statement about infimums in our axiom: it follows already from the statement about supremums, as the following theorem shows.

Theorem 1.10. *Suppose F is an ordered field satisfying the Axiom of Completeness. If $A \subseteq F$ is non-empty and bounded below, then A has an infimum.*

Proof. Suppose $A \subseteq F$ is non-empty and bounded below. Define

$$B = \{-a\colon a \in A\}.$$

Because A is non-empty, B is, too.

We claim that B is also bounded above. Let $x \in F$ be a lower bound for A, which means that $x \leq a$ for every $a \in A$. But this implies $-x \geq -a$ for every $a \in A$, which means $-x \geq b$ for every $b \in B$. This shows that if x is a lower bound for A, then $-x$ is an upper bound for B.

So B is a non-empty set and is bounded above. By the Axiom of Completeness, B has a supremum: let $b = \sup B$. To finish the proof of the theorem, we will show that $-b$ is the infimum of A.

First, note that $-b$ is a lower bound for A because, if this were not the case, then there would be some $a \in A$ with $a < -b$, and then we would have $-a \in B$ and $b < -a$, contradicting the fact that $b = \sup B$. Second, suppose x is some other lower bound for A. As we saw in the second paragraph of the proof, this implies $-x$ is an upper bound for B. But then, $b = \sup B \leq -x$ (by the definition of a supremum), and so $x \leq -b$.

Thus, we see that $-b$ is a lower bound for A, and if x is any lower bound for A, then $x \leq -b$. By definition, $-b = \inf A$. □

A noncomplete ordered field

As mentioned already in Example 1.2, the rational number line $\mathbb{Q}$ is an ordered field. But $\mathbb{R}$ and $\mathbb{Q}$ are very different from each other. (The differences will become more and more evident as we progress in our study of $\mathbb{R}$.) Remembering that our only assumptions about $\mathbb{R}$ are (A1) and (A2), any difference between $\mathbb{R}$ and $\mathbb{Q}$ must arise from (A2). In other words, if everything we prove about $\mathbb{R}$ is proved from (A1) and (A2), and if (A1) is true for $\mathbb{Q}$ as well as $\mathbb{R}$, then for anything we prove about $\mathbb{R}$ that is not true of $\mathbb{Q}$, the proof must involve (A2) in some way or another.

To begin exploring how $\mathbb{Q}$ and $\mathbb{R}$ differ, let us prove that irrational numbers exist, i.e., $\mathbb{Q} \neq \mathbb{R}$. This theorem was first proved in ancient times by the sect of the Pythagoreans. Legend has it that at one time the theorem was an official secret of theirs, and one of their group, Hippasus of Metapontum, was murdered for divulging it to an outsider. Let us proceed with caution.

Theorem 1.11. *There is no rational number x such that $x^2 = 2$.*

Proof. Suppose, aiming for a contradiction, that there is some rational number x with $x^2 = 2$. In other words, we are assuming there are integers p and q, with $q > 0$, such that $x = \frac{p}{q}$. Let us also assume that this fraction is reduced, which means in particular that p and q are not both even.

The equation $x^2 = \frac{p^2}{q^2} = 2$ implies $p^2 = 2q^2$, which means p^2 is an even number. But the square of an odd number is always odd, so this means that p is an even number. In other words, $p = 2r$ for some integer r.

Plugging $p = 2r$ into the equation $\frac{p^2}{q^2} = 2$, we get $\frac{4r^2}{q^2} = 2$, which reduces to $q^2 = 2r^2$. Like in the previous paragraph, this shows that q^2 is an even number, which implies q is an even number.

Thus, p and q are both even numbers, contradicting what we assumed earlier! This contradiction shows that our initial assumption, that there is some rational number x with $x^2 = 2$, must have been wrong. □

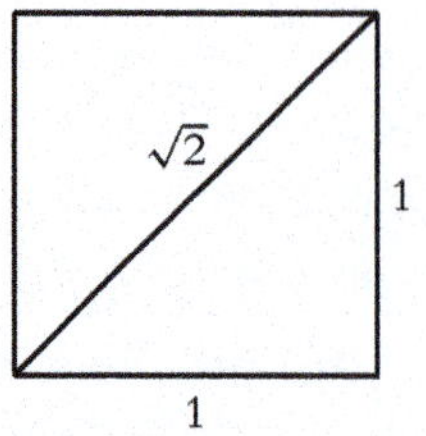

In language a bit closer to their own terminology, the Pythagoreans proved that the diagonal of a square is "incommensurate" with its sides.

Theorem 1.12. *There is a real number $x > 0$, denoted $\sqrt{2}$, such that $x^2 = 2$.*

Proof. Let $A = \{r \in \mathbb{R} \colon r^2 < 2\}$. This set is non-empty, for example, because $0 \in A$, and it is bounded above. (For example, 3 is an upper bound for A because if $r \geq 3$, then $r^2 \geq 3^2 > 2$.) By the Axiom of Completeness, A has a supremum: let $s = \sup A$. Observe that $1 \in A$, which means $s \geq 1 > 0$. To prove the theorem, we wish to show $s^2 = 2$.

Suppose, aiming for a contradiction, that $s^2 > 2$. As $s > 0$, this implies $\frac{s^2-2}{2s} > 0$. Let $t = s - \frac{s^2-2}{2s}$. Then, $t < s$ and

$$t^2 = \left(s - \tfrac{s^2-2}{2s}\right)^2 = s^2 - 2s\tfrac{s^2-2}{2s} + \left(\tfrac{s^2-2}{2s}\right)^2 = 2 + \left(\tfrac{s^2-2}{2s}\right)^2 > 2.$$

Thus, $t > 0$ and $t^2 > 2$. It follows that if $r \geq t$, then $r^2 \geq t^2 > 2$, so $r \notin A$. Hence t is an upper bound for A. But $t < s$, so this contradicts $s = \sup A$.

Again aiming for a contradiction, suppose next that $s^2 < 2$. Then, we have $0 \leq s^2/2 < 1$, which means $s^4/4 = (s^2/2)^2 < s^2/2$. Let $t = s + \frac{2-s^2}{4s}$. Then, $t > s$, and using the fact that $4s \geq 2$,

$$\begin{aligned} t^2 &= \left(s + \tfrac{2-s^2}{4s}\right)^2 = s^2 + 2s\tfrac{2-s^2}{4s} + \left(\tfrac{2-s^2}{4s}\right)^2 = 1 + \tfrac{s^2}{2} + \left(\tfrac{2-s^2}{4s}\right)^2 \\ &\leq 1 + \tfrac{s^2}{2} + \left(\tfrac{2-s^2}{2}\right)^2 = 2 - \tfrac{s^2}{2} + \tfrac{s^4}{4} < 2. \end{aligned}$$

Thus, $t \in A$ and $t > s$, contradicting $s = \sup A$.

Assuming either $s^2 < 2$ or $s^2 > 2$ leads to a contradiction, so $s^2 = 2$. □

Corollary 1.13. *$\mathbb{Q}$ does not satisfy the Axiom of Completeness.*

Proof. Let $A = \{r \in \mathbb{Q} \colon r^2 < 2\}$. This is a non-empty subset of $\mathbb{Q}$ and is bounded above, but we claim that it has no supremum (in $\mathbb{Q}$). To see this, fix $s \in \mathbb{Q}$. We will show s is not the supremum of A.

In the previous proof, we showed that if $s^2 > 2$, then $t = s - \frac{s^2-2}{2s} < s$, and $t^2 > 2$. But notice that if $s \in \mathbb{Q}$, then $t \in \mathbb{Q}$ also. So $t \in \mathbb{Q}$, and t is an upper bound for A (the argument for this is the same as in the previous proof), and $t < s$. This contradicts $s = \sup A$. Similarly, in the previous proof, we showed that if $s^2 < 2$, then $t = s + \frac{2-s^2}{4s} > s$ and $t^2 < 2$. But again, $s \in \mathbb{Q}$ implies $t \in \mathbb{Q}$, while $t > s$ and $t \in A$, also contradicting $s = \sup A$.

This shows that if $s = \sup A$, then we must have $s^2 = 2$. But $s \in \mathbb{Q}$, so this is impossible by Theorem 1.11. □

Less rigorously, but perhaps more clearly, an argument for Corollary 1.13 can be made as follows. We know $\sqrt{2}$ is irrational, but we also know it can be approximated from below by rationals, via its decimal expansion. Because $\sqrt{2} \notin \mathbb{Q}$, the set of these approximations has no supremum in $\mathbb{Q}$.

$$\{1, 1.4, 1.41, 1.414, 1.4142, 1.41421, \dots\}$$

To end this section, we prove that every ordered field contains a copy of $\mathbb{Q}$. In other words, this gives us a precise sense in which we can say $\mathbb{Q}$ is the smallest ordered field. Before proving this, we need to say precisely what it means for an ordered field F to contain a "copy" of $\mathbb{Q}$.

Definition 1.14. Let F and G be ordered fields. A function $\phi : F \to G$ is called an *embedding* if, for every $a, b \in F$,

- $\phi(a + b) = \phi(a) + \phi(b)$
- $\phi(a \cdot b) = \phi(a) \cdot \phi(b)$
- if $a < b$ then $\phi(a) < \phi(b)$

A bijective embedding is called an *isomorphism*, and if there is an isomorphism $F \to G$, then we say that F and G are *isomorphic*.

In this definition, if we wish to be very pedantic, then we should point out that in expressions like $\phi(a + b) = \phi(a) + \phi(b)$, the "+" means something different on the left and on the right. Writing $\phi(a+b)$ means that we add a and b in the field F, and writing $\phi(a)+\phi(b)$ means we add $\phi(a)$ and $\phi(b)$ in the field G. In the context of certain algebraic arguments, it can be confusing if we use the same symbol, +, to denote addition in two different fields. In the present context, however, it is more confusing if we do not.

Observe that for ordered fields, every embedding is injective. Indeed, if $\phi : F \to G$ is an embedding and $a, b \in F$ with $a \neq b$, then either $a < b$ or $a > b$. But then, either $\phi(a) < \phi(b)$ or $\phi(a) > \phi(b)$ in G, which means $\phi(a) \neq \phi(b)$. We may (and do) think of the image of ϕ as a copy of F sitting inside of G.

This definition makes precise what we mean by saying that a field can contain a copy of $\mathbb{Q}$. If there is an embedding $\mathbb{Q} \to F$, then F contains a copy of $\mathbb{Q}$, namely the image of the embedding. Similarly, if F and G are isomorphic, then they are copies of each other, essentially identical.

Theorem 1.15. *For every ordered field F, there is an embedding $\mathbb{Q} \to F$.*

Proof. Let F be an ordered field. Define an embedding $\phi : \mathbb{Q} \to F$ as follows. For each $q \in \mathbb{Q}$, let $q = {}^{m}/_{n}$, where $m \in \mathbb{Z}$, and $n \in \mathbb{N}$, and m and n have no common factors. Then, define

$$\phi(q) = \pm(1_F+1_F+\cdots+1_F)/(1_F+1_F+\cdots+1_F),$$

where the numerator contains $|m|$ copies of 1_F and the denominator contains n copies, and in the numerator we take the $+$ sign if $m \geq 0$, and we take the $-$ sign if $m < 0$.

This defines a function $\phi : \mathbb{Q} \to F$. To finish the proof, we need to argue that ϕ is an embedding. But in fact, we claim this follows from Proposition 1.3, which asserts that arithmetic works as usual in an ordered field.

For example, suppose $m, n \in \mathbb{N}$. The statement "add 1 to itself m times, and then add 1 to itself n times; multiplying the results you get mn, which is the same as 1 added to itself mn times" is certainly true, and is covered by Proposition 1.3 as basic arithmetic. If we replace each 1 with a 1_F, then this is simply what the statement "$\phi(m \cdot n) = \phi(m) \cdot \phi(n)$" asserts. Therefore, this assertion is true in F because Proposition 1.3 is true for F.

Other more complicated assertions of the form $\phi(a+b) = \phi(a) + \phi(b)$ or $\phi(a \cdot b) = \phi(a) \cdot \phi(b)$ or if $a < b$, then $\phi(a) < \phi(b)$ are also expressible as basic assertions of arithmetic if we merely replace all the 1_F's with 1's. All such assertions are true in F by Proposition 1.3. □

We can summarize the previous proof as follows. A given rational number $q = {}^m/_n$ is "built" from the unit 1 by adding 1 to itself m times (or -1, if $m < 0$), then adding 1 to itself n times, then dividing the results. But all of this is just basic arithmetic, so Proposition 1.3 says that we can do this "inside F" and not just in $\mathbb{R}$.

This way of thinking shows the power of Proposition 1.3.

Definition 1.16. Let F be an ordered field. If $q = {}^m/_n$, where $m \in \mathbb{Z}$, and $n \in \mathbb{N}$, and m and n have no common factors, then define

$$q_F = {}^{\pm(1_F+1_F+\cdots+1_F)}/_{(1_F+1_F+\cdots+1_F)},$$

where the numerator contains $|m|$ copies of 1_F, and the denominator contains n copies, and we include the $-$ sign in the numerator if and only if $m < 0$. Let $\mathbb{N}_F = \{n_F \colon n \in \mathbb{N}\}$ and $\mathbb{Q}_F = \{q_F \colon q \in \mathbb{Q}\}$. ☕

The Archimedean property and the density of the rationals

The following theorem and corollary are usually referred to collectively as the *Archimedean property* of the reals. In spite of its importance, both historically and mathematically, the Archimedean property states something fairly obvious about $\mathbb{R}$, namely, that the set $\mathbb{N}$ has no upper bound. However, this "obvious" fact is not part of basic arithmetic because it asserts something not just about how numbers behave with respect to $+$ and $\cdot$ and $<$, but rather something about a *set* of numbers. This kind of assertion is not covered by Proposition 1.3.

One may interpret the Archimedean property as stating that there are no infinitely large or infinitely small real numbers. This is expressed more precisely in the following theorem and corollary.

Theorem 1.17. *For every real number $x \in \mathbb{R}$, there is a natural number $n \in \mathbb{N}$ such that $n > x$. In other words, the set $\mathbb{N}$ is not bounded above.*

Corollary 1.18. *For every real number $x \in \mathbb{R}$ with $x > 0$, there is a natural number $n \in \mathbb{N}$ such that $1/n < x$. In other words, $0 = \inf\{1/n\colon n \in \mathbb{N}\}$.*

Proof of Theorem 1.17. Let us suppose, aiming for a contradiction, that the theorem is false, i.e., there is some $x \in \mathbb{R}$ such that $n \leq x$ for every natural number n. In other words, suppose x is an upper bound for $\mathbb{N}$. By the Axiom of Completeness, this implies $\mathbb{N}$ has a supremum. Let $s = \sup \mathbb{N}$.

Now, consider the number $s - 1$. Because $s - 1 < s$, the definition of "supremum" implies that $s - 1$ is not an upper bound for $\mathbb{N}$. Thus, there is some $N \in \mathbb{N}$ such that $N > s - 1$. But then, $N + 1 \in \mathbb{N}$, and $N + 1 > s$. This is a contradiction since s was supposed to be an upper bound for $\mathbb{N}$. □

This proof uses Proposition 1.3 twice: The assertion that $s - 1 < s$, and the assertion that $N \in \mathbb{N}$ implies $N + 1 \in \mathbb{N}$.

Proof of Corollary 1.18. Let x be a positive real number, and let $y = 1/x$. By Theorem 1.17, there is a natural number n such that $y < n$. But then (because y and n are both positive), this implies $x = 1/y > 1/n$. □

Our next theorem is one of the most important consequences of the Archimedean property. Before stating it, we need a definition and a lemma.

Definition 1.19. Suppose $D \subseteq \mathbb{R}$. We say that D is *dense* in $\mathbb{R}$ if, for every $r, s \in \mathbb{R}$, if $r < s$ then there is some $d \in D$ with $r < d < s$.

In other words, this definition says that given any two (different) real numbers, there is a member of D strictly between them.

Lemma 1.20. *If $S \subseteq \mathbb{Z}$ and S has an infimum, then* $\inf S \in S$.

Proof of Lemma 1.20. Suppose, aiming for a contradiction, that $S \subseteq \mathbb{Z}$ and S has an infimum, but $\inf S \notin S$. Let $b = \inf S$.

Because $b = \inf S$, there is some $s \in S$ with $b \leq s < b + 1$ (since otherwise $b + 1$ would be a larger lower bound than b); and as $b \notin S$, we have $b < s < b + 1$. Applying the same reasoning again, there is some $s' \in S$ such that $b < s' < s$. But then $s, s' \in S$, and because $b < s' < s < b + 1$, we have $0 < s - s' < 1$. This is absurd because $S \subseteq \mathbb{Z}$. □

Theorem 1.21. *$\mathbb{Q}$ is dense in $\mathbb{R}$.*

Proof. Let $r, s \in \mathbb{R}$ with $r < s$. Our goal is to show that there is a rational number q such that $r < q < s$.

By the Archimedean property of the reals (more specifically, by Corollary 1.18), there is some $n \in \mathbb{N}$ such that $1/n < s - r$. The idea of our proof is that if we look at all (not necessarily reduced) fractions with denominator n, then one of these must be between r and s. And looking at the following graph, it is easy to spot a particular fraction of this form that works: the fraction m/n where m is the *least* integer with $m/n > r$.

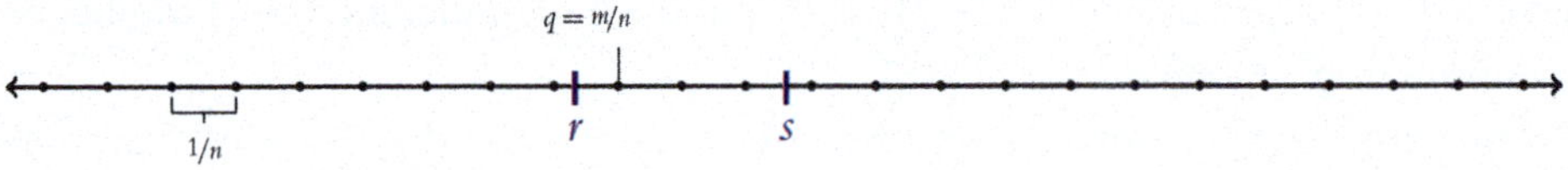

Let $S = \{k \in \mathbb{Z} : k/n > r\}$. Before applying Lemma 1.20, we must show S has an infimum, i.e., we must show S is non-empty and bounded below.

To see that S is non-empty, note that the Archimedean property (specifically Theorem 1.17) states there is some $k \in \mathbb{N}$ such that $k > rn$. But this implies $k/n > r$, which means $k \in S$.

To see that S is bounded below, use the Archimedean property again. By Theorem 1.17, there is some $\ell \in \mathbb{N}$ such that $\ell > -rn$. This implies that $-\ell/n < r$, which implies that if $k < -\ell$, then $k/n < -\ell/n < r$, which means $k \notin S$. In other words, this shows that if $k \in S$, then $k \geq -\ell$, and therefore $-\ell$ is a lower bound for S.

So we see that S is non-empty and has a lower bound. By Lemma 1.20, there is some $m \in S$ with $m = \inf S$. To finish the proof, let us show that the rational number $q = m/n$ is strictly between r and s.

The fact that $r < q$ follows from our definition of S and the fact that $m \in S$. To see that $q < s$, we use the fact that m is the greatest lower bound for S. In particular, $m - 1 \notin S$, which means $(m-1)/n \leq r$. Adding $1/n$ to both sides and recalling $1/n < s - r$, we get

$$\frac{m}{n} = \frac{m-1}{n} + \frac{1}{n} < r + (s - r) = s,$$

and therefore $r < q < s$. $\square$

Observe that our proof of the Archimedean property, Theorem 1.17, uses the Axiom of Completeness in several different places. To end this section, we show that any proof of Theorem 1.17 must use the Axiom of Completeness. The use of (A2) is unavoidable.

Recalling that all our theorems about $\mathbb{R}$ are proved from (A1) and (A2), if we wish to show that (A2) is unavoidable in the proof of the Archimedean property, then we need to show that the Archimedean property does not follow from (A1) alone. To do this, we need to find an example of an ordered field that does not have the Archimedean property.

Definition 1.22. An ordered field F satisfies the *Archimedean property* if, for every $x \in F$, there is some $n \in \mathbb{N}$ such that $n_F > x$, and for every $x \in F$ with $x > 0$, there is some $n \in \mathbb{N}$ such that $n_F^{-1} < x$.

The field $\mathbb{Q}$ clearly does have the Archimedean property. More generally, if F is a field and $F \subseteq \mathbb{R}$, then F has the Archimedean property. (This follows from Theorem 1.17, plus the fact that if $F \subseteq \mathbb{R}$ then $\mathbb{N}_F = \mathbb{N}$.) Thus, in order to show that there is an ordered field without the Archimedean property, we need to describe an ordered field that is more abstract than $\mathbb{Q}$ or $\mathbb{R}$, a field whose members are not real numbers at all.

Example 1.23 (An ordered field without the Archimedean property). A *rational function* is a function of the form $f(x) = {}^{p(x)}/_{q(x)}$, where $p(x)$ and $q(x)$ are both polynomials with real number coefficients and q is not the constant polynomial $x \mapsto 0$. Let F denote the set of all rational functions.

This set F comes equipped with two operations $+$ and $\cdot$ already because the members of F are functions, and we already know what it means to add or multiply two functions together. Observe that if $f(x) = {}^{p(x)}/_{q(x)}$ and $g(x) = {}^{r(x)}/_{s(x)}$ are both in F, then

$$f(x) + g(x) = \frac{p(x)s(x) + r(x)q(x)}{q(x)s(x)} \qquad \text{and} \qquad f(x)g(x) = \frac{p(x)r(x)}{q(x)s(x)}$$

are both in F as well, so $+$ and $\cdot$ are functions $F \times F \to F$. Similarly, the additive and multiplicative inverses of a function $f(x)$, namely, the functions $-f(x)$ and ${}^{1}/_{f(x)}$, are members of F when f is.

The 0_F described in rule 4 of Definition 1.1 is the constant function $x \mapsto 0$. This is a rational function because the constant functions $x \mapsto 0$ and $x \mapsto 1$ are polynomials, and 0_F is just the quotient of these two functions. Similarly, the 1_F described in rule 6 of Definition 1.1 is the constant function $x \mapsto 1$.

Next, we need to define an order $<$ on F, and it is less obvious what to do here. First, observe that if $p(x)$ is a polynomial other than the constant function 0_F, then $p(x)$ has a *leading coefficient*, i.e., the coefficient of the highest power of x appearing in $p(x)$ and that this number is not 0. Given $f(x) = {}^{p(x)}/_{q(x)}$ in F, let us say that f is *positive* if the leading coefficients of $p(x)$ and $q(x)$ have the same sign, either both positive or both negative. Then, given $f(x), g(x) \in F$, define

$$f(x) < g(x) \qquad \text{if and only if} \qquad g(x) - f(x) \text{ is positive.}$$

Thus, F has two operations $+$ and $\cdot$ and the relation $<$. With these operations and this relation, F is an ordered field. Checking this thoroughly entails verifying all 11 rules from Definition 1.1. A few of these are fairly obvious, and the rest are covered in the homework exercises.

We claim that F does not have the Archimedean property. Translated to this particular ordered field F, the Archimedean property states:

1. For every $f \in F$, there is some $n \in \mathbb{N}$ such that $n_F > f$, and
2. for every positive $f \in F$, there is some $n \in \mathbb{N}$ such that $n_F^{-1} < f$.

First, let us try to understand what $\mathbb{N}_F = \{n_F : n \in \mathbb{N}\}$ is for this particular ordered field. Recall that for $n \in \mathbb{N}$, we defined $n_F = 1_F + 1_F + \cdots + 1_F$, where the number of terms in the sum is n. But 1_F is just the constant function $x \mapsto 1$, so this means that n_F is just the constant function $x \mapsto n$.

To see that statement 1 is false, let $f(x) = x$. For any $n \in \mathbb{N}$, $f - n_F$ is the function $x \mapsto x - n = \frac{x-n}{1}$, which is positive, and thus $n_F < f$. In other words, no member of $\mathbb{N}_F$ greater than $f(x) = x$.

To see that statement 2 is false, let $f(x) = 1/x$. This function is positive, so $f > 0_F$. Given $n \in \mathbb{N}$, $n_F^{-1} - f$ is the function $x \mapsto \frac{1}{n} - \frac{1}{x} = \frac{x-n}{nx}$, which is positive. Therefore, $n_F^{-1} > f$, no matter how large n is.

Roughly speaking, the field F from this example is a number system (ordered field) containing infinite numbers larger than every natural number, such as the "number" $f(x) = x$, as well as infinitesimal numbers, smaller than the reciprocal of any natural number, such as $f(x) = 1/x$.

The uniqueness of the real line

The following theorem, one of the most important in this chapter, was first explicitly stated and proved by E. V. Huntington in 1903.

Theorem 1.24 (Huntington's theorem). *Every ordered field satisfying the Axiom of Completeness is isomorphic to the real line* $\mathbb{R}$.

Our proof of this theorem first requires another definition and lemma. The definition is a simple extension of Definition 1.19 to arbitrary ordered fields.

Definition 1.25. Suppose F is an ordered field, with $<$ representing the order relation. A subset $D \subseteq F$ is *dense* if for every $r, s \in F$ with $r < s$, there is some $d \in D$ with $r < d < s$.

Lemma 1.26. *If F is an ordered field satisfying the Axiom of Completeness, then $\mathbb{Q}_F$ is dense in F.*

Proof. In fact, we claim that we proved this already in the previous section, where we showed that $\mathbb{Q}$ is dense in $\mathbb{R}$. If one goes through that proof carefully, including the proofs of the theorems and lemmas leading up to it, one can see that no facts were used about $\mathbb{R}$ except that it is an ordered field satisfying the Axiom of Completeness. Therefore, the conclusions drawn in the previous section are true not just for $\mathbb{R}$, but for every ordered field satisfying the Axiom of Completeness. □

Proof of Theorem 1.24. Let F be an ordered field satisfying the Axiom of Completeness. For each $r \in \mathbb{R}$, let

$$D_r = \{q \in \mathbb{Q}\colon q \leq r\}.$$

(This notation reflects the fact that a set of this form is usually called a *lower Dedekind cut* of the number r.) Then, define $\phi : \mathbb{R} \to F$ by setting $D'_r = \{q_F\colon q \in D_r\}$ and $\phi(r) = \sup D'_r$. That is,

$$\phi(r) = \sup\{q_F\colon q \in D_r\}$$

for all $r \in \mathbb{R}$. As $\{q_F\colon q \in D_r\} \subseteq F$, the supremum here is taken in F.

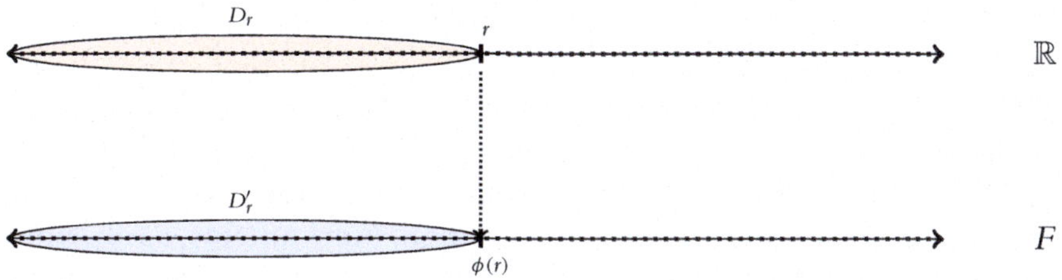

We must check that this function is well-defined. (Recall that not every subset of F must have a supremum, only those that are non-empty and bounded above. So we must check that $\{q_F\colon q \in D_r\}$ is non-empty and has an upper bound in F, for any given $r \in \mathbb{R}$, in order for the definition of ϕ to make sense.) Fix $r \in \mathbb{R}$. Because $\mathbb{Q}$ is dense in $\mathbb{R}$, there are rational numbers a and b such that $r - 1 < a < r$ and $r < b < r + 1$. Thus, $a \in D_r$, and if $q \in D_r$, then $q \leq r < b$. This implies $a_F \in D'_r$ (so $D'_r \neq \emptyset$), and because the mapping $q \mapsto q_F$ is an embedding $\mathbb{Q} \to \mathbb{Q}_F$, if some $q_F \in D'_r$ then $q_F < b_F$, so b_F is an upper bound for D'_r. Therefore, D'_r is non-empty and bounded above. Because F satisfies the Axiom of Completeness, this means that $\sup D'_r$ exists, and ϕ is well-defined.

To finish the proof of the theorem, it remains to prove that ϕ is an isomorphism. First, we need the following claim.

Claim. *If $q \in \mathbb{Q}$ and $r \in \mathbb{R}$, then $q < r$ if and only if $q_F < \phi(r)$.*

Proof of Claim. First, suppose $q < r$. Then (because $\mathbb{Q}$ is dense in $\mathbb{R}$), there is some $q' \in \mathbb{Q}$ with $q < q' < r$. But then, $q_F < q'_F$ and $q'_F \in D'_r$, which implies $q_F < q'_F \leq \sup D'_r = \phi(r)$. Next, suppose $r \leq q$. Then, $s \leq q$ for all $s \in D_r$. Because the map $\mathbb{Q} \to \mathbb{Q}_F$ is order-preserving, this means $s_F \leq q_F$ for all $s_F \in D'_r$, or in other words, q_F is an upper bound for D'_r. Hence, $\phi(r) = \sup D'_r \leq q_F$. ⊣

To see that ϕ is surjective, fix some $x \in F$. Let $E_x = \{q \in \mathbb{Q}\colon q_F \leq x\}$. Using the fact that $\mathbb{Q}_F$ is dense in F, there are some $q, s \in \mathbb{Q}$ such that $x - 1 < q_F < x < s_F < x + 1$. Then, $q \in E_x$, and s is an upper bound for E_x, so E_x has a supremum. Let $r = \sup E_x$. We claim that $\phi(r) = x$.

Suppose, aiming for a contradiction, that $x < \phi(r)$. Then (because $\mathbb{Q}_F$ is dense in F), there is some $q \in \mathbb{Q}$ with $x < q_F < \phi(r)$. But then, our claim implies $q < r$, and (because $r = \sup E_x$) there is some $q' \in E_x$ with $q < q' \leq r$. But then, $x < q_F < q'_F$, contradicting the definition of E_x. Now suppose, again aiming for a contradiction, that $\phi(r) < x$. Then (again using the fact that $\mathbb{Q}_F$ is dense in F), there are $q, s \in \mathbb{Q}$ with $\phi(r) < q_F < s_F < x$. This implies $q < s$, and $q, s \in E_x$, and (applying our claim again) $r \leq q$. This is absurd because $s \in E_x$ and $\sup E_x = r \leq q < s$. Hence, $\phi(r) = x$. And as x was an arbitrary element of F, this shows ϕ is surjective.

Next, let us show ϕ is order-preserving. Fix $r, s \in \mathbb{R}$ with $r < s$. By the density of $\mathbb{Q}$ in $\mathbb{R}$, there is some $q \in \mathbb{Q}$ such that $r < q < s$. Applying our claim twice, $r < q$ implies $\phi(r) \leq q_F$, and $q < s$ implies $q_F < \phi(s)$. Therefore, $r < s$ implies $\phi(r) < \phi(s)$.

Next, let us show that ϕ preserves sums. Fix $r, s \in \mathbb{R}$. Suppose, aiming for a contradiction, that $\phi(r) + \phi(s) < \phi(r+s)$. Because $\mathbb{Q}_F$ is dense in F, there is some $q \in \mathbb{Q}$ with $\phi(r) + \phi(s) < q_F < \phi(r+s)$. Applying the claim, $q < r + s$. But then (using basic arithmetic and the density of $\mathbb{Q}$ in $\mathbb{R}$), there are $a, b \in \mathbb{Q}$ such that $a < r$ and $b < s$ and $a+b = q$. Applying the claim again, we get $a_F < \phi(r)$ and $b_F < \phi(s)$. Because F is an ordered field, this implies $q_F = a_F + b_F < \phi(r) + \phi(s)$, a contradiction. Now suppose, again aiming for a contradiction, that $\phi(r+s) < \phi(r) + \phi(s)$. Because $\mathbb{Q}_F$ is dense in F, there is some $q \in \mathbb{Q}$ with $\phi(r+s) < q_F < \phi(r) + \phi(s)$. Applying the claim, $r + s < q$. But then (again, using basic arithmetic and the density of $\mathbb{Q}$ in $\mathbb{R}$), it is not hard to see that there are $a, b \in \mathbb{Q}$ such that $r < a$ and $s < b$ and $a + b = q$. Applying the claim again, we get $\phi(r) \leq a_F$ and $\phi(s) \leq b_F$. Because F is an ordered field, this implies $\phi(r) + \phi(s) \leq a_F + b_F = q_F$, a contradiction. Thus, $\phi(r+s) = \phi(r) + \phi(s)$.

Finally, let us show that ϕ preserves products. First, observe that the argument from the previous paragraph works just as well for products as it does for sums (just replace every $+$ with a $\cdot$), provided that $r, s > 0$. (The reason for this added assumption on r and s is that we used several times in the previous paragraph that if $a < r$ and $b < s$, then $a + b < r + s$. This is true for sums, regardless of r and s. The analogous fact is true for products if we restrict ourselves to nonnegative numbers, but it is not true for products in general.) Thus, $\phi(r) \cdot \phi(s) = \phi(r \cdot s)$ whenever $r, s > 0$.

Next, observe that $\phi(0) = \sup \{q_F \colon q \leq 0\} = 0_F$. (Note that 0_F is an upper bound for this set, but there cannot be a smaller upper bound because 0_F is in the set.) We already showed ϕ preserves addition, so this implies

$$\phi(-r) + \phi(r) = \phi(-r + r) = \phi(0) = 0_F$$

for any $r \in \mathbb{R}$. But then, this implies

$$\phi(-r) = \phi(-r) + \phi(r) - \phi(r) = (\phi(-r) + \phi(r)) - \phi(r) = 0_F - \phi(r) = -\phi(r).$$

Now suppose $r, s \in \mathbb{R}$ and $r, s < 0$. By the previous two paragraphs,

$$\phi(r \cdot s) = \phi((-r) \cdot (-s)) = \phi(-r) \cdot \phi(-s) = (-\phi(r)) \cdot (-\phi(s)) = \phi(r) \cdot \phi(s).$$

(It is a basic fact of arithmetic that $(-a)\cdot(-b) = a\cdot b$. So, this is true in F by Proposition 1.3.) Next, suppose $r, s \in \mathbb{R}$ and $r < 0 < s$. Then,

$$\phi(r\cdot s) = \phi(-((-r)\cdot s)) = -(\phi((-r)\cdot s)) = -(-\phi(r)\cdot\phi(s)) = \phi(r)\cdot\phi(s).$$

Similarly, exchanging the roles of r and s, if $s < 0 < r$, then $\phi(r\cdot s) = \phi(r)\cdot\phi(s)$. Finally, if $r = 0$, then $\phi(r\cdot s) = \phi(0) = 0_F = 0_F\cdot\phi(s) = \phi(r)\cdot\phi(s)$; and similarly if $s = 0$. This covers all possible cases, and completes the proof that $\phi(r\cdot s) = \phi(r)\cdot\phi(s)$ for all $r, s \in \mathbb{R}$. □

Remember that if two fields are isomorphic, then they are essentially identical. An isomorphism $\phi : F \to G$ may rename the elements of F, but it does not change how they interact, so that all the mathematical structure of F is preserved. In light of this, Huntington's theorem shows that our axioms (A1) and (A2) give a complete description of $\mathbb{R}$: Up to isomorphism, there is only one mathematical structure satisfying these two axioms.

The cardinality of a set

The theorems in the remainder of this chapter reveal another major difference between $\mathbb{Q}$ and $\mathbb{R}$. In a sense that will soon be made precise, $\mathbb{R}$ is much bigger than $\mathbb{Q}$.

How do we determine the number of things in a set A? If A has few enough things in it, we normally just count them. This is done by assigning one number from the set $\{1, 2, 3, \ldots, n\}$ to each member of A. For example, if $A = \{a, b, c, d\}$, we can count the members of A by making the assignments

$$a \mapsto 1 \qquad b \mapsto 2 \qquad c \mapsto 3 \qquad d \mapsto 4.$$

In other words, a set A has size n if and only if there is a bijection between the set A and the set $\{1, 2, \ldots, n\}$.

Now, suppose that we have two sets of size four, $A = \{a, b, c, d\}$ and $B = \{\alpha, \beta, \gamma, \delta\}$. For each of these two sets, we can write down bijections $f_A : A \to \{1, 2, 3, 4\}$ and $f_B : B \to \{1, 2, 3, 4\}$, and then conclude the sets have the same size. But there is a shortcut! The function $f_B^{-1} \circ f_A$ is a bijection $A \to B$, and of course there is a bijection $A \to B$ if and only if A and B have the same size. In other words, a bijection $A \to B$ shows that A and B have the same size, whether or not we bother to count either set and determine precisely what that size is. This observation about finite sets gives us our definition of when two (possibly infinite) sets have the same size.

Definition 1.27. We say that two sets A and B *have the same cardinality* if there is a bijection $A \to B$. This is expressed by writing $|A| = |B|$.

Example 1.28. The function

$$f(n) = \begin{cases} \frac{n}{2} & \text{if } n \text{ is even,} \\ \frac{1-n}{2} & \text{if } n \text{ is odd} \end{cases}$$

is a bijection $\mathbb{N} \to \mathbb{Z}$. Thus, $|\mathbb{N}| = |\mathbb{Z}|$. A bijection $\mathbb{N} \to X$ can be visualized as a kind of "walk" through X. For example, the function f as defined can be pictured as follows:

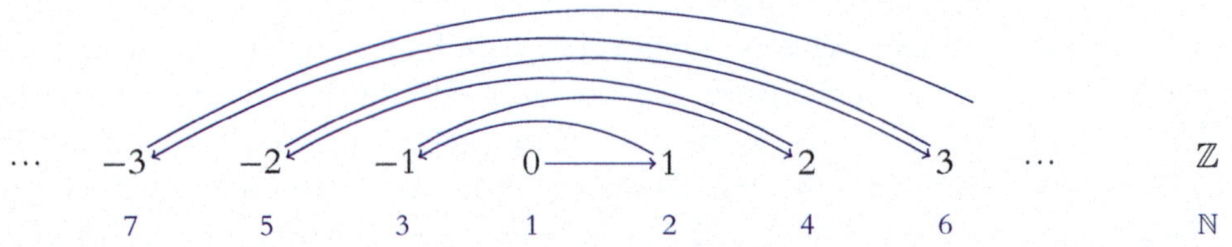

If a finite set A is strictly contained in a finite set B, then A is smaller than B. But this is not always the case for infinite sets. Even though $\mathbb{N}$ is a proper subset of $\mathbb{Z}$, they have the same cardinality.

Definition 1.29. Given two sets A and B, A *has cardinality less than or equal to that of* B if there is an injection $A \to B$. This is denoted $|A| \leq |B|$. If $|A| \leq |B|$, but it is not the case that $|A| = |B|$, then we write $|A| < |B|$, and say A *has smaller cardinality than* B.

Note that we have defined the expressions "$|A| = |B|$" and "$|A| \leq |B|$", but we have not defined what "$|A|$" or "$|B|$" mean by themselves. This is a bit like learning how to use a multi-word phrase in a foreign language without knowing what each word in the phrase means individually. While it is possible to give a formal mathematical meaning to $|A|$ and $|B|$, we do not do so here. It is surprisingly tricky to make precise sense of what the "number" of things in an infinite set really is. The study of such things forms the beginnings of the branch of mathematics known as *set theory*.

Countable and uncountable sets

Definition 1.30. A set A is *countably infinite* if $|\mathbb{N}| = |A|$, or in other words, if there is a bijection $\mathbb{N} \to A$. A set is *countable* if it is either finite or countably infinite. A set is *uncountable* if it is not countable.

Theorem 1.31. *If A is an infinite set, then $|\mathbb{N}| \leq |A|$.*

Proof. Let A be an infinite set. We can obtain an injection $f : \mathbb{N} \to A$ by recursion, as follows. For the base step of the recursion, let $f(1)$ be any element of A. For the recursive step, fix $k \in \mathbb{N}$, and suppose that $f(i)$ are chosen already for each $i < k$. Because A is infinite, $A \setminus \{f(i) : i < k\} \neq \emptyset$. Let $f(k)$ be any element of this non-empty set.

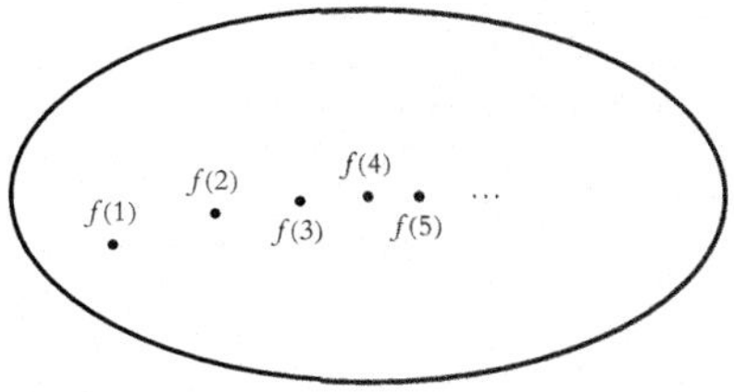

This procedure of recursively choosing new elements of A results in a function $f : \mathbb{N} \to A$. The function is injective because if $i, j \in \mathbb{N}$ and $i < j$, then $f(j)$ was chosen specifically to be different from $f(i)$. □

Corollary 1.32. *If A is uncountable, then $|\mathbb{N}| < |A|$.*

Proof. Suppose A is uncountable. By definition, this means A is infinite, which implies $|\mathbb{N}| \leq |A|$ by the previous theorem. But "uncountable" also means $|\mathbb{N}| \neq |A|$, so $|\mathbb{N}| < |A|$. □

Lemma 1.33. *Every infinite subset of $\mathbb{N}$ is countably infinite.*

Proof. Given $A \subseteq \mathbb{N}$, define a function $\mathbb{N} \to A$ by setting

$$f(n) = \text{the } n^{\text{th}} \text{ element of } A.$$

(The inverse of this function is $f^{-1}(a) = |\{i \in A \colon i \leq a\}|$.) If $A \subseteq \mathbb{N}$ is infinite, then f is a bijection $\mathbb{N} \to A$, which makes A countably infinite. □

Theorem 1.34. *If A is an infinite set and $|A| \leq |\mathbb{N}|$, then $|A| = |\mathbb{N}|$.*

Proof. Let A be an infinite set, and suppose that $f : A \to \mathbb{N}$ is an injection. Because A is infinite, the injectivity of f implies image(f) is infinite. By the previous lemma, $|\text{image}(A)| = |\mathbb{N}|$. But f is a bijection $A \to \text{image}(A)$, so $|A| = |\text{image}(A)| = |\mathbb{N}|$. □

Theorem 1.35. *For an infinite set A, then the following statements are equivalent:*

(1) *A is countably infinite.*
(2) *There is a surjection from a countable set B onto A.*
(3) *There is an injection from A into a countable set B.*

Proof. To prove these three statements equivalent, we show that (1) ⇒ (2), that (2) ⇒ (3), and that (3) ⇒ (1).

To see that (1) ⇒ (2), suppose (1) holds: There is a bijection $f : \mathbb{N} \to A$. But then, f is also a surjection $\mathbb{N} \to A$, so (2) holds.

To see that (2) ⇒ (3), suppose (2) holds, i.e., suppose there is a countable set B and a surjection $f : B \to A$. Because A is infinite, this implies B is also infinite. Thus, $|B| = |\mathbb{N}|$, and there is a bijection $g : \mathbb{N} \to B$. Let $h = f \circ g$, and note that h is a surjection $\mathbb{N} \to A$. For each $a \in A$, $h^{-1}(a)$ is a non-empty subset of $\mathbb{N}$, which implies $h^{-1}(a)$ has a minimum element. Define a function $i : \mathbb{N} \to A$ by setting $i(a) = \min h^{-1}(a)$ for every $a \in A$. This function is an injection because if $a_1, a_2 \in A$ and $a_1 \neq a_2$, then $h(i(a_1)) = a_1 \neq a_2 = h(i(a_2))$, which implies $i(a_1) \neq i(a_2)$. Thus, (3) holds.

To see that (3) $\Rightarrow$ (1), suppose (3) holds: There is a countable set B and an injection $f : A \to B$. Because A is infinite, B is also infinite. Thus, $|B| = |\mathbb{N}|$, i.e., there is a bijection $g : \mathbb{N} \to B$. But then, $h = g \circ f$ is an injection $A \to \mathbb{N}$. By the previous theorem, $|A| = |\mathbb{N}|$, so (1) holds. □

Lemma 1.36. *The set $\mathbb{N} \times \mathbb{N}$, consisting of all ordered pairs (a, b) such that $a, b \in \mathbb{N}$, is countable.*

Proof. We give two proofs of this lemma. The first proof is simply a picture of a bijection $\mathbb{N} \to \mathbb{N} \times \mathbb{N}$, much like the picture in Example 1.28.

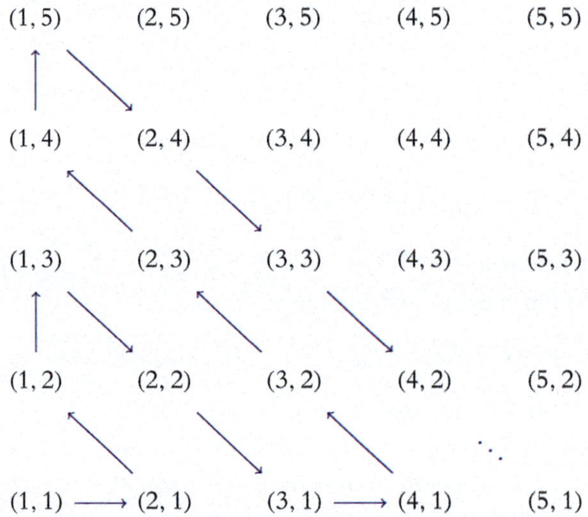

For our second proof, define $f : \mathbb{N} \to \mathbb{N} \times \mathbb{N}$ by setting

$$f(n) = \begin{cases} (a, b) & \text{if } n = 2^a 3^b \text{ for some } a, b \in \mathbb{N}, \\ (1, 1) & \text{if } n \text{ is not of this form.} \end{cases}$$

This function is clearly a surjection $\mathbb{N} \to \mathbb{N} \times \mathbb{N}$, so $\mathbb{N} \times \mathbb{N}$ satisfies statement (3) from Theorem 1.35. Hence, $\mathbb{N} \times \mathbb{N}$ is countably infinite. □

Theorem 1.37. *The set $\mathbb{Q}$ of rational numbers is countable.*

Proof. Let $A = \mathbb{Z} \times \mathbb{N} = \{(m, n) : m \in \mathbb{Z} \text{ and } n \in \mathbb{N}\}$. We claim that A is countable. To see this, first recall that by Example 1.28, there is a bijection $f : \mathbb{N} \to \mathbb{Z}$. Define $g : \mathbb{N} \times \mathbb{N} \to A$

by setting $g(m, n) = (f(m), n)$. It is not difficult to check that g is a bijection, so $|A| = |\mathbb{N} \times \mathbb{N}|$, and $|\mathbb{N} \times \mathbb{N}| = |\mathbb{N}|$ by the previous theorem. Thus, $|A| = |\mathbb{N}|$.

The function $(m, n) \mapsto m/n$ is a surjection $A \to \mathbb{Q}$. Because A is countable, this implies that $\mathbb{Q}$ is countable by Theorem 1.35. □

Instead of using Lemma 1.36 like this to prove $\mathbb{Q}$ is countable, one can show it in a picture similar to the picture from Lemma 1.36. Remember that every $q \in \mathbb{Q}$ is equal to m/n for some $m \in \mathbb{Z}$ and $n \in \mathbb{N}$.

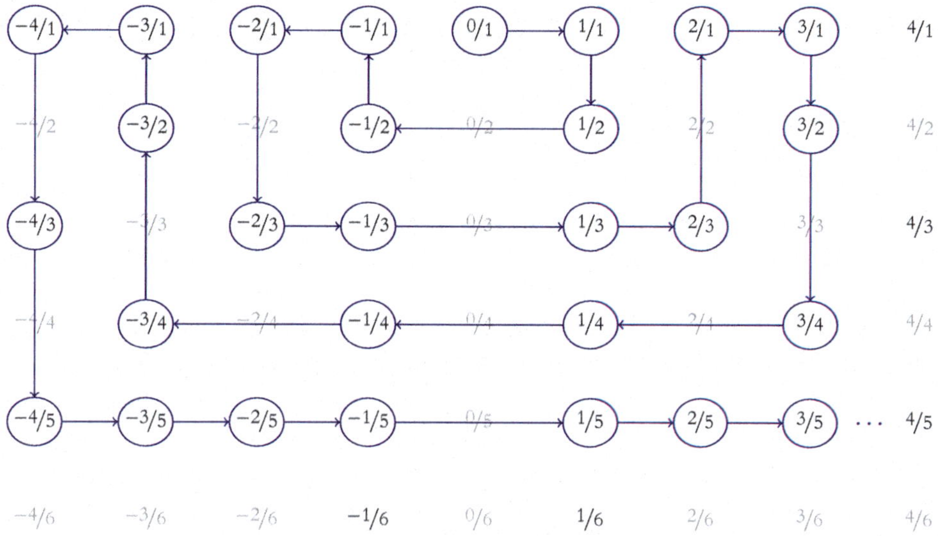

The following theorem can be expressed simply as saying that a countable union of countable sets is countable.

Theorem 1.38. *Suppose $A_1, A_2, A_3, \ldots$ is an infinite sequence of countable sets. Then, $\bigcup_{n=1}^{\infty} A_n$ is countable.*

Proof. For each $n \in \mathbb{N}$, let $B_n = \{n\} \times A_n$.

Because each of the A_n is countable, Theorem 1.35 implies that for each $n \in \mathbb{N}$, there is an injection $f_n : A_n \to \mathbb{N}$. Define a function g from $\bigcup_{n=1}^{\infty} B_n$ to $\mathbb{N} \times \mathbb{N}$ by setting $g(n, a) = (n, f_n(a))$ whenever $(n, a) \in B_n$. This function is injective, and $\mathbb{N} \times \mathbb{N}$ is countable, so $\bigcup_{n=1}^{\infty} B_n$ is countable by Theorem 1.35.

To finish the proof, define a function $h : \bigcup_{n=1}^{\infty} B_n \to \bigcup_{n=1}^{\infty} A_n$ by setting $g(n, a) = a$ whenever $(n, a) \in B_n$. This function is surjective, and $\bigcup_{n=1}^{\infty} B_n$ is countable, so $\bigcup_{n=1}^{\infty} A_n$ is countable by Theorem 1.35. □

The uncountability of the reals

In this section, we prove $\mathbb{R}$ is uncountable. Because $\mathbb{Q}$ is countable, this means $|\mathbb{Q}| < |\mathbb{R}|$. This is a precise sense in which $\mathbb{R}$ is "much bigger" than $\mathbb{Q}$. This theorem is due to the German mathematician Georg Cantor and was first proved in 1874.

Theorem 1.39. *The set $\mathbb{R}$ of real numbers is uncountable.*

Proof. Clearly, $\mathbb{R}$ is an infinite set. To prove $\mathbb{R}$ is uncountable, we must show there is no bijection $\mathbb{N} \to \mathbb{R}$. We will prove the seemingly stronger fact that there is no surjection $\mathbb{N} \to \mathbb{R}$.

Fix a function $f : \mathbb{N} \to \mathbb{R}$. Our goal is to prove f is not a surjection. Define $r_n = f(n)$ for each $n \in \mathbb{N}$. This is just a convenient notation that encourages us to think of the image of f as a sequence $r_1, r_2, r_3, \ldots$ of real numbers. Our proof that f is not a surjection breaks into two cases, according to whether f has the following property:

(∗) For any $a, b \in \mathbb{R}$ with $a < b$, there is some $i \in \mathbb{N}$ such that $a < r_i < b$.

For the first case, suppose f does not have property (∗), i.e., suppose there are some $a, b \in \mathbb{R}$ with $a < b$ such that $r_i \notin (a, b)$ for all i. The interval (a, b) is clearly non-empty, however, and if $r \in (a, b)$, then $r \neq r_i$ for any $n \in \mathbb{N}$. This means r is not in the image of f, so f is not a surjection.

For the second case, suppose f has property (∗). We define two new sequences of numbers $a_1, a_2, a_3, \ldots$ and $b_1, b_2, b_3, \ldots$ via recursion, as follows.

In the base step of the recursion, first let $a_1 = r_1$, and then let b_1 be the first real number in the sequence $r_1, r_2, r_3, \ldots$ bigger than a_1. In other words, $b_1 = r_m$ where $m = \min\{i\colon r_i > a_1\}$. (Note that $\{i\colon r_i > a_1\}$ is non-empty by property (∗): There is some i such that $r_i \in (a, a+1)$.)

In the next step of the recursion, let a_2 be the first real number in the sequence $r_1, r_2, r_3, \ldots$ such that $a_1 < a_2 < b_1$. In other words, $a_2 = r_m$ where $m = \min\{i\colon a_1 < r_i < b_1\}$. Note that $\{i\colon a_1 < r_i < b_1\}$ is non-empty by property (∗), so this definition of a_2 makes sense. Then, let b_2 be the first real number in the sequence $r_1, r_2, r_3, \ldots$ such that $a_2 < b_2 < b_1$. As with a_2, property (∗) implies that some such b_2 must exist.

In general, at stage n of the recursion, we have already chosen some numbers $a_1, \ldots, a_{n-1}$ and $b_1, \ldots, b_{n-1}$ such that

$$a_1 < a_2 < \cdots < a_{n-1} < b_{n-1} < \cdots < b_2 < b_1.$$

Let a_n be the first real number in the sequence $r_1, r_2, r_3, \ldots$ such that $a_{n-1} < a_n < b_{n-1}$, noting that some such number must exist by property (∗). Similarly, let b_n be the first real number in the sequence $r_1, r_2, r_3, \ldots$ such that $a_n < b_n < b_{n-1}$.

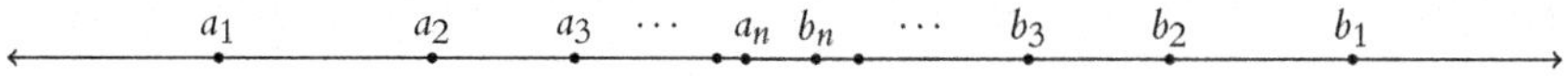

In the end, this recursion produces two infinite sequences $a_1, a_2, a_3, \ldots$ and $b_1, b_2, b_3, \ldots$ in $\mathbb{R}$ such that

$$a_1 < a_2 < a_3 < \cdots < a_n < \cdots < b_n < \cdots < b_3 < b_2 < b_1.$$

Observe that the set $\{a_n \colon n \in \mathbb{N}\}$ is non-empty and bounded above, for example, it is bounded above by b_n for any $n \in \mathbb{N}$. So, this set has a supremum by the Axiom of Completeness. Let $a_\infty = \sup\{a_n \colon n \in \mathbb{N}\}$.

Observe that if $n \in \mathbb{N}$, then $a_n < a_{n+1} \leq a_\infty \leq b_{n+1} < b_n$. In particular, a_∞ is not equal to a_n or to b_n for any n.

We claim a_∞ is not in the image of the function f. Aiming for a contradiction, suppose there is some $N \in \mathbb{N}$ such that $a_\infty = f(N)$, i.e., $a_\infty = r_N$. Because of how a_n and b_n were chosen at each stage of our recursion and because $a_n \neq r_N \neq b_n$ (by the previous paragraph), both of the numbers a_n and b_n must come before r_N in the sequence $r_1, r_2, r_3, \ldots$. This is absurd because only $N-1$ numbers come before r_N in this sequence, whereas there are infinitely many numbers in the sequences of a_n's and b_n's. This contradiction shows a_∞ is not in the image of f.

Therefore, f is not a surjection. As f was an arbitrary function $\mathbb{N} \to \mathbb{R}$, we conclude that there is no surjection, hence no bijection, from $\mathbb{N}$ to $\mathbb{R}$. □

We end this section with a related theorem, also due to Georg Cantor, proved in 1887, which says roughly that there is no largest set.

Definition 1.40. For every set A, the *power set* of A, denoted $\mathcal{P}(A)$, is the set of all subsets of A. That is, $\mathcal{P}(A) = \{X \colon X \subseteq A\}$.

Theorem 1.41. *$|A| < |\mathcal{P}(A)|$ for every set A.*

Proof. Let A be any set. We have two things to prove: $|A| \leq |\mathcal{P}(A)|$, and $|A| \neq |\mathcal{P}(A)|$.

To see that $|A| \leq |\mathcal{P}(A)|$, define $s : A \to \mathcal{P}(A)$ by setting $s(a) = \{a\}$ for all $a \in A$. This is clearly an injective function $A \to \mathcal{P}(A)$, so $|A| \leq |\mathcal{P}(A)|$.

To see that $|A| \neq |\mathcal{P}(A)|$, let f be any function $A \to \mathcal{P}(A)$. We will show that f is not a surjection. To this end, define

$$D = \{a \in A \colon a \notin f(a)\}.$$

Clearly, $D \subseteq A$, which means $D \in \mathcal{P}(A)$. Now, let $x \in A$. If $x \in D$, then this means $x \notin f(x) = D$, which implies $f(x) \neq D$. On the other hand, if $x \notin D$ then this means, $x \in f(x)$, which implies $f(x) \neq D$. Either way, $f(x) \neq D$. Hence, $D \notin \text{image}(f)$, and f is not surjective. □

Exercises

1. The first few exercises are about ordered fields. In Proposition 1.3, we claim that ordered fields obey all the normal rules of arithmetic. Working from the definition of an ordered field (i.e., without appealing to Proposition 1.3), show that the following facts hold in any ordered field F:
   (a) If $a < b$ and $c < 0_F$, then $ac > bc$.
   (b) $a \cdot a > 0$ for every $a \in F$ other than $a = 0_F$.
   (c) If $z \in F$ and $z + a = a$ for every $a \in F$, then $z = 0_F$. (In other words, 0_F is the only additive identity for F.)
   (d) If $a \neq 0_F$, then $a \neq -a$.
   (e) If $a \neq 0_F$, then either $-a < 0_F < a$ or $a < 0_F < -a$.
   (f) Based on the previous part of the question, define
   $$|a| = \begin{cases} a & \text{if } a \geq 0_F, \\ -a & \text{otherwise.} \end{cases}$$
   Prove that if $|a| < |b|$, then $|a^n| < |b^n|$ for all $n \in \mathbb{N}$.
2. Prove that every ordered field is infinite.
3. Prove that there is no possible way of ordering the complex numbers $\mathbb{C}$ such that the ordering makes $\mathbb{C}$ into an ordered field (using the usual $+$ and $\cdot$ operations on $\mathbb{C}$).
4. Suppose $A, B \subseteq \mathbb{R}$, and define $A + B = \{a + b\colon a \in A \text{ and } b \in B\}$. Assuming that A and B are both non-empty and bounded above, prove that $\sup(A + B)$ exists and $\sup(A + B) = \sup A + \sup B$.
5. Suppose $A \subseteq \mathbb{R}$, and suppose c is a positive real number. Define $cA = \{ca\colon a \in A\}$. Assuming that A is non-empty and bounded above, prove that $\sup(cA)$ exists and $\sup(cA) = c \cdot \sup A$. State and prove a similar result for when $c < 0$.
6. Let $A \subseteq \mathbb{R}$, and suppose that A is non-empty and bounded below. Let $B = \{b \in \mathbb{R}\colon b \text{ is a lower bound for } A\}$. Prove that B has a supremum and $\inf A = \sup B$.
7. Suppose $n \in \mathbb{N}$. Prove that unless n is a perfect square (meaning $n = m^2$ for some $m \in \mathbb{N}$), $\sqrt{n}$ is irrational.
8. Using the Axiom of Completeness and basic arithmetic, prove that every nonnegative real number has a square root.
9. Prove that the set of irrational numbers is dense in $\mathbb{R}$.
10. Let F denote the field from Example 1.23. It is easy to see that the first 7 rules in the definition of an ordered field (Definition 1.1) all hold for F. Prove that the last 4 rules also hold.
11. Let F denote the field from Example 1.23. The example shows that F does not have the Archimedean property, and because we proved the Archimedean property from the Axiom of Completeness, it follows that F does not satisfy the Axiom of Completeness. Find an explicit example of a non-empty $A \subseteq F$ such that A is bounded above in F, but does not have a supremum.

12. For each of the following statements, either find a proof or find a counterexample.

(a) If $A \cap B \neq \emptyset$, $A \cap C \neq \emptyset$, and $B \cap C \neq \emptyset$, then $A \cap B \cap C \neq \emptyset$.

(b) Let $A_1, A_2, A_3, \ldots$ be infinite sets with $A_1 \supseteq A_2 \supseteq A_3 \supseteq \ldots$. Then, $\bigcap_{i=1}^{\infty} A_i$ is infinite.

(c) Let $A_1, A_2, A_3, \ldots$ be infinite sets with $A_1 \subseteq A_2 \subseteq A_3 \subseteq \ldots$. Then, $\bigcap_{i=1}^{\infty} A_i$ is infinite.

(d) Let A and B be sets, each a subset of some larger set X. Then, $(X \setminus A) \cup (X \setminus B) = X \setminus (A \cap B)$.

(e) Let A and B be sets, each a subset of some larger set X. Then, $(X \setminus A) \cap (X \setminus B) = X \setminus (A \cup B)$.

13. For each of the following statements, decide whether it is true or false. If it is true, then prove it, and if it is false, then find a counterexample.

(a) Let X and Y be sets, let $f : X \to Y$, and let $A, B \subseteq X$. Then,

i. $f[A \cap B] = f[A] \cap f[B]$,

ii. $f[A \cup B] = f[A] \cup f[B]$, and

iii. $f[X \setminus A] = f[X] \setminus f[A]$.

(b) Let X and Y be sets, let $f : X \to Y$, and let $A, B \subseteq Y$. Then,

i. $f^{-1}[A \cap B] = f^{-1}[A] \cap f^{-1}[B]$,

ii. $f^{-1}[A \cup B] = f^{-1}[A] \cup f^{-1}[B]$, and

iii. $f^{-1}[Y \setminus A] = f^{-1}[Y] \setminus f^{-1}[A]$.

14. Prove that the relation "has the same cardinality as" is an equivalence relation on the class of sets.

15. For every $k \in \mathbb{N}$, let $\mathbb{N}^k$ denote the set of all k-tuples $(n_1, n_2, \ldots, n_k)$ of natural numbers. Prove that $\mathbb{N}^k$ is countable for every k.

16. Let $\mathbb{N}^{\mathbb{N}}$ denote the set of all infinite sequences $(n_1, n_2, n_3, \ldots)$ of natural numbers. Prove that $\mathbb{N}^{\mathbb{N}}$ is uncountable.

17. An *algebraic number* is a number $r \in \mathbb{R}$ that is a root of a polynomial equation in which all of the coefficients are integers. Prove that the set of algebraic numbers is countable.

18. A *transcendental number* is a number that is not algebraic. Prove that the set of transcendental numbers is uncountable.

19. Write down an explicit example of a bijection $[0, 1] \to [0, 1)$.

20. Generalizing the previous exercise, prove that if A is infinite and F is finite, then $|A| = |A \cup F|$.

21. Suppose $a, b \in \mathbb{R}$ and $a < b$. Prove that $|\mathbb{R}| = \big|(a, b)\big| = \big|[a, b)\big| = \big|(a, b]\big| = \big|[a, b]\big|$.

22. Prove that if $B \subseteq A$ and $|A| \leq |B|$, then $|A| = |B|$.
(*Hint:* Suppose $f : A \to B$ is an injection. Your goal is to find a bijection $A \to B$. To do this, consider the sequence of sets $A \setminus B, f[A] \setminus f[B], f^2[A] \setminus f^2[B], \ldots$. Here, the notation $f^n[X]$ means the image of X under the mapping $f \circ f \circ \cdots \circ f$, composed n times.)

23. Prove that for any sets A and C, if $|A| \leq |C|$ and $|C| \leq |A|$, then $|A| = |C|$. This is known as the *Cantor–Schröder–Bernstein Theorem*.
(*Hint:* Use the previous exercise.)

CHAPTER 2

Metric spaces

In this chapter, we learn about *metric spaces*. There are several reasons why metric spaces are important for our study of $\mathbb{R}$. First, $\mathbb{R}$ is itself a metric space, so anything we prove about metric spaces is automatically true of $\mathbb{R}$. Second, like with ordered fields in the previous chapter, we would like to understand how $\mathbb{R}$ fits into this broader class of objects. Third, certain other metric spaces are directly relevant to our study of $\mathbb{R}$. For example, the plane $\mathbb{R}^2$ appears in Chap. 4 when we study continuous functions, and the space $\mathcal{C}(\mathbb{R})$ of all continuous functions $\mathbb{R} \to \mathbb{R}$ appears in Chap. 8.

Fundamentals of metric spaces

Definition 2.1. A *metric space* is a pair (X, d), where X is a set and d is a function $X \times X \to \mathbb{R}$ satisfying the following four rules:

(1) $d(x, x) = 0$ for all $x \in X$.

(2) $d(x, y) > 0$ for all $x, y \in X$ with $x \neq y$.

(3) $d(x, y) = d(y, x)$ for all $x, y \in X$.

(4) $d(x, z) \leq d(x, y) + d(y, z)$ for all $x, y, z \in X$.

A function d with these properties is called a *metric* for X.

Rule (4) is called the *triangle inequality*.

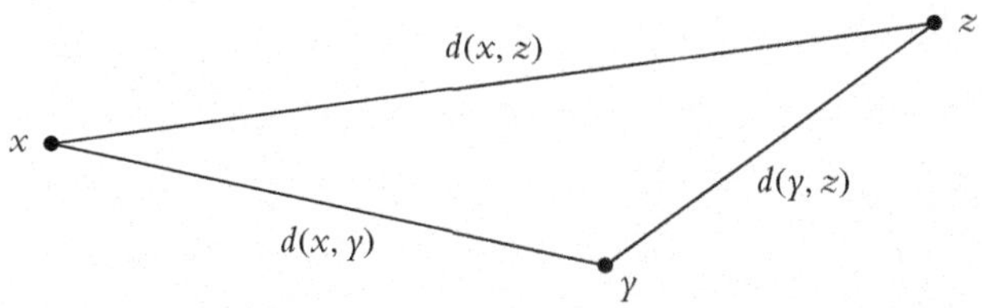

Example 2.2. The function $(x, y) \mapsto |x - y|$ is a metric for $\mathbb{R}$. Thus, the pair consisting of the set $\mathbb{R}$ and this function is a metric space.

In context, we often refer to a set X as a metric space, rather than a pair (X, d). Thus, for example, it is not incorrect to say "$\mathbb{R}$ is a metric space." When one says this, it implies that we are thinking of $\mathbb{R}$ as paired with the function $(x, y) \mapsto |x - y|$.

The function $(x, y) \mapsto |x - y|$ is sometimes called the *standard metric* on $\mathbb{R}$. Unless otherwise specified, any reference to $\mathbb{R}$ as a metric space refers to this particular metric. However, there are also nonstandard metrics on $\mathbb{R}$.

Foundations of Real Analysis
https://doi.org/10.1016/B978-0-44-348527-5.00006-8

Example 2.3. Let X be any set, and define $\delta : X \times X \to \mathbb{R}$ by setting

$$\delta(x, y) = \begin{cases} 0 & \text{if } x = y \\ 1 & \text{otherwise.} \end{cases}$$

This function is a metric for X. It is called the *discrete metric* for X.

So, $(\mathbb{R}, \delta)$ is a metric space, with a metric on $\mathbb{R}$ different than the standard one. Some other examples of metrics on $\mathbb{R}$ can be found in the homework exercises. There is also a standard metric on $\mathbb{R}^2$, the function

$$((x_1, x_2), (y_1, y_2)) \mapsto \sqrt{(x_1 - y_1)^2 + (x_2 - y_2)^2}.$$

There is a similarly defined standard metric for $\mathbb{R}^n$, for any integer $n > 2$.

Definition 2.4. Let X be any set, and suppose $d : X \times X \to \mathbb{R}$ is a metric for X. If $Y \subseteq X$, then $d \restriction Y \times Y$ is a metric for Y, the *restriction* of d to Y.

The metric space $(Y, d \restriction (Y \times Y))$ is called a *subspace* of X. This is normally abbreviated to (Y, d), or just Y.

For example, the function $(x, y) \mapsto |x - y|$ is a metric for $\mathbb{Q}$, and it is the restriction to $\mathbb{Q}$ of the metric from Example 2.2. This function is considered the "standard" metric on $\mathbb{Q}$, so when $\mathbb{Q}$ is referred to as a metric space, then, unless otherwise specified, this is the intended metric.

Definition 2.5. Let (X, d) be a metric space.

- Given $x \in X$ and $r > 0$, define

$$B_r(x) = \{y \in X : d(x, y) < r\}.$$

  This is called the *ball of radius* r or the r*-ball*, with center x.
- A set $U \subseteq X$ is *open* if, for every $x \in U$, there is some $r > 0$ such that $B_r(x) \subseteq U$.
- Given $A \subseteq X$, the *interior* of A, denoted $\text{int}(A)$, is the union of all open subsets of A, i.e., $\text{int}(A) = \bigcup \{U \subseteq A : U \text{ is open}\}$.

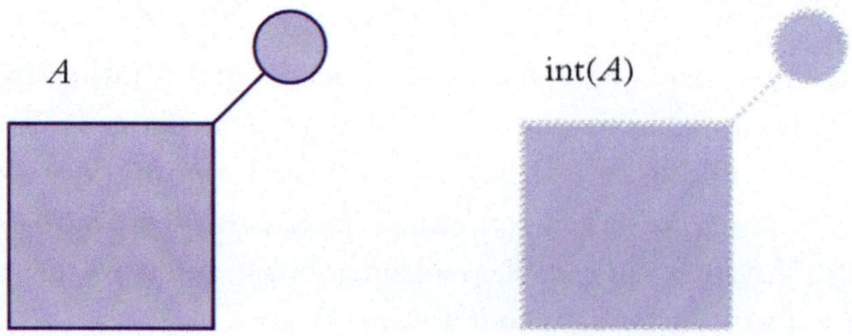

For example, in the metric space $\mathbb{R}$ (with the standard metric), the interval $(1, 2)$ is the ball of radius $1/2$ around the point $3/2$, i.e., $B_{1/2}(3/2)$. The interval $(1, 2)$ is open, but the interval $[1, 2)$ is not open because no ball centered at the point 1 is a subset of $[1, 2)$.

Theorem 2.6. *Let (X, d) be a metric space. For any $x \in X$ and $r > 0$, the ball $B_r(x)$ is open.*

Proof. Fix $x \in X$ and $r > 0$. To show that $B_r(x)$ is open, we must show that for any $y \in B_r(x)$, there is some $\delta > 0$ such that $B_\delta(y) \subseteq B_r(x)$.

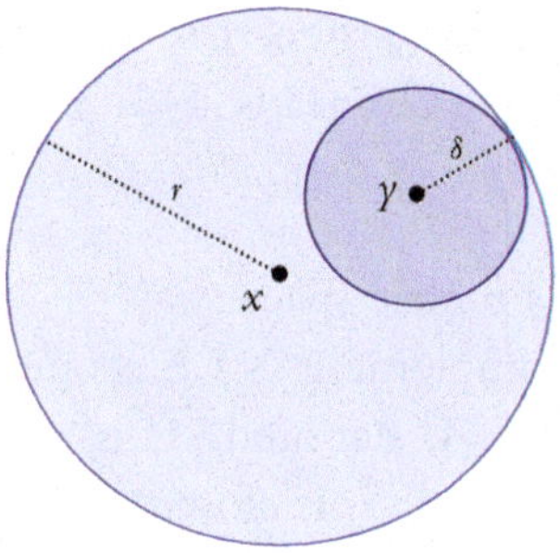

Let $\delta = r - d(x, y)$. Notice that $d(x, y) < r$ because $y \in B_r(x)$, which means $\delta > 0$.

To show that $B_\delta(y) \subseteq B_r(x)$, let $z \in B_\delta(y)$. This means that $d(y, z) < \delta$. By the triangle inequality, this implies

$$\begin{aligned} d(x, z) &\leq d(x, y) + d(y, z) \\ &= r - \delta + d(y, z) = r - (\delta - d(y, z)) < r. \end{aligned}$$

Thus, $z \in B_r(x)$. Because z was an arbitrary member of $B_\delta(y)$, this shows $B_\delta(y) \subseteq B_r(x)$. □

Theorem 2.7. *Let (X, d) be a metric space.*

(1) *$\emptyset$ and X are open subsets of X.*

(2) *If $\mathcal{U}$ is a finite collection of open subsets of X, then $\bigcap \mathcal{U}$ is open.*

(3) *If $\mathcal{U}$ is any collection of open subsets of X, then $\bigcup \mathcal{U}$ is open.*

Proof. Recall that $U \subseteq X$ is open if for every $x \in U$, there is some $r > 0$ such that $B_r(x) \subseteq U$. This is true for $U = \emptyset$ because there are no $x \in U$ for which this is not the case. It is true for $U = X$ because $B_r(x) \subseteq X$ for any $x \in X$ and any $r > 0$. Thus, (1) holds.

For (2), let $\mathcal{U}$ be a finite collection of open subsets of X, and suppose $x \in \bigcap \mathcal{U}$, i.e., $x \in U$ for every $U \in \mathcal{U}$. As each $U \in \mathcal{U}$ is open, for every $U \in \mathcal{U}$, there is some $r_U > 0$ with $B_{r_U}(x) \subseteq U$. Let $r = \min\{r_U : U \in \mathcal{U}\}$. Because $\mathcal{U}$ is finite, $r > 0$. But $B_r(x) \subseteq B_{r_U}(x) \subseteq U$ for every $U \in \mathcal{U}$, so $B_r(x) \subseteq \bigcap \mathcal{U}$. As x was an arbitrary point of $\bigcap \mathcal{U}$, this shows that $\bigcap \mathcal{U}$ is open.

For (3), let $x \in \bigcup \mathcal{U}$, which means $x \in U$ for some $U \in \mathcal{U}$. As U is open, there is some $r > 0$ with $B_r(x) \subseteq U$. But then, $B_r(x) \subseteq U \subseteq \bigcup \mathcal{U}$. As x was an arbitrary point of $\bigcup \mathcal{U}$, this shows that $\bigcup \mathcal{U}$ is open. □

Corollary 2.8. *Let X be a metric space, and let $A \subseteq X$. Then,* $\operatorname{int}(A)$ *is open,* $\operatorname{int}(A) \subseteq A$, *and $A = \operatorname{int}(A)$ if and only if A is open.*

Proof. That $\operatorname{int}(A)$ is open follows from the definition of $\operatorname{int}(A)$ and part (3) of the previous theorem. The fact that $\operatorname{int}(A) \subseteq A$ follows directly from the definition of an interior: Any union of subsets of A is still a subset of A.

It remains to prove that $A = \operatorname{int}(A)$ if and only if A is open. We have already established $\operatorname{int}(A)$ is open, so if $A = \operatorname{int}(A)$, then A is open. Conversely, suppose A is open. Then, $A \in \{U \subseteq A \colon U \text{ is open}\}$, which means $A \subseteq \operatorname{int}(A)$. We already know $\operatorname{int}(A) \subseteq A$, so this implies $A = \operatorname{int}(A)$. □

Definition 2.9. Let (X, d) be a metric space.

- A set $F \subseteq X$ is *closed* if its complement $X \setminus F$ is open.
- Given $A \subseteq X$, the *closure* of A, denoted $\overline{A}$, is the intersection of all closed sets containing A, i.e., $\overline{A} = \bigcap\{F \supseteq A \colon F \text{ is closed}\}$.

Example 2.10. $\mathbb{R}$ is both an open subset of $\mathbb{R}$ (by the previous theorem) and a closed subset (because its complement $\emptyset$ is also open). Thus, unlike doors, a set can be both open and closed at the same time. The set $\mathbb{Q}$ is neither an open nor a closed subset of $\mathbb{R}$. The interval $(0, 1)$ is open but not closed as a subset of $\mathbb{R}$, and the interval $[0, 1]$ is closed but not open.

In the metric space $\mathbb{R}$, the closure of $(0, 1)$ is $[0, 1]$, and the closure of $\mathbb{Q}$ is $\mathbb{R}$. In the metric space $\mathbb{Q}$, the closure of $\mathbb{Q}$ is $\mathbb{Q}$.

Consider the metric spaces $\mathbb{R}$ and $\mathbb{Q}$ (both with the standard metric). The set $\mathbb{Q}$ is not open in the metric space $\mathbb{R}$, but it is open in the metric space $\mathbb{Q}$. Thus, saying that a particular set is "open" or "closed" makes sense only in context when it is considered as a subset of some metric space.

The following theorem and corollary, whose proofs we leave as a homework exercise, highlight the close relationship between open and closed sets.

Theorem 2.11. *Let (X, d) be a metric space.*

(1) *$\emptyset$ and X are closed subsets of X.*
(2) *If $\mathcal{F}$ is a finite collection of closed subsets of X, then $\bigcup \mathcal{F}$ is closed.*
(3) *If $\mathcal{F}$ is any collection of closed subsets of X, then $\bigcap \mathcal{F}$ is closed.*

Corollary 2.12. *Let X be a metric space, and let $A \subseteq X$. Then, $\overline{A}$ is closed, $A \subseteq \overline{A}$, and $A = \overline{A}$ if and only if A is closed.*

Theorem 2.13. *Let X be a metric space, let $x \in X$, and let $A \subseteq X$. Then, $x \in \overline{A}$ if and only if $B_r(x) \cap A \neq \emptyset$ for every $r > 0$.*

Proof. For the "only if" direction of the theorem, we prove the contrapositive. Suppose $x \in X$, and there is some $r > 0$ such that $B_r(x) \cap A = \emptyset$. In other words, $B_r(x) \subseteq X \setminus A$. Because $B_r(x)$ is open, this implies $B_r(x) \subseteq \operatorname{int}(X \setminus A)$, and in particular $x \in \operatorname{int}(X \setminus A)$. Because $\operatorname{int}(X \setminus A)$ is open, its complement $F = X \setminus \operatorname{int}(X \setminus A)$ is closed, and $x \notin F$. Also, $A \subseteq F$, because $X \setminus A \supseteq \operatorname{int}(X \setminus A)$. Thus, $x \notin \bigcap\{G \supseteq A : G \text{ is closed}\} = \overline{A}$. By contraposition, if $x \in \overline{A}$, then $B_r(x) \cap A \neq \emptyset$ for every $r > 0$.

For the "if" direction, suppose $B_r(x) \cap A \neq \emptyset$ for every $r > 0$. Let F be a closed set with $A \subseteq F$. This implies $X \setminus F$ is open and $A \cap (X \setminus F) = \emptyset$. But then, $x \notin X \setminus F$ because (as $X \setminus F$ is open) this would mean there is some $r > 0$ with $B_r(x) \subseteq X \setminus F$, which would mean $B_r(x) \cap A = \emptyset$. Thus, $x \in F$ whenever F is a closed set containing A. Hence $x \in \overline{A}$. □

Definition 2.14. Let X be a metric space, and let $A \subseteq X$. A point $x \in X$ is a *limit point* of A if $B_r(x) \cap A$ is infinite for every $r > 0$. A point x is an *isolated point* of A if $x \in A$, but x is not a limit point of A.

For example, let $A = \{1/n : n \in \mathbb{N}\}$. Every point of this set is an isolated point. It has a limit point, namely, 0, but this limit point is not in A.

For another example, the interval $(0, 1)$ has no isolated points, and every point of the interval $[0, 1]$ is a limit point of $(0, 1)$.

Proposition 2.15. *Let X be a metric space, and let $A \subseteq X$. A point x is an isolated point of A if and only if there is some $r > 0$ with $B_r(x) \cap A = \{x\}$.*

Proof. If $B_r(x) \cap A = \{x\}$, then $x \in A$ and is not a limit point of A; so, x is an isolated point of A. For the reverse implication, suppose that x is an isolated point of A. Then, $x \in A$, and there is some $r_0 > 0$ such that $B_{r_0}(x) \cap A$ is finite. Let $r = \min\{d(x, a) : a \in A \cap B_{r_0}(x) \text{ and } a \neq x\}$. This is a finite set of positive real numbers, so $r > 0$. And $B_r(x) \cap A = \{x\}$. □

Theorem 2.16. *Let X be a metric space and let $A \subseteq X$. Then,*

$$\overline{A} = A \cup \{x \in X : x \text{ is a limit point of } A\}.$$

Proof. We already proved that $A \subseteq \overline{A}$ for any set A. If x is a limit point of A, then $B_r(x) \cap A \neq \emptyset$ for all $r > 0$, and this implies $x \in \overline{A}$ by Theorem 2.13. Thus, $A \cup \{x \in X : x \text{ is a limit point of } A\} \subseteq \overline{A}$.

To prove the reverse inclusion, suppose $x \in \overline{A}$. By Theorem 2.13, this means $B_r(x) \cap A \neq \emptyset$ for every $r > 0$. If $B_r(x) \cap A$ is infinite for all $r > 0$, then x is a limit point of A. If not, fix $r_0 > 0$ such that $B_{r_0}(x) \cap A$ is finite. As in the previous proof, let $r = \min\{d(x, a) : a \in A \cap B_{r_0}(x) \text{ and } a \neq x\}$, and then observe that $r > 0$ and $B_r(x) \cap A \subseteq \{x\}$. But, $B_r(x) \cap A \neq \emptyset$ for every $r > 0$, so this implies $B_r(x) \cap A = \{x\}$. In particular, $x \in A$. Thus, either x is a limit point of A or $x \in A$. □

Continuous functions

Definition 2.17. Suppose (X, d_X) and (Y, d_Y) are metric spaces, f is a function $X \to Y$, and $x \in X$. We say f is *continuous at* x if, for every $\varepsilon > 0$, there is some $\delta > 0$ such that if $z \in X$ and $d_X(x, z) < \delta$, then $d_Y(f(x), f(z)) < \varepsilon$. We say f is *continuous* if it is continuous at every $x \in X$.

Example 2.18. Many familiar functions $\mathbb{R} \to \mathbb{R}$ are continuous: polynomials; the sine and cosine functions; exponential functions; and all sums, differences, and compositions of these (to name a few).

A simple example of a discontinuous function $f : \mathbb{R} \to \mathbb{R}$ is

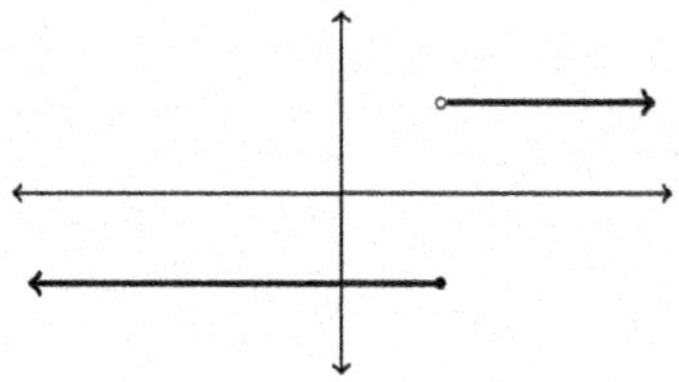

$$f(x) = \begin{cases} -1 & \text{if } x \leq 1, \\ 1 & \text{if } x > 1. \end{cases}$$

This function and others like it are sometimes called "step functions." It is continuous at x for every $x \neq 1$, but it is not continuous at 1.

Example 2.19. Not every discontinuity in a function $\mathbb{R} \to \mathbb{R}$ involves a "jump" like the function in the previous example. The function

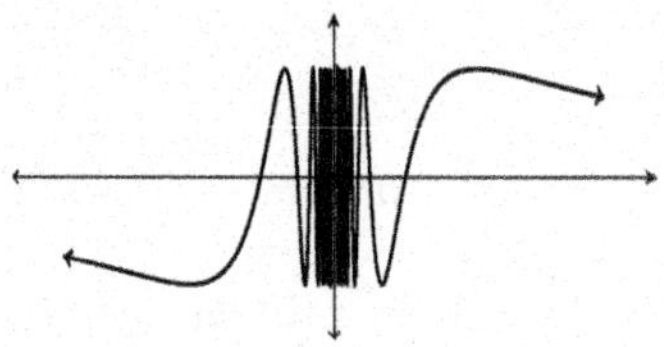

$$f(x) = \begin{cases} \sin(1/x) & \text{if } x \neq 0, \\ 0 & \text{if } x = 0 \end{cases}$$

is not continuous at the point $x = 0$, though it is continuous at every other point in $\mathbb{R}$.

These examples show how a function can be discontinuous at a single point. In the homework exercises, there is an example of a function $[0, 1] \to \mathbb{R}$ that is continuous at x if and only if x is irrational.

Theorem 2.20. *Let X and Y be metric spaces, and suppose f is a function $X \to Y$. The following statements are equivalent:*
(1) *f is continuous.*
(2) *For every open $U \subseteq Y$, $f^{-1}[U]$ is an open subset of X.*

Proof. To see that (1) $\Rightarrow$ (2), suppose $f : X \to Y$ is continuous, and let $U \subseteq Y$ be open. Fix $x \in f^{-1}[U]$. Because U is open and $f(x) \in U$, there is some $\varepsilon > 0$ such that $B_\varepsilon(f(x)) \subseteq U$. Because f is continuous at x, there is some $\delta > 0$ such that, if $z \in X$ and $d_X(x, z) < \delta$, then $d_Y(f(x), f(z)) < \varepsilon$. Or, rephrasing, if $z \in B_\delta(x)$, then $f(z) \in B_\varepsilon(f(x)) \subseteq U$. Thus, $B_\delta(x) \subseteq f^{-1}[U]$. As x was an arbitrary element of $f^{-1}[U]$, this shows that $f^{-1}[U]$ is open.

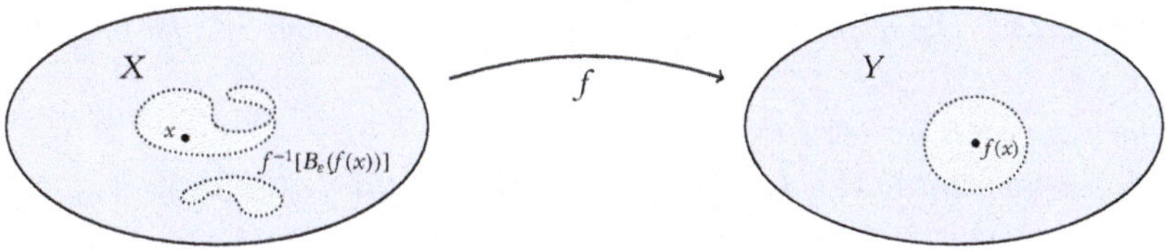

To see that (2) $\Rightarrow$ (1), suppose $f : X \to Y$ is a function with the property that $f^{-1}[U]$ is an open subset of X for every open $U \subseteq Y$. Fix $x \in X$, and let $\varepsilon > 0$. By Theorem 2.6, $B_\varepsilon(f(x))$ is an open subset of Y. Thus, $f^{-1}[B_\varepsilon(f(x))]$ is open in X. As $x \in f^{-1}[B_\varepsilon(f(x))]$, this implies there is some $\delta > 0$ such that $B_\delta(x) \subseteq f^{-1}[B_\varepsilon(f(x))]$. Or, rephrasing this, if $z \in X$ and $d_X(x, z) < \delta$, then $f(z) \in B_\varepsilon(f(x))$, which means $d_Y(f(x), f(z)) < \varepsilon$. □

In Chap. 4, we will study continuous functions further, and prove several results concerning how to combine continuous functions into new continuous function (by adding or subtracting them, multiplying them, etc.). For now, we prove just one result along these lines.

Corollary 2.21. *Let X, Y, and Z be metric spaces. If $f : X \to Y$ and $g : Y \to Z$ are continuous functions, then $g \circ f$ is also continuous.*

Proof. If U is an open subset of Z, then $g^{-1}[U]$ is an open subset of Y (because g is continuous), and then $f^{-1}[g^{-1}[U]] = (g \circ f)^{-1}[U]$ is an open subset of X. Thus, $g \circ f$ satisfies condition (2) of the previous theorem and is therefore continuous. □

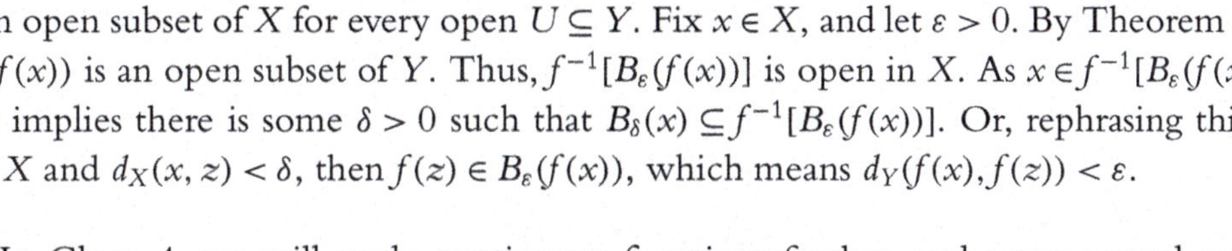

When are two metric spaces the same?

In Chap. 1, we saw that two ordered fields can be isomorphic: essentially the same, because there is a bijection between them preserving all the mathematical structure of an ordered field. Roughly, an isomorphism is a relabeling of the points that preserves all the structure. With metric spaces, there is a similar notion, a way of identifying two spaces and saying they are essentially the same thing. Only, with metric spaces, we call this an "isometry" rather than an isomorphism.

Definition 2.22. Suppose (X, d_X) and (Y, d_Y) are metric spaces. An *isometry* is a bijection $\phi : X \to Y$ such that $d_X(a, b) = d_Y(\phi(a), \phi(b))$ for every $a, b \in X$. If there is an isometry $X \to Y$, then we say (X, d_X) and (Y, d_Y) are *isometric.* ☕

Proposition 2.23. *The relation "is isometric to" is an equivalence relationship on the class of metric spaces, i.e.,*

1. *Every metric space is isometric to itself.*
2. *If (X, d_X) is isometric to (Y, d_Y), then (Y, d_Y) is isometric to (X, d_X).*
3. *If (X, d_X) is isometric to (Y, d_Y) and (Y, d_Y) is isometric to (Z, d_Z), then (X, d_X) is isometric to (Z, d_Z).*

Proof. For (1), observe that for any given metric space (X, d_X), the identity map id_X is an isometry $X \to X$. For (2), note that if $\phi : X \to Y$ is an isometry, then the inverse mapping $\phi^{-1} : Y \to X$ is also an isometry. For (3), suppose $\phi : X \to Y$ and $\psi : Y \to Z$ are isometries; then, the composite mapping $\psi \circ \phi : X \to Z$ is an isometry as well. □

Proposition 2.24. *Isometries are continuous functions.*

Proof. Suppose (X, d_X) and (Y, d_Y) are metric spaces and $\phi : X \to Y$ is an isometry. To see that ϕ is continuous, fix some $\varepsilon > 0$. Let $\delta = \varepsilon$. If $a, b \in X$ and $d_X(a, b) < \delta$, then $d_Y(\phi(a), \phi(b)) = d_X(a, b) < \delta = \varepsilon$. Thus, ϕ is continuous. □

Example 2.25. If two metric spaces are isometric, then they have the same cardinality (because isometries are bijections). For example, $\mathbb{Q}$ and $\mathbb{R}$ are not isometric because $|\mathbb{Q}| < |\mathbb{R}|$. ☕

Example 2.26. The real line $\mathbb{R}$ is not isometric to the plane $\mathbb{R}^2$. To see this, observe that $\mathbb{R}^2$ contains three points all at distance 1 from each other (the corners of a unit equilateral triangle), but $\mathbb{R}$ does not contain any such triple of points. This shows the two spaces are not isometric because it is a "metric" property the two spaces do not share, and isometries preserve all such properties. More precisely, if $\phi : \mathbb{R}^2 \to \mathbb{R}$ were an isometry, then taking a, b, c to be the corners of a unit equilateral triangle in $\mathbb{R}^2$, their $\phi(a)$, $\phi(b)$, $\phi(c)$ would be three points in $\mathbb{R}$ with $|\phi(a) - \phi(b)| = |\phi(a) - \phi(c)| = |\phi(b) - \phi(c)| = 1$, which is absurd. ☕

Example 2.27. The metric spaces $(0,1)$ and $(0,2)$, considered as subspaces of $\mathbb{R}$ with the usual metric, are not isometric. To see this, let $\phi:(0,2)\to(0,1)$ be any function. Now, $1/2, 3/2 \in (0,2)$, and $d(1/2, 3/2) = |1/2 - 3/2| = 1$. But, of course, $d(a,b) < 1$ for all $a, b \in (0,1)$. Thus, $d(1/2, 3/2) > d(\phi(1/2), \phi(3/2))$, and this means ϕ is not an isometry. ☕

In this last example, although $(0,2)$ and $(0,1)$ are not isometric, they certainly seem very similar. Informally, one can think of $(0,2)$ as just a stretched out copy of $(0,1)$, different, but very similar. The function $x \mapsto \frac{1}{2}x$ is not an isometry $(0,2)\to(0,1)$, but it is a bijection that seems to preserve a lot of significant features between these two metric spaces.

With metric spaces, we have not just one, but two useful notions of when two spaces are "essentially the same." One is the notion of isometry, which we have seen already and which preserves the metric completely from one space to the other. The other, weaker notion is called a homeomorphism, and it preserves open/closed sets.

Definition 2.28. Suppose (X, d_X) and (Y, d_Y) are metric spaces. A *homeomorphism* is a bijection $\phi: X \to Y$ such that both ϕ and ϕ^{-1} are continuous. If there is a homeomorphism $X \to Y$, then we say (X, d_X) and (Y, d_Y) are *homeomorphic*. ☕

Proposition 2.29. *Every isometry is a homeomorphism.*

Proof. Suppose $\phi: X \to Y$ is an isometry. Then, ϕ is a bijection by definition and is continuous by Proposition 2.24. As pointed out in the proof of Proposition 2.23(2), ϕ^{-1} is also an isometry, so ϕ^{-1} is also continuous, again by Proposition 2.24. □

Example 2.30. Every isometry is a homeomorphism, but not every homeomorphism is an isometry. The metric spaces $(0,2)$ and $(0,1)$, with their usual metrics, are not isometric (Example 2.27). Nonetheless, the map $x \mapsto \frac{1}{2}x$ is a homeomorphism $(0,2)\to(0,1)$. ☕

This proposition and example express what was meant by saying the notion of homeomorphism is "weaker" than that of isometry. Two spaces being isometric implies they are homeomorphic, but being homeomorphic does not imply being isometric. This is like the notions of similarity and congruence in geometry: Congruent figures are always similar, but two figures can be similar without being congruent. Essentially, every geometric property of a shape is preserved under congruence, but many important properties are also preserved by similarity. In the same way, we shall see that essentially every property of a metric space is preserved under isometry, and many important properties are preserved by homeomorphisms.

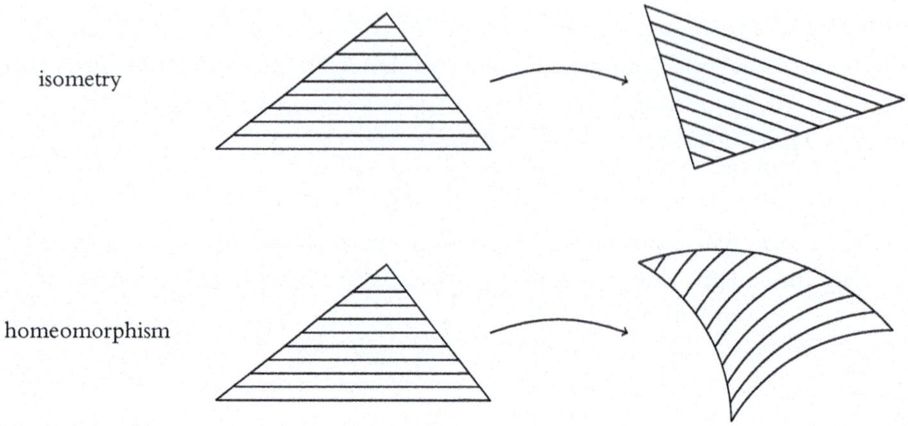

Just like isometries give rise to an equivalence relation on the class of metric spaces, homeomorphisms do, too.

Proposition 2.31. *The relationship "is homeomorphic to" is an equivalence relationship on the class of metric spaces. That is,*

1. *Every metric space is homeomorphic to itself.*
2. *If (X, d_X) is homeomorphic to (Y, d_Y), then (Y, d_Y) is homeomorphic to (X, d_X).*
3. *If (X, d_X) is homeomorphic to (Y, d_Y) and (Y, d_Y) is homeomorphic to (Z, d_Z), then (X, d_X) is homeomorphic to (Z, d_Z).*

Proof. The proof of this proposition is essentially identical to the proof of Proposition 2.23: (1) holds because id_X is always a homeomorphism from a metric space X to itself, (2) holds because the inverse of a homeomorphism is a homeomorphism, and (3) holds because the composition of two homeomorphisms is a homeomorphism. □

Connectedness and the Intermediate Value Theorem

In Chap. 1, we introduced the Axiom of Completeness, one of our two fundamental assumptions about $\mathbb{R}$, the one that distinguishes it from $\mathbb{Q}$ and says, roughly, that $\mathbb{R}$ has no holes in it. In this section, we develop an alternative version of this intuition that $\mathbb{R}$ has no holes. This second version says $\mathbb{R}$ is *connected.* This will allow us to prove the Intermediate Value Theorem, the main result of this section, which expresses the intuition that the graph of a continuous function has no holes in it.

Definition 2.32. Suppose (X, d) is a metric space, and $A \subseteq X$. A is *disconnected* if there are two open sets $U, V \subseteq X$ such that

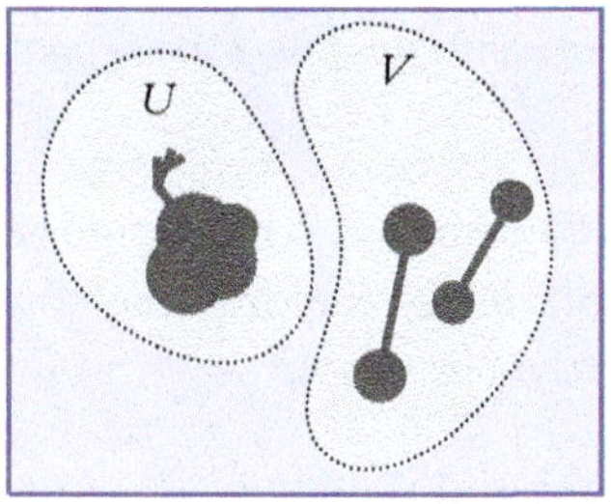

- $U \cap A \neq \emptyset$ and $V \cap A \neq \emptyset$,
- $U \cup V \supseteq A$, and
- $U \cap V \cap A = \emptyset$.

In this case, we say that the pair (U, V) is a *separation* of A or that the sets U and V form a separation of A. On the other hand, A is *connected* if it is not disconnected. ☕

Taking $A = X$ in this definition tells us what it means to say a metric space X is connected or disconnected. Be warned that Definition 2.32 is a slightly unusual way of defining when a subset of a metric space is disconnected. The usual definition first says what it means for a space X to be connected (the $A = X$ case of our definition) and then states that $A \subseteq X$ is connected if it is a connected space when viewed as a subspace of X. We prove in the homework exercises that these two ways of defining the connectedness of a subspace of a metric space are equivalent.

Example 2.33. In the metric space $\mathbb{R}$, the set $\mathbb{Q}$ is disconnected. For example, it is not hard to check that the two sets

$$U = (-\infty, \sqrt{2}) \qquad \text{and} \qquad V = (\sqrt{2}, \infty)$$

are a separation of $\mathbb{Q}$. Of course, there are other separations of $\mathbb{Q}$. In general, a disconnected set may have many different separations showing it is disconnected. ☕

The following theorem can be paraphrased as saying "continuous functions map connected sets to connected sets."

Theorem 2.34. *Suppose X and Y are metric spaces and f is a continuous function $X \to Y$. If A is a connected subset of X, then $f[A]$ is a connected subset of Y.*

Proof. Let us prove the contrapositive. Fix $A \subseteq X$, and suppose that $f[A]$ is a disconnected subset of Y. Let U and V be open subsets of Y forming a separation of $f[A]$. Looking at the three parts of the definition of "separation" in Definition 2.32, and taking inverse images, we get (1) $f^{-1}[U] \cap A \neq \emptyset$ and $f^{-1}[V] \cap A \neq \emptyset$, (2) $f^{-1}[U] \cap f^{-1}[V] \cap A = \emptyset$, and (3) $f^{-1}[U] \cup f^{-1}[V] \supseteq A$. And, because f is continuous, $f^{-1}[U]$ and $f^{-1}[V]$ are open subsets of X. Thus, $f^{-1}[U]$ and $f^{-1}[V]$ form a separation of A, so that A is disconnected. □

Corollary 2.35. *If X and Y are homeomorphic metric spaces, then either they are both connected or both not.*

Proof. Homeomorphisms are continuous, so if X and Y are homeomorphic, then each is a continuous image of the other. By the previous theorem, this implies that if one of them is connected, then the other is, too. □

This corollary can be paraphrased as saying that connectedness is preserved by homeomorphisms. A property that is preserved by homeomorphisms is sometimes called a "topological" property, and connectedness is one example. This corollary shows that connectedness is preserved by isometries, as well, because isometries are special kinds of homeomorphisms.

Definition 2.36. Given $x, y, z \in \mathbb{R}$, we say that z is *between* x and y if either $x < z < y$ or $y < z < x$. A set $I \subseteq \mathbb{R}$ is *convex* if, for any $x, y \in I$, all the points between x and y are also in I. A convex subset of $\mathbb{R}$ is sometimes also called an *interval*, with the convex sets of size 0 or 1 (either $\emptyset$ or a singleton) being referred to as *degenerate* intervals. ☕

There are 11 varieties of convex subsets of $\mathbb{R}$: There are nine different kinds of non-degenerate intervals (depending on whether the interval is bounded closed, bounded open, or unbounded, on the right or on the left) plus the two kinds of degenerate intervals. Specifically, the convex subsets of $\mathbb{R}$ are

- bounded intervals: $[a, b]$, $[a, b)$, $(a, b]$, and (a, b) are convex for every $a, b \in \mathbb{R}$ with $a < b$.
- unbounded intervals: $[a, \infty)$, (a, ∞), $(-\infty, a]$, and $(-\infty, a)$ are convex for every $a \in \mathbb{R}$, and $\mathbb{R}$ itself is convex.
- degenerate intervals: $\emptyset$ is convex, and $\{a\}$ is convex for every $a \in \mathbb{R}$.

Theorem 2.37. *A subset of $\mathbb{R}$ is connected if and only if it is convex.*

Proof. Suppose $A \subseteq X$ and A is not convex. This means that there are points $x, y, z \in \mathbb{R}$ such that z is between x and y, and $x, y \in A$, but $z \notin A$. Let $U = (-\infty, z)$ and $V = (z, \infty)$. Then, $U \cap A$ and $V \cap A$ are both non-empty (one contains x and the other y), $U \cap V = \emptyset$, and $U \cup V = \mathbb{R} \setminus \{z\} \supseteq A$. Thus, U and V form a separation of A, which means A is disconnected.

For the reverse implication, we prove the contrapositive: If A is disconnected, then it is not convex. Suppose $A \subseteq \mathbb{R}$ is disconnected, and let U and V be a separation of A. Pick $x \in U$ and $y \in V$ and, without loss of generality, let us assume $x < y$. Note that $U \cap [x, y]$ is non-empty (it contains x) and bounded above (by y). Let $z = \sup(U \cap [x, y])$. Clearly, $x \leq z \leq y$.

Observe $y \notin U$ because $y \in V \cap A$ and $U \cap V \cap A = \emptyset$. We claim $z \notin U$. Aiming for a contradiction, suppose instead that $z \in U$. Then, $z \neq y$, which means $z < y$. Because

U is open, there is some $\varepsilon > 0$ with $(z-\varepsilon, z+\varepsilon) \subseteq U$. As $y \notin U$, this implies $y \geq z+\varepsilon$, and therefore $z+\varepsilon/2 \in U \cap [x, y]$. This contradicts our definition of z.

We now consider two cases: $z \notin V$ and $z \in V$. Suppose first that $z \notin V$. In this case, $z \neq y$ (because $y \in V$), and the previous paragraph shows $z \notin U$. Hence, $z \notin U \cup V \supseteq A$, so $z \notin A$. Since $x, y \in A$, this means $z \neq x, y$, so $x < z < y$. So z is between x, and y and $z \notin A$, which shows A is not convex.

Suppose next that $z \in V$. Because V is open, there is some $\varepsilon > 0$ such that $(z-\varepsilon, z+\varepsilon) \subseteq V$. Because $z = \sup(U \cap [x, y])$, there is some $z' \in U \cap [x, y]$ with $z - \varepsilon < z' \leq z$. Then, $z' \in U \cap V$, which implies $z' \notin A$ (because $U \cap V \cap A = \emptyset$). Furthermore, $x \notin V$ (because $U \cap V \cap A = \emptyset$ and $x \in U \cap A$) and $x \leq z$; since $(z-\varepsilon, z+\varepsilon) \subseteq V$, this implies $x \leq z - \varepsilon < z'$. Also, $z' \in U$ and $z \notin U$, so $z' < z \leq y$. Thus, $x < z' < y$ and $z' \notin A$, which shows A is not convex. □

The following result, which follows fairly easily from Theorem 2.37, is known as the *Intermediate Value Theorem*, or the *IVT* for short.

Theorem 2.38 (Intermediate Value Theorem). *Suppose $a, b \in \mathbb{R}$ with $a < b$, and suppose f is a continuous function $[a, b] \to \mathbb{R}$. For every $y \in \mathbb{R}$ that is between $f(a)$ and $f(b)$, there is some x between a and b such that $f(x) = y$.*

Proof. Suppose f is a continuous function $[a, b] \to \mathbb{R}$, and let y be between $f(a)$ and $f(b)$. Because $[a, b]$ is connected, Theorem 2.34 implies that the image of $[a, b]$ under f, namely, $A = f\big[[a, b]\big]$, is also connected. By the previous theorem, this means that A is convex. Because $f(a), f(b) \in A$ and y is between $f(a)$ and $f(b)$, the convexity of A implies $y \in A$. By definition, this means there is some $x \in [a, b]$ such that $f(x) = y$. □

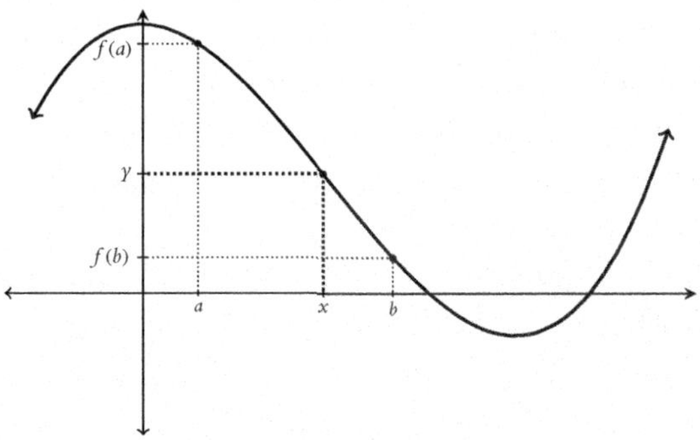

Example 2.39. Recall our painstaking proof from the previous chapter that there is a real number, $\sqrt{2}$, whose square is 2. This kind of thing can be deduced immediately from the IVT.

Later, in Chap. 4, when we study continuous functions $\mathbb{R} \to \mathbb{R}$ more closely, we shall see that all polynomial functions are continuous. For the moment, let us take this for granted. Consider the function $y = x^2$.

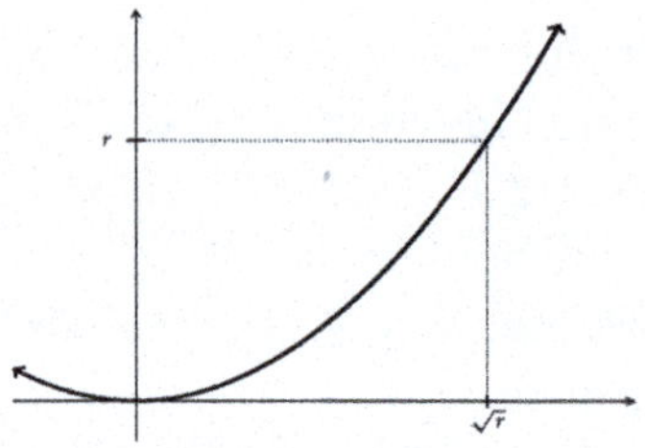

This is a polynomial function, hence continuous. Fix $r > 0$. Let $a = 0$, and let b be some number large enough that $b^2 > r$ (e.g., take $b \in \mathbb{N}$ with $b > r$, by the Archimedean property). Then, f is continuous on $[a, b]$, and r is between $f(0) = 0$ and $f(b) = b^2 > r$. By the IVT, it follows that there is some $x \in (a, b)$ such that $x^2 = r$. Because $r > 0$ was arbitrary, this shows that every positive number has a square root.

Using similar arguments, the IVT can be used to show that every positive real number has a 4th root, a 6th root, etc., and every real number has a 3rd root, a 5th root, etc.

Compactness

We now turn our attention to the notion of compactness, which is the topic of the rest of this chapter. The definition of compactness is often difficult to understand at first, but it is worth the effort. The idea of compactness plays an important role in analysis, and it will reappear in almost every subsequent chapter.

Definition 2.40. Let X be a metric space and let $A \subseteq X$.

- If $\mathcal{U}$ is a collection of subsets of X, we say that $\mathcal{U}$ *covers* A if $\bigcup \mathcal{U} \supseteq A$.
- An *open cover* of A is a collection of open subsets of X that covers A.
- Suppose $\mathcal{U}$ is a cover of A. A *finite subcover* of $\mathcal{U}$ is a finite $\mathcal{V} \subseteq \mathcal{U}$ such that $\mathcal{V}$ covers A. If some such $\mathcal{V}$ exists, we say that $\mathcal{U}$ *has a finite subcover.*
- A is *compact* if every open cover of A has a finite subcover.

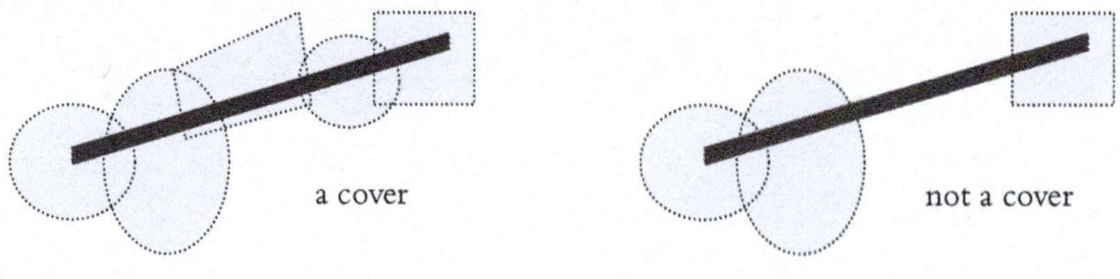

Just like with connectedness, taking $A = X$ in the previous definition tells us what it means to say that a metric space X is compact.

Example 2.41. Let $X = \mathbb{R}$, $A = [0, 1)$, and $\mathcal{U} = \{U \subseteq \mathbb{R} : U \text{ is open}\}$. Then, $\mathcal{U}$ is an open cover of A because $\bigcup \mathcal{U} = \mathbb{R} \supseteq A$. Notice that $\mathcal{U}$ has a finite subcover: For example, $\mathcal{V} = \{(-1, 1)\}$ and $\mathcal{W} = \{(-\frac{1}{2}, \frac{1}{2}), (0, \frac{2}{3}), (0, 2)\}$ are two different finite subcovers of $\mathcal{U}$.

However, this does not mean that A is compact. Compactness requires that *every* open cover of A to have a finite subcover. In fact, A is not compact: $\mathcal{U} = \{(-1, 1 - \frac{1}{n}) : n \in \mathbb{N}\}$ is also an open cover of A, and this open cover does not have a finite subcover.

Example 2.42. Let $X = \mathbb{R}$, and let $A = \mathbb{N}$. We claim that A is not compact. To see this, let $\mathcal{U} = \{(-1, n) : n \in \mathbb{N}\}$. Note that $\mathcal{U}$ is an open cover for A, but it has no finite subcover.

Proposition 2.43. *Every finite subset of a metric space is compact.*

Proof. Let X be a metric space, and let $F \subseteq X$ be finite. Let $\mathcal{U}$ be an open cover of F. This means that for every $x \in F$, there are some members of $\mathcal{U}$ containing x. For each $x \in F$, choose one $U_x \in \mathcal{U}$ with $x \in U_x$. Then, $\{U_x : x \in F\}$ is a finite subset of $\mathcal{U}$, and it covers F. Thus, $\mathcal{U}$ has a finite subcover. As $\mathcal{U}$ was arbitrary, this shows F is compact. □

Example 2.44. For an example of an infinite compact set, let $X = \mathbb{R}$ and let $A = \{\frac{1}{n} : n \in \mathbb{N}\} \cup \{0\}$.

To see that this set is compact, let $\mathcal{U}$ be an arbitrary open cover of A. There is some $U \in \mathcal{U}$ such that $0 \in U$. But because U is open and $U \in \mathcal{U}$, there is some $\varepsilon > 0$ such that $B_\varepsilon(0) = (-\varepsilon, \varepsilon) \subseteq U$. By the Archimedean property, there is some $N \in \mathbb{N}$ such that $1/N < \varepsilon$. For each $n \leq N$, choose some $U_{1/n} \in \mathcal{U}$ such that $1/n \in U_{1/n}$. Then, $\{U_{1/n} : n \leq N\} \cup \{U\}$ is an open cover for A. As $\mathcal{U}$ was arbitrary, this shows A is compact.

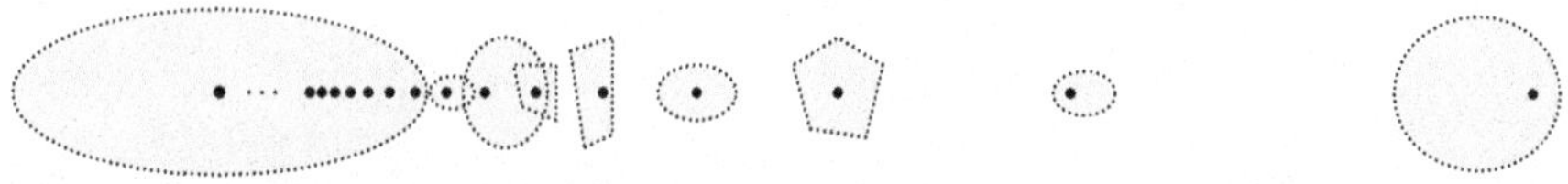

The next theorem can be summarized as saying "closed subsets of compact sets are compact."

Theorem 2.45. *Let X be a metric space, and suppose that A is a compact subset of X. If B is closed and $B \subseteq A$, then B is also compact.*

Proof. Let $\mathcal{U}$ be an open cover of $\mathcal{B}$. Because B is closed, $X \setminus B$ is open. Let $\mathcal{U}^+ = \mathcal{U} \cup \{X \setminus \mathcal{B}\}$. (It is possible that $X \setminus B \in \mathcal{U}$ already, in which case we would have $\mathcal{U}^+ = \mathcal{U}$.) Then, $\mathcal{U}^+$ is a collection of open sets, and

$$\bigcup \mathcal{U}^+ = (X \setminus B) \cup \bigcup \mathcal{U} \supseteq (X \setminus B) \cup B = X \supseteq A.$$

Thus, $\mathcal{U}^+$ is an open cover of A. Because A is compact, there is a finite $\mathcal{V} \subseteq \mathcal{U}^+$ such that $\bigcup \mathcal{V} \supseteq A$. Let $\mathcal{V}^- = \mathcal{V} \setminus \{X \setminus B\}$. (It is possible that $X \setminus B \notin \mathcal{V}$ already, in which case we would have $\mathcal{V}^- = \mathcal{V}$.)

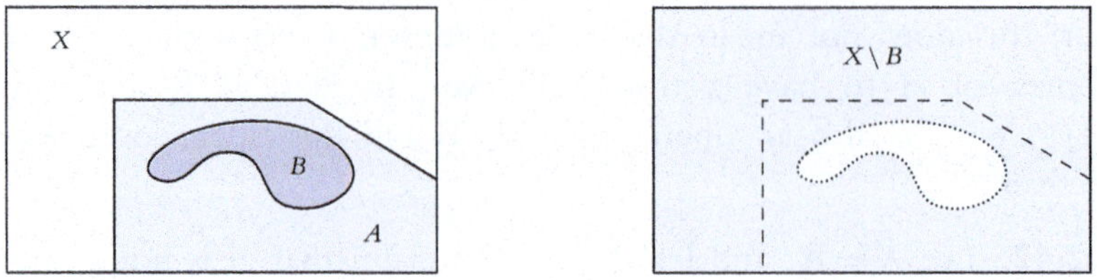

Notice that $\mathcal{V}^- \subseteq \mathcal{U}$. Also,

$$\bigcup \mathcal{V}^- \supseteq \left(\bigcup \mathcal{V}\right) \setminus (X \setminus B) = \left(\bigcup \mathcal{V}\right) \cap B = B,$$

so $\mathcal{V}^-$ covers B. Therefore, $\mathcal{U}$ has a finite subcover. As $\mathcal{U}$ was an arbitrary open cover of B, this shows that B is compact. □

The next theorem can be summarized as saying "continuous images of compact sets are compact." This is similar to Theorem 2.34, where we proved that continuous images of connected sets are connected.

Theorem 2.46. *Let X and Y be metric spaces, and $f : X \to Y$ continuous. If A is a compact subset of X, then $f[A]$ is a compact subset of Y.*

Proof. Let $\mathcal{U}$ be an open cover of $f[A]$. Define $\mathcal{U}^{\leftarrow} = \{f^{-1}[U] : U \in \mathcal{U}\}$. Because $\mathcal{U}$ is a collection of open subsets of Y and f is continuous, $\mathcal{U}^{\leftarrow}$ is a collection of open subsets of X. Furthermore, if $x \in A$, then there is some $U \in \mathcal{U}$ such that $f(x) \in U$, which implies $x \in f^{-1}[U]$. Hence, $A \subseteq \bigcup \mathcal{U}^{\leftarrow}$.

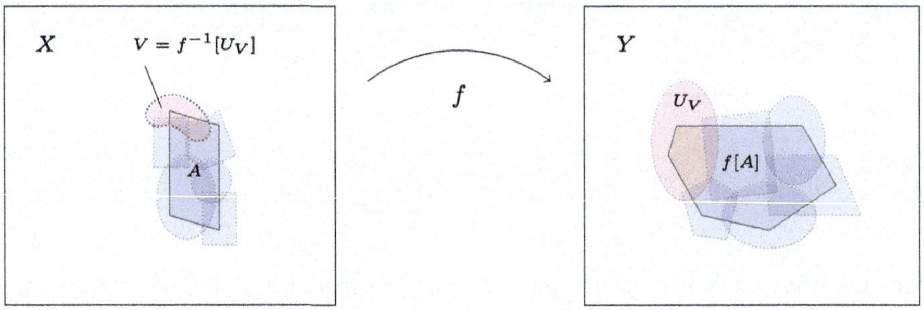

Because A is compact, this implies there is a finite $\mathcal{V} \subseteq \mathcal{U}^{\leftarrow}$ such that $A \subseteq \bigcup \mathcal{V}$. For each $V \in \mathcal{V}$, there is some $U_V \in \mathcal{U}$ such that $V = f^{-1}[U_V]$. (If f is surjective, then there is exactly one set with this property, but this need not be true for nonsurjective functions.) Let $\mathcal{V}^{\rightarrow} = \{U_V : V \in \mathcal{V}\}$.

By design, $\mathcal{V}^{\rightarrow} \subseteq \mathcal{U}$. Furthermore, $\mathcal{V}^{\rightarrow}$ is finite: It has the same size as $\mathcal{V}$. Finally, if $y \in f[A]$ then there is some $x \in A$ such that $y = f(x)$. This implies there is some $V \in \mathcal{V}$ such that $x \in V$, which implies $y \in U_V$. Therefore, $\mathcal{V}^{\rightarrow}$ covers $f[A]$, which shows that $\mathcal{U}$ has a finite subcover. As $\mathcal{U}$ was an arbitrary open cover of $f[A]$, this shows A is compact. □

Corollary 2.47. *If X and Y are homeomorphic metric spaces, then either they are both compact or both not. In other words, compactness is preserved by homeomorphisms.*

Proof. Homeomorphisms are continuous, so if X and Y are homeomorphic, then each is a continuous image of the other. By the previous theorem, this implies that if one of them is compact, then the other is, too. □

This corollary shows that compactness is preserved by isometries because isometries are just special kinds of homeomorphisms.

Definition 2.48. Let X be a metric space, and let $A \subseteq X$. We say that A is *bounded* if there is some $x \in X$ and some $r > 0$ such that $A \subseteq B_r(x)$.

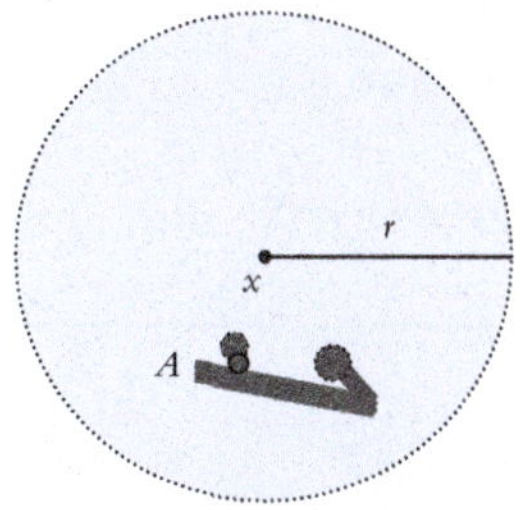

The next proposition shows that if we replace "some" with "every" in Definition 2.48, it does not change the meaning.

Proposition 2.49. *Let X be a metric space. A set $A \subseteq X$ is bounded if and only if for every $x \in X$ there is some $r > 0$ such that $A \subseteq B_r(x)$.*

Proof. The "if" direction of the theorem follows immediately from the definition of boundedness. For the "only if" direction, suppose $A \subseteq X$ is bounded. This means there is some particular $x_0 \in X$ and some $r_0 > 0$ such that $A \subseteq B_{r_0}(x_0)$. Let $x \in X$, and let $r = r_0 + d(x, x_0)$. If $a \in A$, then

$$d(x, a) \leq d(x, x_0) + d(x_0, a) \leq d(x, x_0) + r_0 = r,$$

and thus $a \in B_r(x)$. Thus, $A \subseteq B_r(x)$. □

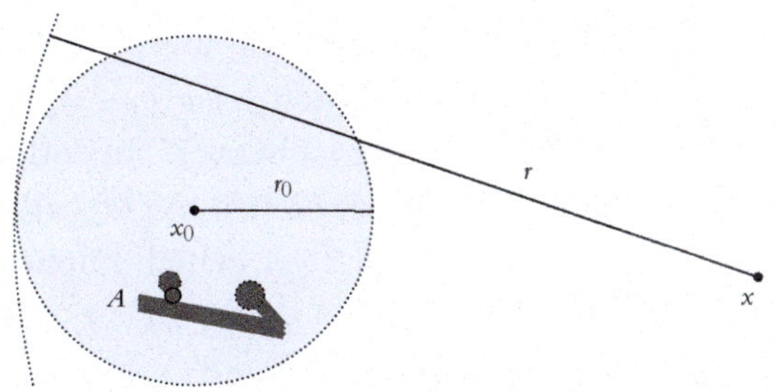

Theorem 2.50. *Every compact subset of a metric space is bounded.*

Proof. Let X be a metric space, and suppose $A \subseteq X$ is compact. Fix a point $x \in X$, and let $\mathcal{U} = \{B_r(x) : r > 0\}$. Because A is compact, $\mathcal{U}$ has a finite subcover, say $\mathcal{V}$. In other words, there are some $r_1, r_2, \ldots, r_n > 0$ such that $\mathcal{V} = \{B_{r_1}(x), B_{r_2}(x), \ldots, B_{r_n}(x)\}$ covers A. Let $R = \max\{r_1, r_2, \ldots, r_n\}$. Then, $B_{r_i}(x) \subseteq B_R(x)$ for all $i \leq n$, which implies $\bigcup \mathcal{V} \subseteq B_R(x)$. Because $\mathcal{V}$ covers A, this means $A \subseteq B_R(x)$, and therefore A is bounded. □

Theorem 2.51. *Every compact subset of a metric space is closed.*

Proof. Let X be a metric space, and suppose $A \subseteq X$ is compact. To show A is closed, we will prove that its complement $X \setminus A$ is open.

Let $x \in X \setminus A$. For each $r > 0$, define

$$U_r(x) = \{y \in X : d(x, y) > r\}.$$

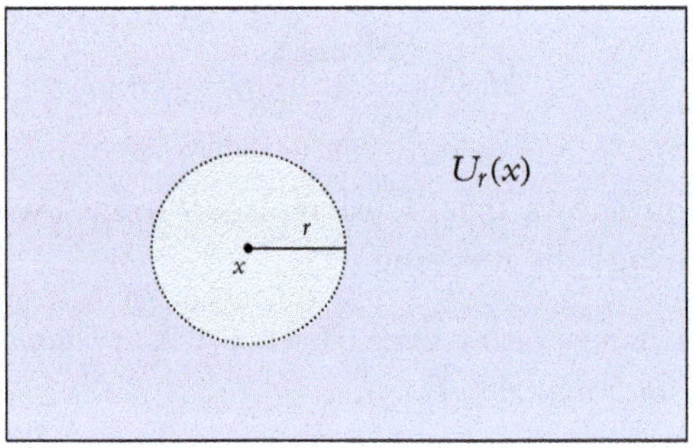

We claim that $U_r(x)$ is open for every $r > 0$. To see this, fix $r > 0$ and let $y \in U_r(x)$. Let $\delta = d(x, y) - r$. If $z \in B_\delta(y)$, then

$$d(x, z) \geq d(x, y) - d(y, z) \geq d(x, y) - \delta = r,$$

which means $z \in U_r(x)$. Thus, $B_\delta(y) \subseteq U_r(x)$, and as y was an arbitrary point of $U_r(x)$, this shows $U_r(x)$ is open.

Let $\mathcal{U} = \{U_r(x) : r > 0\}$. By the previous paragraph, this is a collection of open subsets of X. Furthermore, $\mathcal{U}$ covers A because $x \notin A$ and $\bigcup \mathcal{U} = \{y \in X : d(x, y) > 0\} = X \setminus \{x\}$.

Because A is compact, $\mathcal{U}$ has a finite subcover. In other words, there are some $r_1, r_2, \ldots, r_n > 0$ such that $\mathcal{V} = \{U_{r_1}(x), U_{r_2}(x), \ldots, U_{r_n}(x)\}$ covers A. Let $R = \min\{r_1, r_2, \ldots, r_n\}$. Then, $U_R(x) \supseteq U_{r_i}(x)$ for all $i \leq n$, which implies $U_R(x) \supseteq \bigcup \mathcal{V} \supseteq A$.

Taking complements, this shows that $B_R(x) \subseteq X \setminus U_R(x) \subseteq X \setminus A$. Because x was arbitrary, this shows that $X \setminus A$ is open, and A is closed. □

The Heine–Borel Theorem and the Extreme Value Theorem

In this section we prove two important theorems: the *Heine–Borel Theorem* and the *Extreme Value Theorem* (or the EVT, for short).

In the previous section, we proved that compact sets are always closed and bounded. The Heine–Borel Theorem asserts that for subsets of $\mathbb{R}$, the converse is also true: Sets that are closed and bounded are compact. Thus, the Heine–Borel Theorem characterizes the compact subsets of $\mathbb{R}$ as precisely those sets that are both closed and bounded. It is important to note that while "closed and bounded" is equivalent to compactness in $\mathbb{R}$, this is not true for every metric space.

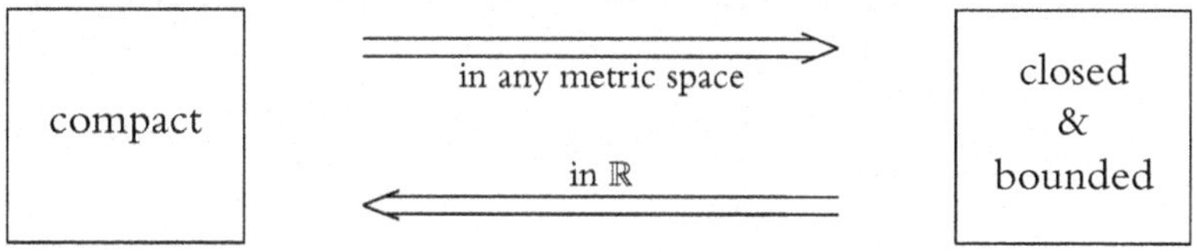

Example 2.52. Let $X = \mathbb{R} \setminus \{0\}$, and for any $x, y \in X$ let $d(x, y) = |x - y|$ (the usual metric). Then, (X, d) is a metric space. In this metric space, there are sets that are both closed and bounded, but are not compact. For example, let $A = (0, 1]$. This is a closed and bounded subset of X. (It is not closed as a subset of $\mathbb{R}$, but it is closed in X.) $\mathcal{U} = \left\{(\frac{1}{n}, 2) : n \in \mathbb{N}\right\}$ is an open cover of A that does not have a finite subcover, so A is not compact.

For a more extreme example, let $X = \mathbb{R}$, and for any $x, y \in X$, let

$$\delta(x, y) = \begin{cases} 0 & \text{if } x = y, \\ 1 & \text{if } x \neq y \end{cases}$$

(the discrete metric). Then, (X, δ) is a metric space. In this metric space, every subset is both closed and bounded! But the only compact subsets of this space are the finite sets.

Some of what we proved in the previous section, Theorems 2.45, 2.50, and 2.51, could be counted as steps toward a proof of the Heine–Borel Theorem. The next theorem is the last major step in this direction. It also gives us our first example of an uncountable compact subset of $\mathbb{R}$.

Theorem 2.53. *If $a, b \in \mathbb{R}$ and $a < b$, then $[a, b]$ is compact.*

Proof. Let $\mathcal{U}$ be an open cover of A. Define

$$C = \left\{x \in [a, b] : \text{There is some finite } \mathcal{V} \subseteq \mathcal{U} \text{ that covers } [a, x]\right\}.$$

Our goal is to show that $b \in C$.

Because $\mathcal{U}$ covers $[a, b]$, there is some $U \in \mathcal{U}$ with $a \in U$. This implies that $a \in C$ (which can be seen by taking $\mathcal{V} = \{U\}$ in the definition of C). In particular, $C \neq \emptyset$. Also C is bounded above because $C \subseteq [a, b]$. By the Axiom of Completeness, C has a supremum: Let $c = \sup C$.

We claim that $c \in C$ and $c = b$. To see this, first note that $C \subseteq [a, b]$ implies $\sup C \in [a, b]$. Because $\mathcal{U}$ covers $[a, b]$, this means there is some $V \in \mathcal{U}$ with $c \in V$. As V is open, there is some $\varepsilon > 0$ such that $B_\varepsilon(c) = (c - \varepsilon, c + \varepsilon) \subseteq V$. Because $c - \varepsilon < c = \sup C$, $c - \varepsilon$ is not an upper bound for C, hence there is some $x \in C$ with $x > c - \varepsilon$. This means there is some finite $\mathcal{V} \subseteq \mathcal{U}$ that covers $[a, x]$. But then, $\mathcal{V} \cup \{V\}$ is also a finite subset of $\mathcal{U}$, and it covers $[a, x] \cup (c - \varepsilon, c + \varepsilon) = [a, c + \varepsilon)$. It follows that every $x \in [a, b] \cap [a, c + \varepsilon)$ is in C. Because $c = \sup C$, this implies $c = b$ and $b \in C$.

Because $b \in C$, $\mathcal{U}$ has a finite subcover. As $\mathcal{U}$ was an arbitrary open cover of $[a, b]$, this shows that $[a, b]$ is compact. □

Theorem 2.54 (Heine–Borel Theorem). *A subset of $\mathbb{R}$ is compact if and only if it is both closed and bounded.*

Proof. The "only if" direction of the theorem was proved already as Theorems 2.50 and 2.51. For the "if" direction, suppose $A \subseteq \mathbb{R}$ is closed and bounded. Because A is bounded, there is some $R > 0$ and some $x \in \mathbb{R}$ such that $A \subseteq B_R(x) = (x - R, x + R)$. In particular, $A \subseteq [x - R, x + R]$. The set $[x - R, x + R]$ is compact by the previous theorem, and closed subsets of compact sets are compact by Theorem 2.45. Thus, A is compact. □

The Extreme Value Theorem is proved using the ideas of compactness and connectedness working together. The EVT can be summarized as saying "a continuous function defined on a compact set attains both its minimum and its maximum."

Lemma 2.55. *Suppose $C \subseteq \mathbb{R}$ is closed. If* $\sup C$ *exists, then* $\sup C \in C$. *Similarly, if* $\inf C$ *exists, then* $\inf C \in C$.

Proof. Suppose C is non-empty and bounded above, so that $\sup C$ exists. Let $\varepsilon > 0$. By the definition of a supremum, $\sup C - \varepsilon$ is not an upper bound for C. Thus, there is some $c \in C$ with $c > \sup C - \varepsilon$. Therefore, every ε-ball around $\sup C$ contains points of C, which means $\sup C \in \overline{C} = C$ by Theorem 2.13. The argument for $\inf C$ is essentially the same. □

Theorem 2.56 (Extreme Value Theorem). *Suppose A is a non-empty compact subset of some metric space X, and f is a continuous function $X \to \mathbb{R}$. Then,* $\inf f[A]$ *and* $\sup f[A]$ *both exist, and furthermore, there are $s, t \in A$ such that $f(s) = \inf f[A]$ and $f(t) = \sup f[A]$.*

Proof. Recall that the continuous image of a compact set is compact, by Theorem 2.46. Therefore, $f[A]$ is a compact subset of $\mathbb{R}$. By the results in the previous section, namely Theorems 2.50 and 2.51, $f[A]$ is closed and bounded. Because $f[A]$ is bounded, it has a supremum and an infimum. Because $f[A]$ is closed, Lemma 2.55 implies $\inf f[A], \sup f[A] \in f[A]$. By the definition of $f[A]$, this means there are $s, t \in A$ such that $f(s) = \inf f[A]$ and $f(t) = \sup f[A]$. □

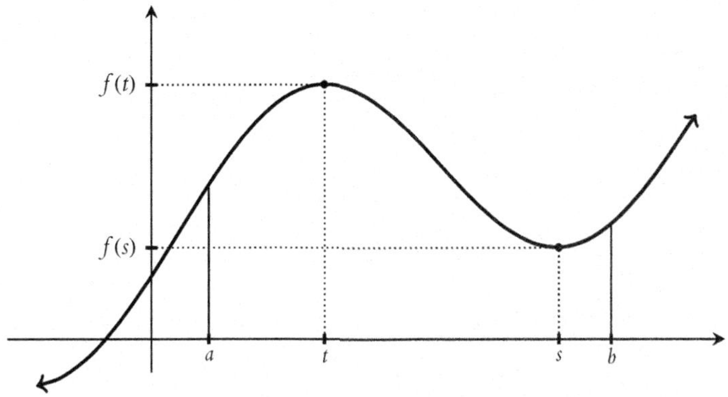

The Cantor space and the Cantor set

The final section of this chapter is devoted to describing an interesting metric space known as the Cantor space, and a related subset of $\mathbb{R}$ known as the Cantor set. This subset of $\mathbb{R}$ is uncountable, closed, and bounded (hence compact), yet it contains no intervals.

Definition 2.57. Let $2^{\mathbb{N}}$ denote the set of all functions $\mathbb{N} \to \{0, 1\}$ (which one can think of as sequences of 0's and 1's). Given $f, g \in 2^{\mathbb{N}}$, define

$$d(f, g) = \begin{cases} 1/2^n & \text{if } f(i) = g(i) \text{ for all } i < n \text{ but } f(n) \neq g(n), \\ 0 & \text{if } f = g. \end{cases}$$

In other words, if $f \neq g$, then f and g must disagree at some $n \in \mathbb{N}$, and $d(f, g)$ is defined as $1/2^n$ for the least such value of n.

It is not difficult to check that d is a metric on $2^{\mathbb{N}}$ (and doing so is a homework exercise). The metric space $(2^{\mathbb{N}}, d)$ is called the *Cantor space.*

This metric d is the "usual" metric on $2^{\mathbb{N}}$ (in the same sense that there is a "usual" metric on $\mathbb{R}$ or $\mathbb{Q}$). When we refer to $2^{\mathbb{N}}$ as a metric space, this is always the metric that is meant.

One may think of each $f \in 2^{\mathbb{N}}$ as corresponding to a branch through the following infinite tree, where the value of $f(n)$ determines whether our branch turns left ($f(n) = 0$) or right ($f(n) = 1$) at level n of the tree.

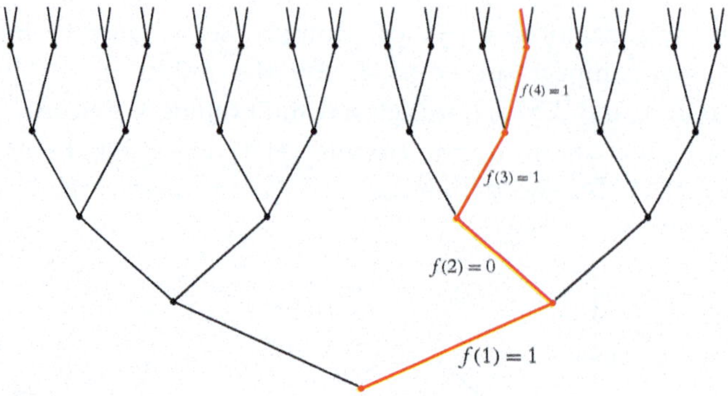

If $f \in 2^{\mathbb{N}}$ and $n \in \mathbb{N}$, then

$$B_{1/2^n}(f) = \left\{g \in 2^{\mathbb{N}} : g(i) = f(i) \text{ for all } i < n\right\}.$$

In other words, the $1/2^n$-ball around f consists of all the functions in $2^{\mathbb{N}}$ that agree with f on $\{1, 2, \ldots, n-1\}$. (This follows straightaway from our definition of d.)

The Cantor space looks rather different from the familiar Euclidean spaces $\mathbb{R}$ and $\mathbb{R}^n$. For example, every point of the ball just described is at its center, i.e., if $g \in B_{1/2^n}(f)$, then $B_{1/2^n}(g) = B_{1/2^n}(f)$. Furthermore, $B_r(f) = B_{1/2^n}(f)$ whenever $\frac{1}{2^{n-1}} < r \leq \frac{1}{2^n}$.

For each $n \in \mathbb{N}$ and each length-n sequence s of 0's and 1's (meaning that s is a function $\{1, 2, \ldots, n\} \to \{0, 1\}$), define

$$[\![s]\!] = \left\{g \in 2^{\mathbb{N}} : g(i) = s(i) \text{ for all } i \leq n\right\}.$$

Each set of this form is an open subset of $2^{\mathbb{N}}$: It is equal to $B_{1/2^{n-1}}(f)$ for any $f \in [\![s]\!]$. In our tree picture of $2^{\mathbb{N}}$, each node s of the tree represents a finite 0-1 sequence, and $[\![s]\!]$ is represented by all the branches of the tree passing through s. Observe that a node t is above a node s in the tree if and only if $[\![t]\!] \subseteq [\![s]\!]$.

To define the Cantor set, we first define an infinite sequence of sets $K_0, K_1, K_2, \ldots$, with each of these sets smaller than the previous ones. Afterward, the Cantor set is defined as the intersection $\bigcap_{n \in \mathbb{N}} K_n$. The sequence of K_n's is defined by recursion as follows.

To begin, let $K_0 = [0, 1]$. Then, K_1 is obtained from K_0 by removing the open interval $(\frac{1}{3}, \frac{2}{3})$ from K_0, so $K_1 = [0, \frac{1}{3}] \cup [\frac{2}{3}, 1]$. Next, K_2 is obtained from K_1 by removing the middle third of each of the two intervals comprising K_1, so $K_2 = [0, \frac{1}{9}] \cup [\frac{2}{9}, \frac{1}{3}] \cup [\frac{2}{3}, \frac{7}{9}] \cup [\frac{8}{9}, 1]$. Generally speaking, at stage n, suppose we have defined K_n already and that it is a disjoint union of 2^n closed intervals, each of length $\frac{1}{3^n}$. Then, K_{n+1} is formed from K_n by removing an open interval of length $\frac{1}{3^{n+1}}$ from the exact middle of each of the 2^n intervals in K_n, leaving 2^{n+1} closed intervals, each of length $\frac{1}{3^{n+1}}$.

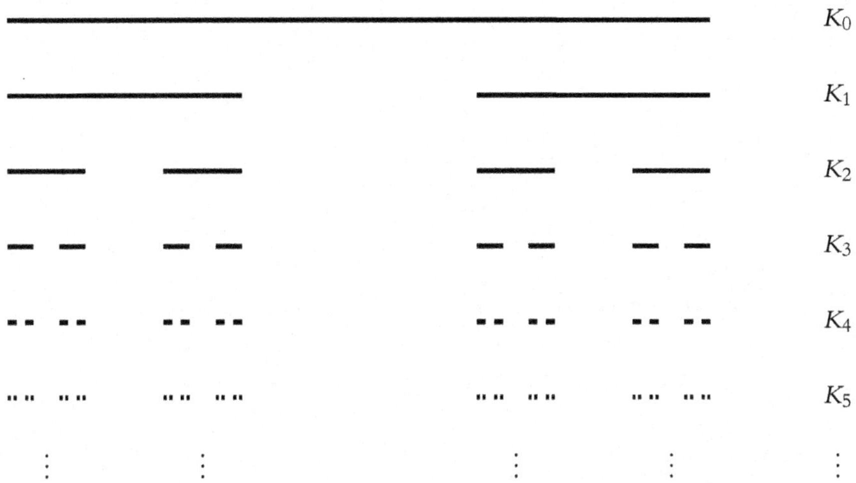

The Cantor set is $K = \bigcap_{n=0}^{\infty} K_n$. An alternative description of K is that it is the set of all $x \in [0, 1]$ such that x can be written in ternary decimal notation using just the digits 0 and 2, but not 1. For example, $\frac{1}{3} \in K$ and (as a ternary decimal) $\frac{1}{3} = .1 = .02222222\ldots$. (Roughly, this describes K because K_n is formed from K_{n-1} precisely by removing those members of K_{n-1} that require a 1 as the nth digit of their ternary decimal representation.)

Proposition 2.58. *The Cantor set K is closed and bounded, hence compact.*

Proof. It is clear that K is bounded since $K \subseteq K_0 = [0, 1]$. It is also not difficult to see that K_n is closed for every n because K_n is a union of finitely many closed intervals. Thus, K is the intersection of a collection of closed sets, and therefore closed. □

Proposition 2.59. *The Cantor set contains no non-empty open intervals. Equivalently, $\mathbb{R} \setminus K$ is dense in $\mathbb{R}$, i.e., if $a, b \in \mathbb{R}$ and $a < b$, there is some $c \in \mathbb{R} \setminus K$ with $a < c < b$.*

Proof. Fix $a, b \in \mathbb{R}$ with $a < b$. There is some $n \in \mathbb{N}$ such that $\frac{1}{3^n} < b - a$. The set K_n consists of 2^n disjoint intervals, each of length $\frac{1}{3^n}$. Therefore, $(a, b) \nsubseteq K_n$, and it follows that $(a, b) \nsubseteq K$. □

Our next goal is to prove that the Cantor set K (with the metric it inherits as a subspace of $\mathbb{R}$) is homeomorphic to the Cantor space $2^{\mathbb{N}}$. The idea of the proof is suggested by the similarity between the tree picture of $2^{\mathbb{N}}$ and the recursive construction of the sets $K_0, K_1, \ldots$ defining K.

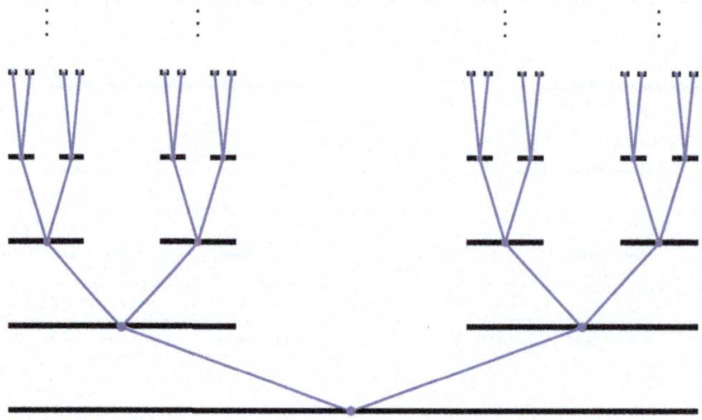

To prove K and $2^{\mathbb{N}}$ are homeomorphic, we make use of the *Nested Interval Property* of the reals, a close relative of the Heine–Borel Theorem.

Theorem 2.60 (the Nested Interval Property). *Let $I_0, I_1, I_2, I_3, \ldots$ be a sequence of non-empty closed bounded intervals in $\mathbb{R}$, with the property that $I_0 \supseteq I_1 \supseteq I_2 \supseteq I_3 \supseteq \ldots$. Then, $\bigcap_{n=1}^{\infty} I_n \neq \emptyset$.*

Proof. Suppose, aiming for a contradiction, that $\bigcap_{n=1}^{\infty} I_n = \emptyset$. For each n, let $U_n = \mathbb{R} \setminus I_n$. Notice that U_n is open because I_n is closed, and that $\bigcap_{n=1}^{\infty} I_n = \emptyset$ implies $\bigcup_{n=1}^{\infty} U_n = \mathbb{R}$ (by De Morgan's laws). Thus, $\mathcal{U} = \{U_n \colon n \in \mathbb{N}\}$ is an open cover of I_0. By the Heine–Borel Theorem, I_0 is compact. Consequently, there is some finite set $\{U_{n_1}, U_{n_2}, \ldots, U_{n_k}\} \subseteq \{U_n \colon n \in \mathbb{N}\}$ such that $\bigcup_{i=1}^{k} U_{n_i} \supseteq I_0$. Let $N = \max\{n_1, n_2, \ldots, n_k\}$, and note that $U_n \supseteq U_{n_i}$ for all $i \leq k$. Therefore, $I_0 \subseteq \bigcup_{i=1}^{k} U_{n_i} \subseteq U_N$. But this is nonsense because $\emptyset \neq I_N \subseteq I_0$ and U_N is the complement of I_N. □

Theorem 2.61. *The Cantor set is homeomorphic to the Cantor space.*

Proof. We begin by describing a bijection $\phi : 2^{\mathbb{N}} \to K$. The idea for this bijection is more or less captured by the picture, showing the correspondence between the tree representation of $2^{\mathbb{N}}$ and the recursive construction of K. That is, just as every $f \in 2^{\mathbb{N}}$ can be viewed as describing a path through our tree, picking out exactly one of the 2^n nodes at level n of the tree, f can also be viewed as describing a path through the K_n's, picking out exactly one of the 2^n intervals comprising each K_n.

Specifically, let us label the intervals of $K_1, K_2, K_3, \ldots$ via recursion as follows. At step 1, we have $K_1 = \left[0, \frac{1}{3}\right] \cup \left[\frac{2}{3}, 1\right]$, and we define

$$I_{\langle 0 \rangle} = \left[0, \tfrac{1}{3}\right] \qquad \text{and} \qquad I_{\langle 1 \rangle} = \left[\tfrac{2}{3}, 1\right].$$

Now, suppose that at step n of the recursion, we labeled each of the 2^n intervals comprising K_n as I_s, where s is some length-n sequence of 0's and 1's. There are 2^n such sequences, and we suppose our labeling assigns every sequence to one and only one interval. Given some length-n sequence s, the set K_{n+1} contains exactly two intervals that are subsets of I_s, one on the left and one on the right. Let us call these L_s and R_s. Then, define

$$I_{s^\frown 0} = L_s \qquad \text{and} \qquad I_{s^\frown 1} = R_s,$$

where $s^\frown i$ denotes the length-$(n+1)$ sequence obtained by putting an i at the end of the sequence s. For example, at step 2 of the recursion, we get

$$I_{\langle 0,0\rangle} = [0, \tfrac{1}{9}], \qquad I_{\langle 0,1\rangle} = [\tfrac{2}{9}, \tfrac{1}{3}], \qquad I_{\langle 1,0\rangle} = [\tfrac{2}{3}, \tfrac{7}{9}], \qquad I_{\langle 1,1\rangle} = [\tfrac{8}{9}, 1].$$

Having labeled all the intervals in all the K_n's this way, we are ready to define our bijection $\phi : 2^{\mathbb{N}} \to K$. Given $f \in 2^{\mathbb{N}}$, define $\phi(f)$ to be the unique point in the set $\bigcap_{n\in\mathbb{N}} I_{f\restriction\{1,2,\ldots,n\}}$. (Here, as before, we are interpreting the function $f \restriction \{1, 2 \ldots, n\}$ as a length-n sequence of 0's and 1's.)

First, let us check that ϕ is well-defined, i.e., that $\bigcap_{n\in\mathbb{N}} I_{f\restriction\{1,2,\ldots,n\}}$ contains exactly one point for each $f \in 2^{\mathbb{N}}$. Fix $f \in 2^{\mathbb{N}}$. The fact that $\bigcap_{n\in\mathbb{N}} I_{f\restriction\{1,2,\ldots,n\}} \neq \emptyset$ follows from Theorem 2.60 (the Nested Interval Property) because $I_{f\restriction\{1\}} \supseteq I_{f\restriction\{1,2\}} \supseteq \ldots$. On the other hand, $\bigcap_{n\in\mathbb{N}} I_{f\restriction\{1,2,\ldots,n\}}$ cannot contain two distinct points x and y because, when N is large enough (when $|x-y| > 1/3^N$), the interval $I_{f\restriction\{1,2,\ldots,N\}}$ cannot contain both x and y. Thus, $\bigcap_{n\in\mathbb{N}} I_{f\restriction\{1,2,\ldots,n\}}$ contains exactly one point, which we call $\phi(f)$.

Next, let us check that ϕ is injective. If $f, g \in 2^{\mathbb{N}}$ and $f \neq g$, then there is some n such that $f(n) \neq g(n)$. This implies $I_{f\restriction\{1,2,\ldots,n\}} \neq I_{g\restriction\{1,2,\ldots,n\}}$, which means $I_{f\restriction\{1,2,\ldots,n\}} \cap I_{g\restriction\{1,2,\ldots,n\}} = \emptyset$. This implies $\phi(f) \neq \phi(g)$.

Next, let us check that ϕ is surjective. Let $x \in K$. For each $n \in \mathbb{N}$, there is one and only one length-n sequence s_n of 0's and 1's such that $x \in I_s$. Furthermore, if $m < n$ then clearly $s_m = s_n\restriction\{1, 2, \ldots, m\}$, as otherwise we would have $I_{s_m} \cap I_{s_n} = \emptyset$. Thus, there is a function $f : \mathbb{N} \to \{0, 1\}$ such that $f\restriction\{1, 2, \ldots, n\} = s_n$ for every n. But then, $\phi(f) = x$.

Next, let us check that ϕ is continuous. Fix $f \in 2^{\mathbb{N}}$, and let $\varepsilon > 0$. Fix some $n \in \mathbb{N}$ large enough that $1/3^n < \varepsilon$, and then let $\delta = 1/2^n$. Let $s = f\restriction\{1, 2, \ldots, n\}$, and note that

$B_\delta(f) = [\![s]\!]$ in $2^{\mathbb{N}}$. So if $g \in 2^{\mathbb{N}}$ and $d(f,g) < \delta$, then $g \in [\![s]\!]$ also. But then, $\phi(f)$ and $\phi(g)$ are both members of the same interval in K_n, namely, I_s. Because all the intervals comprising K_n have length $1/3^n$, this means $|\phi(f) - \phi(g)| \le 1/3^n < \varepsilon$.

Finally, let us check that ϕ^{-1} is continuous. Fix $x \in K$, and let $\varepsilon > 0$. Fix some $n \in \mathbb{N}$ large enough that $1/2^n < \varepsilon$, and then let $\delta = 1/3^n$. There is some $s : \{1, 2, \dots, n\} \to \{0, 1\}$ such that $x \in I_s$. Note that our choice of δ implies $B_\delta(x) \subseteq I_s$, and, in particular, if $y \in K$ and $|x - y| < \delta$, then $y \in I_s$. But, $x, y \in I_s$ means that $\phi^{-1}(x)$ and $\phi^{-1}(y)$ are both in $[\![s]\!]$. This implies $d(\phi^{-1}(x), \phi^{-1}(y)) \le 1/2^n < \varepsilon$. □

Corollary 2.62. $|K| = |2^{\mathbb{N}}| = |\mathcal{P}(\mathbb{N})|$. *In particular, K is uncountable.*

Proof. That $|K| = |2^{\mathbb{N}}|$ follows from the previous theorem. To see that $|2^{\mathbb{N}}| = |\mathcal{P}(\mathbb{N})|$, we use the idea of characteristic functions. Given $A \subseteq \mathbb{N}$, the *characteristic function* of A is the function $\chi_A \in 2^{\mathbb{N}}$ defined by

$$\chi_A(n) = \begin{cases} 1 & \text{if } n \in A, \\ 0 & \text{if } n \notin A. \end{cases}$$

It is not difficult to see that the mapping $A \mapsto \chi_A$ is a bijection $\mathcal{P}(\mathbb{N}) \to 2^{\mathbb{N}}$, so this mapping shows that $|2^{\mathbb{N}}| = |\mathcal{P}(\mathbb{N})|$ as claimed. Recalling from the end of the previous chapter that $|\mathcal{P}(\mathbb{N})| > |\mathbb{N}|$, this means that $2^{\mathbb{N}}$ and K are uncountable. □

Exercises

1. Which of the following functions are metrics for the set $\mathbb{R}$? For each function that is not a metric, prove that it is not by showing that it fails to satisfy one of the properties in our definition of a metric.
   (a) $d(x, y) = \max\{|x|, |y|\}$
   (b) $d(x, y) = x - y$
   (c) $d(x, y) = (x - y)^2$
   (d) $d(x, y) = 2|x - y|$
   (e) $d(x, y) = \sqrt{x^2 + y^2}$
   (f) $d(x, y) = |x - y|$ if $|x - y| < 1$, and $d(x, y) = 1$, otherwise.
   (g) $d(x, y) = \sin(|x - y|)$
2. Prove Theorem 2.11 and Corollary 2.12.
3. Let X be a metric space, and let $A \subseteq X$. Prove that $X \setminus \text{int}(A) = \overline{X \setminus A}$.
4. The *Dirichlet function* is the function $f : \mathbb{R} \to \mathbb{R}$ defined by

$$f(x) = \begin{cases} 1 & \text{if } x \in \mathbb{Q}, \\ 0 & \text{if } x \notin \mathbb{Q}. \end{cases}$$

   Prove that this function is not continuous at x for any $x \in \mathbb{R}$.

5. *Thomae's function* is the function $f : \mathbb{R} \to \mathbb{R}$ defined by

$$f(x) = \begin{cases} 1/q & \text{if } x = p/q \text{ for some integers } p \text{ and } q \text{ with } \gcd(p, q) = 1, \\ 0 & \text{if } x \notin \mathbb{Q}. \end{cases}$$

Prove that this function is continuous at x if and only if x is irrational.

6. Let X and Y be metric spaces, and let $f : X \to Y$. Prove that f is continuous if and only if $f^{-1}[C]$ is closed in X for every closed $C \subseteq Y$.
7. Let X and Y be metric spaces, and let $f : X \to Y$ be a bijection. Prove that f is a homeomorphism if and only if, for every $A \subseteq X$, A is open in X if and only if $f[A]$ is open in Y.
8. Prove that $\mathbb{R}$ and $(0, 1)$ are homeomorphic, but not isometric.
9. Give an example of two metric spaces X and Y and a continuous bijection $f : X \to Y$ that is not a homeomorphism.
10. For each $a \in \mathbb{R}$, prove that the functions $x \mapsto a + x$ and $x \mapsto a - x$ are isometries from $\mathbb{R}$ to itself. Furthermore, prove that every isometry from $\mathbb{R}$ to itself is a function of this form.
11. Let ρ be the metric on the plane $\mathbb{R}^2$ defined by

$$\rho\big((x_1, y_1), (x_2, y_2)\big) = |x_1 - x_2| + |y_1 - y_2|.$$

This is called the "taxicab metric" on $\mathbb{R}^2$.

(a) Prove that $(\mathbb{R}^2, \rho)$ is homeomorphic to $\mathbb{R}^2$ with its usual metric.

(b) Prove that $(\mathbb{R}^2, \rho)$ is not isometric to $\mathbb{R}^2$ with its usual metric.

12. Suppose (X, d_X) and (Y, d_Y) are metric spaces. Define a metric σ on $X \times Y$ by setting

$$\sigma\big((x_1, y_1), (x_2, y_2)\big) = \max\{d_X(x_1, x_2), d_Y(y_1, y_2)\}.$$

This is called the "sup metric" on $X \times Y$. Prove that this is a metric.

13. Let ρ denote the taxicab metric on $\mathbb{R}^2$, and let σ denote the sup metric on $\mathbb{R}^2$. Prove that $(\mathbb{R}^2, \rho)$ and $(\mathbb{R}^2, \sigma)$ are isometric.
14. Prove that a metric space X is disconnected if and only if there is some $E \subseteq X$ such that $E \neq \emptyset, X$ and E is both open and closed.
15. Prove that a subspace A of a metric space X is disconnected (as defined in Definition 2.32) if and only if when A is viewed as a subspace of X, there is a subset $E \subseteq A$ such that $E \neq \emptyset, A$ and E is both open and closed in A.
16. Suppose X is a countable set containing at least two points. Prove that if d is any metric on X, then (X, d) is disconnected.
17. (The fixed point property for closed bounded intervals.) Let $a, b \in \mathbb{R}$ with $a < b$, and suppose $f : [a, b] \to [a, b]$ is continuous. Prove there is some $x \in [a, b]$ such that $f(x) = x$.

18. Find an example showing that boundedness is not always preserved by homeomorphisms. On the other hand, prove that boundedness is preserved by isometries.

19. Suppose that X is a metric space and $A \subseteq X$ is bounded. Prove that $\overline{A}$ is also bounded.

20. Let X and Y be metric spaces and $f : X \to Y$. Describe a counterexample showing that the following statement is not true: If $f^{-1}[C]$ is a compact subset of X for every compact $C \subseteq Y$, then f is continuous.

21. Let X be a metric space and $A \subseteq X$. Prove that A is compact if and only if for every collection $\mathcal{C}$ of closed subsets of X, if $A \cap \bigcap \mathcal{C}' \neq \emptyset$ for every finite $\mathcal{C}' \subseteq \mathcal{C}$, then $A \cap \bigcap \mathcal{C} \neq \emptyset$.

22. Let X be a metric space and let $A \subseteq X$. Prove that if $A \subseteq X$ is compact, then every infinite subset of A has a limit point in A.

23. Prove that if $A \subseteq \mathbb{R}$ and $\mathcal{U}$ is an open cover of A, then $\mathcal{U}$ has a countable subcover. (*Hint:* There are only countably many intervals (p, q) with $p, q \in \mathbb{Q}$. For each such interval that is contained in some member of $\mathcal{U}$, choose some $U_{(p,q)} \in \mathcal{U}$ containing it.)

24. Let $A \subseteq \mathbb{R}$. Using the previous exercise, prove that if every infinite subset of A has a limit point in A, then A is compact. Combining this with Exercise 22, conclude that a subset A of $\mathbb{R}$ is compact if and only if every infinite subset of A has a limit point in A.

25. Prove that the function d from Definition 2.57 is a metric.

26. Suppose C is a non-empty, compact subset of $\mathbb{R}$ that contains no intervals and has no isolated points. Prove that C is homeomorphic to the Cantor space.

27. Prove that for every uncountable closed $A \subseteq \mathbb{R}$, there is some $C \subseteq A$ that is homeomorphic to $2^{\mathbb{N}}$. Conclude that all uncountable closed subsets of $\mathbb{R}$ have the same cardinality.
(*Hint:* Use the previous exercise and Exercise 23.)

CHAPTER 3

Completeness

In this chapter, we learn about *completeness* in metric spaces. What does it mean for a metric space to be complete? Somewhat confusingly, it does not mean that it satisfies the Axiom of Completeness. In fact, the Axiom of Completeness talks about upper bounds and supremums, and these words simply have no meaning in the context of general metric spaces. These words make sense for ordered sets, but many metric spaces are not ordered in any natural or meaningful way. (What would be an order for the plane $\mathbb{R}^2$? Or for an arbitrary set X equipped with the discrete metric?) Nevertheless, the idea of metric completeness is the same, in some sense, as the idea expressed by the Axiom of Completeness: There are no holes. Our definition of completeness in arbitrary metric spaces expresses this idea in a language that makes sense in any metric space, rather than just for ordered spaces like $\mathbb{R}$. Before saying what a complete metric space is, we must discuss sequences and convergence.

Sequences and subsequences

Definition 3.1. Suppose X is a metric space and $\langle x_n : n \in \mathbb{N} \rangle$ is a sequence of points in X. This sequence *converges* to a point $p \in X$ if, for every $\varepsilon > 0$, there is some $N \in \mathbb{N}$ such that if $n \geq N$ then $d(p, x_n) < \varepsilon$.

This definition may feel similar to our definition of a continuous function in the previous chapter. The two definitions are closely related, as the following theorem shows.

Theorem 3.2. *Suppose X and Y are metric spaces, f is a function $X \to Y$, and $x \in X$. Then, f is continuous at x if and only if whenever $\langle x_n : n \in \mathbb{N} \rangle$ is a sequence of points in X that converges to x, the sequence $\langle f(x_n) : n \in \mathbb{N} \rangle$ converges to $f(x)$ in Y.*

Proof. For the "only if" direction, suppose f is continuous at some $x \in X$. Let $\langle x_n : n \in \mathbb{N} \rangle$ be a sequence of points in X that converges to x. We wish to show $\langle f(x_n) : n \in \mathbb{N} \rangle$ converges to $f(x)$. Let $\varepsilon > 0$. By the definition of continuity, there is some $\delta > 0$ such that for any $x' \in X$, if $d(x', x) < \delta$, then $d(f(x'), f(x)) < \varepsilon$. Because $\langle x_n : n \in \mathbb{N} \rangle$ converges to x, there is some $N \in \mathbb{N}$ such that if $n \geq N$, then $d(x_n, x) < \delta$. It follows that if $n \geq N$, then $d(f(x_n), f(x)) < \varepsilon$, and this means that $\langle f(x_n) : n \in \mathbb{N} \rangle$ converges to $f(x)$.

Foundations of Real Analysis
https://doi.org/10.1016/B978-0-44-348527-5.00007-X

For the "if" direction, we prove the contrapositive. Suppose f is not continuous at x. This means that for some particular $\varepsilon > 0$, given any $\delta > 0$ (no matter how small), there is some $x' \in X$ with $d(x', x) < \delta$ and $d(f(x'), f(x)) \geq \varepsilon$. In particular, for any $n \in \mathbb{N}$, we may take $\delta = 1/n$ and find some $x_n \in X$ with $d(x_n, x) < 1/n$ and $d(f(x_n), f(x)) \geq \varepsilon$. Now, the sequence $\langle x_n : n \in \mathbb{N} \rangle$ converges to x: This is because for any $\eta > 0$ there is some $N \in \mathbb{N}$ such that $1/N < \eta$, which implies that $d(x_n, x) < 1/n \leq 1/N < \eta$ for all $n \geq N$. However, the sequence $\langle f(x_n) : n \in \mathbb{N} \rangle$ does not converge to $f(x)$: This is because $d(f(x_n), f(x)) \geq \varepsilon$ for all $n \in \mathbb{N}$. □

Definition 3.3. Suppose X is a metric space and $\langle x_n : n \in \mathbb{N} \rangle$ is a sequence of points in X. If $\langle x_n : n \in \mathbb{N} \rangle$ converges to some $x \in X$, then we denote this by writing $\lim_{n\to\infty} x_n = x$.

The notation in this definition is a bit unusual, and we must take care! To write "$\lim_{n\to\infty} x_n = x$" is really to make two assertions, not one: that the sequence $\langle x_n : n \in \mathbb{N} \rangle$ converges, and also that it converges to x (not some other point). Thus, the negation of the assertion "$\lim_{n\to\infty} x_n = x$" is: *Either* the sequence $\langle x_n : n \in \mathbb{N} \rangle$ does not converge, *or* it does converge but not to x. Because of the either/or nature of this negation, we do not usually write $\lim_{n\to\infty} x_n \neq x$ to mean the negation of $\lim_{n\to\infty} x_n = x$ (because writing $\lim_{n\to\infty} x_n \neq x$ suggests that $\lim_{n\to\infty} x_n$ exists and $\lim_{n\to\infty} x_n = y$ for some $y \neq x$).

Proposition 3.4. *Suppose X is a metric space and $\langle x_n : n \in \mathbb{N} \rangle$ is a sequence of points in X. Then, $\langle x_n : n \in \mathbb{N} \rangle$ converges to at most one point.*

Proof. Suppose, aiming for a contradiction, that $\langle x_n : n \in \mathbb{N} \rangle$ is a sequence of points in a metric space X that converges both to x and to y, where $x, y \in X$ and $x \neq y$. Let $\varepsilon = \frac{1}{2} d(x, y)$. Because $\langle x_n : n \in \mathbb{N} \rangle$ converges to x, there is some $N_x \in \mathbb{N}$ such that if $n \geq N_x$ then $d(x_n, x) < \varepsilon$. Similarly, because $\langle x_n : n \in \mathbb{N} \rangle$ converges to x, there is some $N_y \in \mathbb{N}$ such that if $n \geq N_y$ then $d(x_n, y) < \varepsilon$. Pick some $n \in \mathbb{N}$ large enough that $n \geq N_x$ and $n \geq N_y$. Then, $d(x_n, x) < \varepsilon$ and $d(x_n, y) < \varepsilon$, which implies

$$d(x, y) \leq d(x, x_n) + d(x_n, y) < \varepsilon + \varepsilon = d(x, y),$$

which is, of course, nonsense. □

Proposition 3.5. *Suppose X is a metric space, $A \subseteq X$ and $x \in X$. Then, $x \in \overline{A}$ if and only if there is a sequence of points in A that converges to x.*

Proof. For the "only if" direction, suppose $x \in \overline{A}$. This implies that for every $\varepsilon > 0$, there is some $a \in A$ such that $d(a, x) < \varepsilon$. In particular, for every $n \in \mathbb{N}$ there is some $a_n \in A$ such that $d(a_n, x) < 1/n$. Then, $\langle a_n : n \in \mathbb{N} \rangle$ is a sequence of points in A that converges to x.

For the "if" direction, suppose $x \in X$ and there is a sequence $\langle a_n : n \in \mathbb{N} \rangle$ of points in A that converges to x. Let $\varepsilon > 0$. Because $\langle a_n : n \in \mathbb{N} \rangle$ converges to x, there is some

$N \in \mathbb{N}$ such that if $n \geq N$, then $d(a_n, x) < \varepsilon$. In particular, $B_\varepsilon(x)$ contains members of the sequence $\langle a_n \colon n \in \mathbb{N} \rangle$, which are all points of A. Thus, $B_\varepsilon(x) \cap A \neq \emptyset$ for all $\varepsilon > 0$, which implies $x \in \overline{A}$. □

Corollary 3.6. *Suppose X is a metric space and A is a closed subset of X. If a sequence of points in A converges, then it converges to a point in A.*

Proof. If A is closed, then $A = \overline{A}$, so the previous proposition asserts that if some sequence of points in A converges to x, then $x \in A$. □

Definition 3.7. Suppose $\langle x_n \colon n \in \mathbb{N} \rangle$ is an infinite sequence. If $\langle n_k \colon k \in \mathbb{N} \rangle$ is a sequence in $\mathbb{N}$ that is strictly increasing (i.e., $k < \ell$ implies $n_k < n_\ell$), then $\langle x_{n_k} \colon k \in \mathbb{N} \rangle$ is called a *subsequence* of $\langle x_n \colon n \in \mathbb{N} \rangle$.

In other words, a subsequence is obtained from an infinite sequence by deleting some entries, or skipping over them, and listing only some of the members of the original sequence.

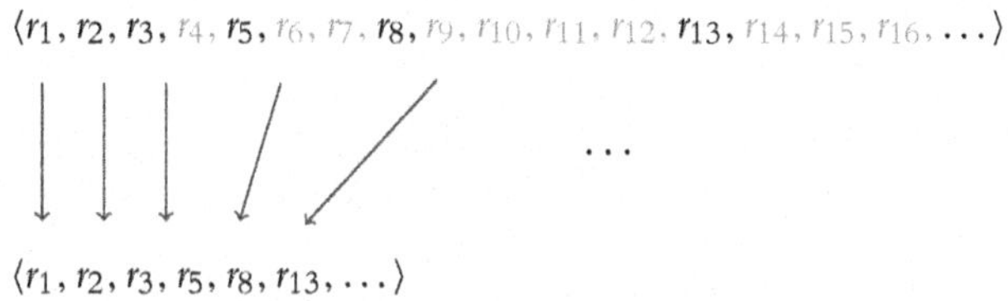

Example 3.8. Even if a sequence of points in a metric space does not converge, subsequences of that sequence may nonetheless converge. For example, $\langle 0, 1, 0, 1, 0, 1, \ldots \rangle$ is a sequence of points in $\mathbb{R}$ that does not converge. But $\langle 0, 0, 0, \ldots \rangle$ and $\langle 1, 1, 1, \ldots \rangle$ are both subsequences of this sequence, and they both converge.

Definition 3.9. Suppose $\langle x_n \colon n \in \mathbb{N} \rangle$ is an infinite sequence in a metric space X. If some subsequence of this sequence converges to some $x \in X$, then x is called a *subsequential limit* of $\langle x_n \colon n \in \mathbb{N} \rangle$.

For example, the sequence from Example 3.8 does not converge, but it has two subsequential limits (0 and 1). It is possible for a sequence to have many subsequential limits.

Example 3.10. Even though a sequence contains only countably many points, it is possible for a sequence to have uncountably many subsequential limits. For example, we know that the set $\mathbb{Q}$ of all rational numbers is countable, which means there is a bijection $\phi \colon \mathbb{N} \to \mathbb{Q}$. Letting $q_n = \phi(n)$ for all n, the sequence $\langle q_n \colon n \in \mathbb{N} \rangle$ lists all the rational numbers. In particular, every open subset of $\mathbb{R}$ contains infinitely many members of this sequence. This implies (using the soon-to-be-proved Theorem 3.11) that every real number is a subsequential limit of this sequence.

The following picture shows a sequence of points in the plane $\mathbb{R}^2$ with uncountably many subsequential limits.

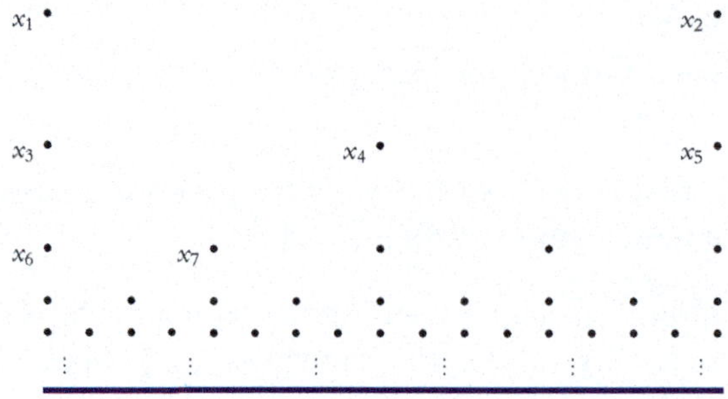

Theorem 3.11. *Suppose X is a metric space, $x \in X$, and $\langle x_n : n \in \mathbb{N}\rangle$ is a sequence of points in X.*

(1) *Then, x is a subsequential limit of $\langle x_n : n \in \mathbb{N}\rangle$ if and only if for every $\varepsilon > 0$, there are infinitely many $n \in \mathbb{N}$ with $x_n \in B_\varepsilon(x)$.*

(2) *Furthermore, $x = \lim_{n\to\infty} x_n$ if and only if for every $\varepsilon > 0$, there are only finitely many $n \in \mathbb{N}$ with $x_n \notin B_\varepsilon(x)$.*

Proof. To prove the "only if" direction of (1), suppose x is a subsequential limit of $\langle x_n : n \in \mathbb{N}\rangle$. This means there is an increasing sequence $\langle n_k : k \in \mathbb{N}\rangle$ of natural numbers such that $\langle x_{n_k} : k \in \mathbb{N}\rangle$ converges to x. This means that for any $\varepsilon > 0$, there is some $K \in \mathbb{N}$ such that if $k \geq K$, then $d(x_{n_k}, x) < \varepsilon$. In other words, $x_n \in B_\varepsilon(x)$ whenever $n = n_k$ for some $k \geq K$, and this means $x_n \in B_\varepsilon(x)$ for infinitely many n.

To prove the "if" direction of (1), suppose that for any $\varepsilon > 0$ there are infinitely many $n \in \mathbb{N}$ with $x_n \in B_\varepsilon(x)$. Then, we can recursively define a sequence $\langle n_k : k \in \mathbb{N}\rangle$ of natural numbers as follows. First, choose n_1 so that $d(x_{n_1}, x) < 1$. Then, given $k \in \mathbb{N}$ and supposing that n_{k-1} has been chosen already, chose n_k so that

- $n_k > n_{k-1}$, and
- $d(x_{n_k}, x) < 1/k$.

Some such n_k must exist because by assumption there are infinitely many integers satisfying the second condition, and this implies that there are some integers bigger than n_k satisfying it.

This recursive definition of the n_k's results in a subsequence $\langle x_{n_k} : k \in \mathbb{N}\rangle$ of our original sequence that converges to x: because, for any $\varepsilon > 0$, there is some K with $1/K < \varepsilon$, which implies $d(x_{n_k}, x) < 1/k < \varepsilon$ for all $k \geq K$.

To prove the "only if" direction of (2), suppose $x = \lim_{n\to\infty} x_n$. Then, for any $\varepsilon > 0$, there is some $N \in \mathbb{N}$ such that if $n \geq N$, then $d(x_n, x) < \varepsilon$. In other words, if $x_n \notin B_\varepsilon(x)$ then $n < N$, and this means that $x_n \notin B_\varepsilon(x)$ for only finitely many values of n.

To prove the "if" direction of (2), suppose that for any $\varepsilon > 0$ there are only finitely many $n \in \mathbb{N}$ with $x_n \notin B_\varepsilon(x)$. Given some $\varepsilon > 0$, let $N = \max\{n \in \mathbb{N} : x_n \notin B_\varepsilon(x)\}$. Then, $n \geq N + 1$ implies $d(x_n, x) < \varepsilon$. As ε was arbitrary, this shows that $\langle x_n : n \in \mathbb{N}\rangle$ converges to x. $\square$

Corollary 3.12. *Suppose* $\langle x_n : n \in \mathbb{N} \rangle$ *is a sequence of points in a metric space* X, *and* $x \in X$. *Then,* $x = \lim_{n\to\infty} x_n$ *if and only if every subsequence of* $\langle x_n : n \in \mathbb{N} \rangle$ *converges to* x.

Proof. For the "if" direction, suppose every subsequence of $\langle x_n : n \in \mathbb{N} \rangle$ converges to x. But, $\langle x_n : n \in \mathbb{N} \rangle$ is a subsequence of itself (just set $n_k = k$ for all $k \in \mathbb{N}$), so this means $\langle x_n : n \in \mathbb{N} \rangle$ converges to x.

For the "only if" direction, suppose $x = \lim_{n\to\infty} x_n$. By the previous theorem, this means that for any $\varepsilon > 0$, there are only finitely many $n \in \mathbb{N}$ with $x_n \notin B_\varepsilon(x)$. So, if $\langle x_{n_k} : k \in \mathbb{N} \rangle$ is a subsequence of $\langle x_n : n \in \mathbb{N} \rangle$, there are only finitely many k with $x_{n_k} \notin B_\varepsilon(x)$, and thus $x = \lim_{k\to\infty} x_{n_k}$. □

Completeness, compactness, and the real line

We are nearly ready to define what a complete metric space is. Roughly, the idea is that a metric space is complete when sequences that "should" converge actually do. The following definition is meant to describe those sequences that "should" converge.

Definition 3.13. Let X be a metric space. An infinite sequence $\langle x_n : n \in \mathbb{N} \rangle$ of points in X is *Cauchy* if for every $\varepsilon > 0$, there is some $N \in \mathbb{N}$ such that if $m, n \geq N$, then $d(x_m, x_n) < \varepsilon$.

Theorem 3.14. *Every convergent sequence (in any metric space) is Cauchy.*

Proof. Let X be a metric space, and let $\langle x_n : n \in \mathbb{N} \rangle$ be a sequence of points in X converging to some $x \in X$. Let $\varepsilon > 0$. By the definition of convergence, there is some $N \in \mathbb{N}$ such that if $n \geq N$, then $d(x_n, x) < \varepsilon/2$. This implies that if $m, n \geq N$, then

$$d(x_m, x_n) \leq d(x_m, x) + d(x, x_n) < \varepsilon/2 + \varepsilon/2 = \varepsilon.$$

Thus, $\langle x_n : n \in \mathbb{N} \rangle$ is Cauchy. □

If all convergent sequences are Cauchy, is it also the case that all Cauchy sequences are convergent? It turns out that this depends on what metric space we are talking about.

Example 3.15. Let $X = \mathbb{Q}$, and let d denote the usual metric on $\mathbb{Q}$, i.e., $d(x, y) = |x - y|$. For each $n \in \mathbb{N}$, there is some rational number q_n such that $\sqrt{2} - \frac{1}{n} < q_n < \sqrt{2}$. Then, it is not difficult to see that the sequence $\langle q_n : n \in \mathbb{N} \rangle$ is a Cauchy sequence in X, but it does not converge. (This sequence does converge to $\sqrt{2}$ in the larger metric space $\mathbb{R}$, but it does not converge in the metric space $X = \mathbb{Q}$.)

Example 3.16. Suppose X is a finite metric space. Then, every Cauchy sequence $\langle x_n : n \in \mathbb{N} \rangle$ in X is eventually constant, which means that there is some $x \in X$ and $N \in \mathbb{N}$ such that $x_n = x$ for all $n \geq N$. (To see this, just set $\varepsilon = \min\{d(x, y) : x, y \in X \text{ and } x \neq y\}$ in the definition of a Cauchy sequence.) Eventually, constant sequences obviously converge, therefore every Cauchy sequence in X converges.

Thus, the statement "every Cauchy sequence converges" is true of some metric spaces but not others. In other words, it can be taken to define a property that a metric space may or may not have.

Definition 3.17. A metric space (X, d) is *complete* if every Cauchy sequence of points in X converges.

We prove next that, unlike $\mathbb{Q}$, the real line $\mathbb{R}$ is a complete metric space. Along the way we prove another important result: That every compact metric space is complete. First we need a few lemmas.

Lemma 3.18. *If a Cauchy sequence has a subsequential limit, then the sequence converges to that subsequential limit.*

Proof. Suppose $\langle x_n : n \in \mathbb{N} \rangle$ is a Cauchy sequence in a metric space X, and suppose a subsequence of this sequence converges to some $x \in X$. Let $\varepsilon > 0$. Because $\langle x_n : n \in \mathbb{N} \rangle$ is Cauchy, there is some $N \in \mathbb{N}$ such that if $m, n \geq N$ then $d(x_m, x_n) < \varepsilon/2$. Because a subsequence of $\langle x_n : n \in \mathbb{N} \rangle$ converges to x, there are infinitely many $n \in \mathbb{N}$ such that $d(x, x_n) < \varepsilon/2$; fix some particular $k \geq N$ such that $d(x, x_k) < \varepsilon/2$. Consequently, if $n \geq N$, then

$$d(x_n, x) \leq d(x_n, x_k) + d(x_k, x) < \varepsilon/2 + \varepsilon/2 = \varepsilon.$$

As ε was arbitrary, this shows $\langle x_n : n \in \mathbb{N} \rangle$ converges to x. □

Lemma 3.19. *If X is a metric space and $A \subseteq X$ is compact, then every sequence of points in A has a subsequential limit.*

Proof. We prove the contrapositive: If some sequence in A has no subsequential limit, then A is not compact.

Let X be a metric space, let $A \subseteq X$, and let $\langle x_n : n \in \mathbb{N} \rangle$ be a sequence of points in A with no convergent subsequences. By Theorem 3.11, this implies that for every $x \in A$, there is some $\varepsilon_x > 0$ such that $\{n \in \mathbb{N} : x_n \in B_{\varepsilon_x}(x)\}$ is finite.

Let $\mathcal{U} = \{B_{\varepsilon_x}(x) : x \in A\}$. Then, $\mathcal{U}$ is a collection of open subsets of X. Furthermore, because $x \in B_{\varepsilon_x}(x)$ for every $x \in A$, $\mathcal{U}$ covers A. If $\mathcal{V}$ is a finite subset of $\mathcal{U}$, then (as each member of $\mathcal{V}$ contains only finitely many members of our sequence) there are only finitely many $n \in \mathbb{N}$ such that $x_n \in \bigcup \mathcal{V}$. But, $x_n \in A$ for every n, so this means $\mathcal{V}$ does not cover A. Thus, $\mathcal{U}$ is an open cover of A with no finite subcover, which means A is not compact. □

Theorem 3.20. *Every compact metric space is complete.*

Proof. Let X be a compact metric space. By Lemma 3.19, every Cauchy sequence in X has a subsequential limit. By Lemma 3.18, this implies that every Cauchy sequence in X converges. □

Definition 3.21. A sequence $\langle x_n \colon n \in \mathbb{N}\rangle$ of points in a metric space X is *bounded* if the corresponding set $\{x_n \colon n \in \mathbb{N}\}$ is bounded.

Lemma 3.22. *Every Cauchy sequence (in any metric space) is bounded.*

Proof. Let X be a metric space, and let $\langle x_n \colon n \in \mathbb{N}\rangle$ be a Cauchy sequence of points in X. This implies there is some $N \in \mathbb{N}$ such that if $m, n \geq N$, then $d(x_m, x_n) < 1$. In particular, if $n \geq N$, then $d(x_N, x_n) < 1$. Let

$$R = \max\{1, d(x_0, x_N), d(x_1, x_N), \ldots, d(x_{N-1}, x_N)\}.$$

This choice of R ensures that $x_n \in B_R(x_N)$ for every $n \in \mathbb{N}$, which means that $\{x_n \colon n \in \mathbb{N}\}$ is a bounded subset of X. □

Theorem 3.23. $\mathbb{R}$ *is a complete metric space.*

Proof. Suppose $\langle x_n \colon n \in \mathbb{N}\rangle$ is a Cauchy sequence in $\mathbb{R}$. By Lemma 3.22, this sequence is bounded, i.e., there is some $x \in \mathbb{R}$ and some $R > 0$ such that $x_n \in B_R(x) = (x - R, x + R)$ for every $n \in \mathbb{N}$. Let $A = [x - R, x + R]$. Then, A is a closed and bounded set containing every point of our sequence. By the Heine–Borel Theorem, A is compact. By Lemma 3.19, this implies that $\langle x_n \colon n \in \mathbb{N}\rangle$ has a subsequential limit in A. Therefore, by Lemma 3.18, $\langle x_n \colon n \in \mathbb{N}\rangle$ converges. □

Limits of sequences and the extended real line

The following theorem shows that limits taken in $\mathbb{R}$ behave well with respect to the field operations defined on $\mathbb{R}$.

Theorem 3.24 (Algebraic Limit Theorem). *Let $\langle s_n \colon n \in \mathbb{N}\rangle$, $\langle t_n \colon n \in \mathbb{N}\rangle$ be convergent sequences in $\mathbb{R}$, and let $s = \lim_{n\to\infty} s_n$ and $t = \lim_{n\to\infty} t_n$.*

(1) *For every $c \in \mathbb{R}$, $\lim_{n\to\infty} cs_n = cs$.*
(2) $\lim_{n\to\infty} s_n + t_n = s + t$.
(3) $\lim_{n\to\infty} s_n t_n = st$.
(4) *If $s_n \neq 0$ for all n and if $s \neq 0$, then $\lim_{n\to\infty} {}^1\!/\!_{s_n} = {}^1\!/\!_s$.*

Proof. In each of these four parts of the theorem, it is important to remember that we are asserting not one fact but two: first that the limit exists, and also what it is equal to.

(1) If $c = 0$, then (1) just says that the constant 0 sequence converges to 0, which is obvious. So, assume $c \neq 0$. Fix $\varepsilon > 0$. Because $s = \lim_{n\to\infty} s_n$, there is some $N \in \mathbb{N}$ such that if $n \geq N$, then $|s_n - s| < \varepsilon/|c|$. But then,

$$|cs_n - cs| = |c||s_n - s| < |c|\frac{\varepsilon}{|c|} = \varepsilon,$$

for all $n \geq N$, which shows that $\lim_{n\to\infty} cs_n = cs$.

(2) Fix $\varepsilon > 0$. Because $s = \lim_{n\to\infty} s_n$, there is some $N_s \in \mathbb{N}$ such that if $n \geq N_s$, then $|s_n - s| < \varepsilon/2$. Similarly, there is some $N_t \in \mathbb{N}$ such that if $n \geq N_t$, then $|t_n - t| < \varepsilon/2$. Let $N = \max\{N_s, N_t\}$. Then,

$$|(s_n + t_n) - (s + t)| = |(s_n - s) + (t_n - t)| \leq |s_n - s| + |t_n - t| < \varepsilon/2 + \varepsilon/2 = \varepsilon$$

for all $n \geq N$, which shows that $\lim_{n\to\infty} s_n + t_n = s + t$.

(3) Fix $\varepsilon > 0$. Because $\lim_{n\to\infty} s_n = s$, there is some $N_s \in \mathbb{N}$ such that $|s_n - s| < \varepsilon/2(|t|+1)$ whenever $n \geq N_s$. Recall that convergent sequences are bounded, so there is some $M > 0$ such that $|s_n| < M$ for all $n \in \mathbb{N}$. Because $\lim_{n\to\infty} t_n = t$, there is some $N_t \in \mathbb{N}$ such that $|t_n - t| < \varepsilon/2M$ whenever $n \geq N_t$. Let $N = \max\{N_s, N_t\}$. Then,

$$\begin{aligned} |s_n t_n - st| &= |s_n t_n - s_n t + s_n t - st| \leq |s_n t_n - s_n t| + |s_n t - st| \\ &= |s_n||t_n - t| + |t||s_n - s| < M\tfrac{\varepsilon}{2M} + |t|\tfrac{\varepsilon}{2(|t|+1)} < \varepsilon \end{aligned}$$

for all $n \geq N$, which shows that $\lim_{n\to\infty} s_n t_n = s + t$.

(4) Because $s = \lim_{n\to\infty} s_n \neq 0$, there is some $N_0 \in \mathbb{N}$ such that if $n \geq N_0$, then $|s_n - s| < |s|/2$, which implies $|s_n| > |s|/2$. Fix $\varepsilon > 0$. Because $s = \lim_{n\to\infty} s_n$, there is some $N_1 \in \mathbb{N}$ such that if $n \geq N$, then $|s_n - s| < \varepsilon|s|^2/2$. Let $N = \max\{N_0, N_1\}$. Then,

$$\left|\frac{1}{s_n} - \frac{1}{s}\right| = \left|\frac{s - s_n}{s s_n}\right| = \frac{|s_n - s|}{|s||s_n|} < \frac{\varepsilon|s|^2/2}{|s||s|/2} = \varepsilon$$

for all $n \geq N$, which shows that $\lim_{n\to\infty} 1/s_n = 1/s$. □

Definition 3.25. The *extended real line* consists of the set $\mathbb{R} \cup \{-\infty, \infty\}$. We extend the $<$ relation to this set in the natural way: $a < \infty$ for all $a \in \mathbb{R} \cup \{-\infty\}$, and $-\infty < a$ for all $a \in \mathbb{R} \cup \{\infty\}$. We also partially extend the two operations $+$ and $\cdot$ to $\mathbb{R} \cup \{\pm\infty\}$ (by "partially extend" we mean that $a + b$ and $a \cdot b$ do not always exist). We define

$$a + \infty = \infty + a = \infty \qquad \text{and} \qquad a - \infty = -\infty + a = -\infty$$

for every $a \in \mathbb{R}$. Furthermore, $\infty + \infty = \infty$ and $-\infty - \infty = -\infty$, but the expressions $-\infty + \infty$ and $\infty - \infty$ are left undefined. Similarly,

$$a \cdot \infty = \infty \cdot a = \begin{cases} \infty & \text{if } a < 0 \\ -\infty & \text{if } a > 0 \end{cases} \qquad a \cdot (-\infty) = -\infty \cdot a = \begin{cases} \infty & \text{if } a < 0 \\ -\infty & \text{if } a > 0 \end{cases}$$

for all $a \in \mathbb{R}$, and $\infty \cdot \infty = -\infty \cdot (-\infty) = \infty$, and $-\infty \cdot \infty = \infty \cdot (-\infty) = -\infty$. But the expressions $\pm\infty \cdot 0$ and $0 \cdot \pm\infty$ are left undefined. Additionally, though ∞ and $-\infty$ do not have multiplicative inverses in the extended reals, we define $\frac{a}{\infty} = \frac{a}{-\infty} = 0$ for all $a \in \mathbb{R}$, though expressions like $\frac{\pm\infty}{\infty}$ are undefined. ☕

The extended real line is useful for generalizing the notion of convergent sequences in $\mathbb{R}$, as in the following definition and theorem.

Definition 3.26. Let $A \subseteq \mathbb{R}$. If A is non-empty and not bounded above, then we write $\sup A = \infty$. If A is non-empty and not bounded below, then we write $\inf A = -\infty$.

Definition 3.27. A sequence $\langle x_n \colon n \in \mathbb{N} \rangle$ of real numbers *diverges to* ∞ if for every $r \in \mathbb{R}$, there is some $N \in \mathbb{N}$ such that if $n \geq N$, then $x_n > r$. This is denoted $\lim_{n\to\infty} x_n = \infty$. Similarly, a sequence $\langle x_n \colon n \in \mathbb{N} \rangle$ of real numbers *diverges to* $-\infty$ if for every $r \in \mathbb{R}$, there is some $N \in \mathbb{N}$ such that if $n \geq N$, then $x_n < r$. This is denoted $\lim_{n\to\infty} x_n = -\infty$. If a sequence either converges in the usual sense, or diverges to $\pm\infty$, then we say the sequence *converges in the extended reals.*

With these new notions in place, we can expand the Algebraic Limit Theorem to the context of the extended reals:

Theorem 3.28 (Extended Algebraic Limit Theorem). *Let* $\langle s_n \colon n \in \mathbb{N} \rangle$ *and* $\langle t_n \colon n \in \mathbb{N} \rangle$ *be sequences in* $\mathbb{R}$ *that converge in the extended reals, and let* $s = \lim_{n\to\infty} s_n$ *and* $t = \lim_{n\to\infty} t_n$.

(1) *For every nonzero* $c \in \mathbb{R}$, $\lim_{n\to\infty} cs_n = cs$.

(2) $\lim_{n\to\infty} s_n + t_n = s + t$, *provided that this sum is well-defined.*

(3) $\lim_{n\to\infty} s_n t_n = st$, *provided that this product is well-defined.*

(4) *If* $s_n \neq 0$ *for all* n *and if* $s \neq 0$, *then* $\lim_{n\to\infty} 1/s_n = 1/s$.

Proof. (1) The Algebraic Limit Theorem already covers the case where $s \in \mathbb{R}$, so we only need to prove the case where $s = \pm\infty$. Let us address the case where $s = \infty$ and $c > 0$. Fix $r \in \mathbb{R}$. Because $\lim_{n\to\infty} s_n = \infty$, there is some $N \in \mathbb{N}$ such that if $n \geq N$, then $s_n > r/c$. But this means $cs_n > r$ whenever $n \geq N$, which shows $\lim_{n\to\infty} cs_n = \infty = cs$. The three remaining cases (where $s = \infty$ and $c < 0$, where $s = -\infty$ and $c > 0$, and where $s = -\infty$ and $c < 0$) are proved by nearly identical arguments.

(2) Once again, the Algebraic Limit Theorem covers the case where $s, t \in \mathbb{R}$. There are several remaining cases to address, but like with (1), they are all proved in more or less the same way, so we will just write down the proof for one case. Suppose $s \in \mathbb{R}$ and $t = \infty$. Fix $r \in \mathbb{R}$. Because $\lim_{n\to\infty} s_n = s \in \mathbb{R}$, the sequence $\langle s_n \colon n \in \mathbb{N} \rangle$ is bounded; fix $M > 0$ such that $|s_n| < M$ for all n. Because $\lim_{n\to\infty} t_n = \infty$, there is some $N \in \mathbb{N}$ such that if $n \geq N$, then $t_n > r + M$. But then, $s_n + t_n > -M + r + M = r$ for all $n \geq N$, which shows $\lim_{n\to\infty} s_n + t_n = \infty = s + t$.

(3) Like with (1) and (2), we only need to consider when s or t or both are equal to $\pm\infty$. This leads to four nearly identical cases, and we write down the proof of just one. Suppose $s \in \mathbb{R}$ and $t = \infty$. Assume $s \neq 0$, since otherwise the product st is not well-defined. Because $\lim_{n\to\infty} s_n = s \in \mathbb{R} \setminus \{0\}$, there is some $N_s \in \mathbb{N}$ such that if $n \geq N$, then $|s_n - s| < |s|/2$, which implies $|s_n| > |s|/2$. Fix $r \in \mathbb{R}$. Because $\lim_{n\to\infty} t_n = \infty$, there is some $N_t \in \mathbb{N}$ such that if $n \geq N_t$, then $t_n > 2r/|s|$. Let $N = \max\{N_s, N_t\}$. Then, $s_n t_n > \frac{|s|}{2}\frac{2r}{|s|} = r$ for all $n \geq N$, which shows $\lim_{n\to\infty} s_n t_n = \infty = st$.

(4) Here there are only two cases not covered by the Algebraic Limit Theorem: $s = \infty$ and $s = -\infty$. The proofs of these two cases are nearly identical, so we consider only the first case. Suppose $\lim_{n\to\infty} s_n = \infty$, and fix $\varepsilon > 0$. There is some $N \in \mathbb{N}$ such that if $n \geq N$, then $s_n > 1/\varepsilon$. But then, $1/s_n < \varepsilon$ for all $n \geq N$, which means $\lim_{n\to\infty} 1/s_n = 0 = 1/\infty = 1/s$. □

The basic idea of the extended Algebraic Limit Theorem is that limits behave nicely with respect to the $+$ and $\cdot$ operations, even when using the extended reals. The basic idea of the next theorem is similar: Limits also behave nicely with respect to the $<$ relationship.

Theorem 3.29. *Suppose $\langle r_n\colon n \in \mathbb{N}\rangle$ and $\langle t_n\colon n \in \mathbb{N}\rangle$ are sequences in $\mathbb{R}$ that converge in the extended reals; let $r = \lim_{n\to\infty} r_n$ and $t = \lim_{n\to\infty} t_n$.*

(1) *If $r_n \leq t_n$ for all but finitely many values of n, then $r \leq t$.*

(2) *Suppose $r = t$, and let $\langle s_n\colon n \in \mathbb{N}\rangle$ be another sequence of reals such that $r_n \leq s_n \leq t_n$ for all but finitely many n. Then, $\lim_{n\to\infty} s_n$ exists, and $\lim_{n\to\infty} s_n = r = t$.*

Proof. (1) If $r = t$, then (1) is true, so let us suppose $r \neq t$.

First, suppose $r, t \in \mathbb{R}$. Let $\varepsilon = |r-t|/2$. Because $r_n \leq t_n$ for all but finitely many n, there is some $N_0 \in \mathbb{N}$ such that $r_n \leq t_n$ for all $n \geq N_0$. Furthermore, there is some $N_r \in \mathbb{N}$ such that $|r_n - r| < \varepsilon$ whenever $n \geq N_r$, and there is some $N_t \in \mathbb{N}$ such that $|t_n - t| < \varepsilon$ whenever $n \geq N_t$. Let $N = \max\{N_0, N_r, N_t\}$. Then, $n \geq N$ implies $|r_n - r| < |r-t|/2$, $|t_n - t| < |r-t|/2$, and $r_n \leq t_n$. These three things together imply $r \leq t$.

Now, suppose that one or both of r and t are equal to $\pm\infty$. If $r = -\infty$ or if $t = \infty$, then $r \leq t$ automatically. We must consider only what happens if $t = -\infty$ or if $r = \infty$. Suppose $t = -\infty$. For any $s \in \mathbb{R}$, there is some $N_s \in \mathbb{N}$ such that $n \geq N_s$ implies $t_{N_s} < s$. Because $r_n \leq t_n$ for all but finitely many n, there is some $N_0 \in \mathbb{N}$ such that $r_n \leq t_n$ for all $n \geq N_0$. Setting $N = \max\{N_0, N_s\}$, if $n \geq N$, then $r_n \leq t_n \leq s$. As s was arbitrary, this shows $\lim_{n\to\infty} r_n = -\infty$, that is, $r = -\infty = t$. A nearly identical argument shows that if $r = \infty$, then $t = \infty$ as well.

(2) We consider two cases: $r = t \in \mathbb{R}$ and $r = t = \pm\infty$. First, suppose $r = t \in \mathbb{R}$. Fix $\varepsilon > 0$. There is some $N_r \in \mathbb{N}$ such that $|r_n - r| < \varepsilon/2$ whenever $n \geq N_r$, there is some $N_t \in \mathbb{N}$ such that $|t_n - t| < \varepsilon/2$ whenever $n \geq N_t$, and there is some $N_s \in \mathbb{N}$ such that $r_n \leq s_n \leq t_n$ whenever $n \geq N_s$. Let $N = \max\{N_r, N_s, N_t\}$. Then, $n \geq N$ implies

$$r - \varepsilon/2 < r_n \leq s_n \leq t_n < t + \varepsilon/2.$$

Recalling that $r = t$, this implies $|r - s_n| = |t - s_n| < \varepsilon$. This shows $r = t = \lim_{n\to\infty} s_n$.

Suppose next that $r = t = \infty$. Fix $a \in \mathbb{R}$. There is some $N_r \in \mathbb{N}$ such that $r_n > a$ whenever $n \geq N_r$, and there is some $N_s \in \mathbb{N}$ such that $r_n \leq s_n \leq t_n$ whenever $n \geq N_s$. Then, $n \geq \max\{N_r, N_s\}$ implies $r_n > a$. Hence, $\lim_{n\to\infty} s_n = \infty$. The remaining case, where $r = t = -\infty$, is proved in exactly the same way. □

Definition 3.30. A sequence $\langle r_n : n \in \mathbb{N} \rangle$ of real numbers is *nondecreasing* if $r_m \leq r_n$ whenever $m \leq n$; it is *nonincreasing* if $r_m \geq r_n$ whenever $m \leq n$. A sequence that is either nondecreasing or nonincreasing is called *monotone*.

Theorem 3.31. *Every monotone sequence of reals converges in the extended reals. A nondecreasing sequence* $\langle r_n : n \in \mathbb{N} \rangle$ *converges to* $\sup\{r_n : n \in \mathbb{N}\}$, *and a nonincreasing sequence* $\langle r_n : n \in \mathbb{N} \rangle$ *converges to* $\inf\{r_n : n \in \mathbb{N}\}$.

Proof. Suppose $\langle r_n : n \in \mathbb{N} \rangle$ is a nondecreasing sequence of reals. Because $\sup\{r_n : n \in \mathbb{N}\}$ cannot be $-\infty$, we consider two cases, according to whether this supremum is ∞ or not.

For the first case, suppose $\sup\{r_n : n \in \mathbb{N}\} = r \in \mathbb{R}$. Fix $\varepsilon > 0$. Observe that $r - \varepsilon$ is not an upper bound for $\{r_n : n \in \mathbb{N}\}$ since it is less than the supremum. Thus, there is some $N \in \mathbb{N}$ such that $r_N > r - \varepsilon$. If $n \geq N$, then because the sequence is nondecreasing, $a - \varepsilon < r_N \leq r_n \leq r$; so $n \geq N$ implies $|r_n - r| < \varepsilon$. Hence, $\lim_{n \to \infty} r_n = r$.

For the second case, suppose $\sup\{r_n : n \in \mathbb{N}\} = \infty$. Fix $a \in \mathbb{R}$. Observe that a is not an upper bound for $\{r_n : n \in \mathbb{N}\}$ since, if it were, then $\{r_n : n \in \mathbb{N}\}$ would have a supremum in $\mathbb{R}$ (at most a). Thus, there is some $N \in \mathbb{N}$ such that $r_N > a$. If $n \geq N$, then because the sequence is nondecreasing, $a < r_N \leq r_n$; so $n \geq N$ implies $r_n > a$. Hence, $\lim_{n \to \infty} r_n = \infty$.

The argument for nonincreasing sequences is essentially the same. □

Infinite series

Informally, an infinite series is an infinitely long summation problem, like

$$1 + \tfrac{1}{4} + \tfrac{1}{16} + \tfrac{1}{64} + \tfrac{1}{256} + \tfrac{1}{1024} + \dots$$

Such a problem may have a solution, as the following picture shows.

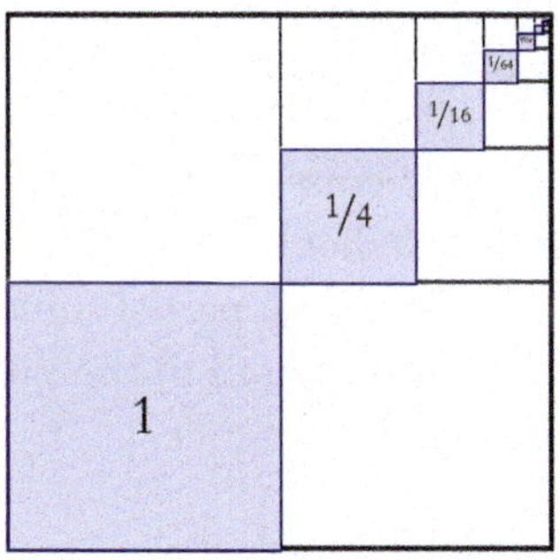

$$1 + \tfrac{1}{4} + \tfrac{1}{16} + \tfrac{1}{64} + \tfrac{1}{256} + \dots$$
$$= \tfrac{1}{3}(2 \cdot 2) = 4/3$$

Other infinite series have no solution: for example, it seems impossible to assign any sensible value, not even $\pm\infty$, to a series like $1-2+3-4+5-6+\ldots$. In this section we aim to understand some of the strange features of infinite series, and how they can behave rather differently from finite sums.

Definition 3.32. Let $\langle a_i \colon i \in \mathbb{N}\rangle$ be a sequence of real numbers. The sum $\sum_{i=1}^{n} a_i$ is the nth *finite partial sum* of this sequence. The notation $\sum_{i=1}^{\infty} a_i$ denotes the limit of the sequence $\langle \sum_{i=1}^{n} a_i \colon n \in \mathbb{N}\rangle$ of finite partial sums:

$$\sum_{i=1}^{\infty} a_i = \lim_{n\to\infty} \sum_{i=1}^{n} a_i.$$

If this limit is a real number r, we say the infinite series $\sum_{i=1}^{\infty} a_i$ *converges to* r. If, instead, the limit is $\pm\infty$, we say $\sum_{i=1}^{\infty} a_i$ *diverges to* $\pm\infty$. If the limit does not converge in the extended reals at all, then the infinite series $\sum_{i=1}^{\infty} a_i$ *diverges by oscillation.* ☕

If the terms of an infinite sequence $\langle a_n \colon n \in \mathbb{N}\rangle$ are all defined by some formula, say $a_n = f(n)$ for all n, then we can denote the corresponding infinite series simply by writing $\sum_{n=1}^{\infty} f(n)$, without needing to explicitly write down the sequence first. For example, $\sum_{n=1}^{\infty} \frac{1}{n^2}$ denotes the infinite series defined from the sequence $\langle 1, \frac{1}{4}, \frac{1}{9}, \frac{1}{16}, \ldots\rangle$. If the formula $f(n)$ is easy to guess, then we sometimes denote an infinite series just by listing the first few terms, e.g., writing $1 + \frac{1}{4} + \frac{1}{9} + \frac{1}{16} + \ldots$ instead of $\sum_{n=1}^{\infty} \frac{1}{n^2}$.

As another notational flexibility, we can write an infinite series where the indexing begins at a number other than 1. Formally, for any $m \in \mathbb{Z}$, $\sum_{n=m}^{\infty} f(n)$ denotes the infinite series $\sum_{n=1}^{\infty} f(m+1-n)$. For example,

$$\sum_{n=0}^{\infty} \tfrac{1}{n!} = 1 + 1 + \tfrac{1}{2} + \tfrac{1}{6} + \ldots, \qquad \sum_{n=3}^{\infty} \tfrac{1}{2^n} = \tfrac{1}{8} + \tfrac{1}{16} + \tfrac{1}{32} + \tfrac{1}{64} + \ldots.$$

Example 3.33. The infinite series $\sum_{n=1}^{\infty} \frac{1}{2^n}$ converges to 1. It is not too hard to see that for any $n \in \mathbb{N}$,

$$\tfrac{1}{2} + \tfrac{1}{4} + \tfrac{1}{8} + \cdots + \tfrac{1}{2^n} = 1 - \tfrac{1}{2^n}.$$

Consequently, $\sum_{n=1}^{\infty} \frac{1}{2^n} = \lim_{n\to\infty} 1 - \frac{1}{2^n} = 1$. ☕

Proposition 3.34 (The Cauchy criterion). *An infinite series $\sum_{i=1}^{\infty} a_i$ converges if and only if, for all $\varepsilon > 0$, there is some $N \in \mathbb{N}$ such that if $n \geq m \geq N$, then $\left|\sum_{i=m}^{n} a_i\right| < \varepsilon$.*

Proof. By Theorems 3.14 and 3.23, a sequence of reals converges if and only if it is Cauchy. Combining this with Definition 3.32, this means that the infinite series $\sum_{i=1}^{\infty} a_i$ converges if and only if the sequence $\langle \sum_{i=1}^{n} a_i \colon n \in \mathbb{N}\rangle$ of its finite partial sums is Cauchy. If $n \geq m$, then $\sum_{i=1}^{n} a_i - \sum_{i=1}^{m-1} a_i = \sum_{i=m}^{n} a_i$, so the condition stated in the proposition is equivalent to the assertion that $\langle \sum_{i=1}^{n} a_i \colon n \in \mathbb{N}\rangle$ is a Cauchy sequence. □

Proposition 3.35. *If $\sum_{n=1}^{\infty} a_n$ is a convergent series, then* $\lim_{n\to\infty} a_n = 0$.

Proof. Suppose $\sum_{n=1}^{\infty} a_n$ is a convergent series, and let $\varepsilon > 0$. By the previous proposition (with $m = n$), there is some $N \in \mathbb{N}$ such that if $n \geq N$, then $\left|\sum_{i=n}^{n} a_i\right| = |a_n| < \varepsilon$. Thus, $\lim_{n\to\infty} a_n = 0$. □

It is important to understand that the converse of this proposition is not true. That is, there are infinite series $\sum_{n=1}^{\infty} a_n$ such that $\lim_{n\to\infty} a_n = 0$, and yet the series does not converge. One such series is described in the following example.

Example 3.36 (The harmonic series). Consider the series $\sum_{n=1}^{\infty} \frac{1}{n}$, which is known as the *harmonic series*. This series diverges to ∞. To see this, group the terms of the series into blocks as follows:

$$1 + \tfrac{1}{2} + \underbrace{\tfrac{1}{3} + \tfrac{1}{4}}_{>1/2} + \underbrace{\tfrac{1}{5} + \tfrac{1}{6} + \tfrac{1}{7} + \tfrac{1}{8}}_{>1/2} + \underbrace{\tfrac{1}{9} + \cdots + \tfrac{1}{16}}_{>1/2} + \underbrace{\tfrac{1}{17} + \cdots + \tfrac{1}{32}}_{>1/2} + \ldots.$$

To put it another way, within each of these groups the sum $\sum_{i=2^n+1}^{2^{n+1}} \frac{1}{n}$ contains 2^n terms all of size $\geq \frac{1}{2^{n+1}}$, and so $\sum_{i=2^n+1}^{2^{n+1}} \frac{1}{n} \geq 2^n \frac{1}{2^{n+1}} = \frac{1}{2}$. Applying Proposition 3.34, $\sum_{n=1}^{\infty} \frac{1}{n}$ does not converge.

On the other hand, because every term of this infinite series is positive, the sequence of finite partial sums is nondecreasing. By Theorem 3.31, the sequence of finite partial sums converges in the extended reals to its supremum, which by the previous paragraph is ∞.

In the following two theorems, we use some of our results concerning limits of sequences to conclude that various infinite series converge.

Theorem 3.37. *Suppose $\sum_{i=1}^{\infty} a_i$ and $\sum_{i=1}^{\infty} b_i$ are infinite series that converge in the extended reals.*

(1) *Given $c \in \mathbb{R}$, the series $\sum_{i=1}^{\infty} ca_i$ converges in the extended reals, and $\sum_{i=1}^{\infty} ca_i = c\sum_{i=1}^{\infty} a_i$ (unless $c = 0$ and $\sum_{i=1}^{\infty} a_i = \pm\infty$).*

(2) *If $m > 1$, the series $\sum_{i=m}^{\infty} a_i$ converges in the extended reals, and $\sum_{i=1}^{\infty} a_i = a_1 + \cdots + a_{m-1} + \sum_{i=m}^{\infty} a_i$. If $m < 1$, the series $\sum_{i=m}^{\infty} a_i$ converges in the extended reals, and $a_1 + \cdots + a_{m-1} + \sum_{i=1}^{\infty} a_i = \sum_{i=m}^{\infty} a_i$.*

(3) *If $\sum_{i=1}^{\infty} a_i + \sum_{i=1}^{\infty} b_i$ is well-defined, then $\sum_{i=1}^{\infty} a_i + b_i$ converges in the extended reals, and $\sum_{i=1}^{\infty} a_i + b_i = \sum_{i=1}^{\infty} a_i + \sum_{i=1}^{\infty} b_i$.*

Proof. (1) For any n, the nth finite partial sums of these two series are equal: $\sum_{i=1}^{n} ca_i = c\sum_{i=1}^{n} a_i$. Thus, by the extended Algebraic Limit Theorem,

$$\sum_{i=1}^{\infty} ca_i = \lim_{n\to\infty} \sum_{i=1}^{n} ca_i = \lim_{n\to\infty} c\sum_{i=1}^{n} a_i = c\sum_{i=1}^{\infty} a_i.$$

(2) In the case $m > 1$, the nth finite partial sum of the new series is $a_1 + \cdots + a_{m-1} + \sum_{i=1}^{n} a_i$. Thus,

$$\sum_{i=m}^{\infty} a_i = \lim_{n\to\infty} a_1 + \cdots + a_{m-1} \sum_{i=m}^{n} a_i = a_1 + \cdots + a_{m-1} + \sum_{i=1}^{\infty} a_i.$$

The case with $m < 1$ is handled similarly.

(3) For any n, the nth finite partial sums of these two series are equal: $\sum_{i=1}^{n} a_i + b_i = \sum_{i=1}^{n} a_i + \sum_{i=1}^{n} b_i$. By the extended Algebraic Limit Theorem,

$$\begin{aligned}\sum_{i=1}^{\infty} a_i + b_i &= \lim_{n\to\infty} \sum_{i=1}^{n} a_i + b_i = \lim_{n\to\infty} \left(\sum_{i=1}^{n} a_i + \sum_{i=1}^{n} b_i\right) \\ &= \lim_{n\to\infty} \sum_{i=1}^{n} a_i + \lim_{n\to\infty} \sum_{i=1}^{n} b_i = \sum_{i=1}^{\infty} a_i + \sum_{i=1}^{\infty} b_i,\end{aligned}$$

provided that this last sum is well-defined. □

Theorem 3.38 (Comparison test). *Suppose $\sum_{i=1}^{\infty} a_i$ and $\sum_{i=1}^{\infty} b_i$ are infinite series, and $|a_i| \le b_i$ for all $i \in \mathbb{N}$. If $\sum_{i=1}^{\infty} b_i$ converges (to a real number), then $\sum_{i=1}^{\infty} a_i$ converges also, and $\sum_{i=1}^{\infty} a_i \le \sum_{i=1}^{\infty} b_i$.*

Proof. Observe that $|a_i| \le b_i$ for all i implies $a_i \ge 0$ for all i. Fix $\varepsilon > 0$. Using the Cauchy criterion, there is some $N \in \mathbb{N}$ such that if $n \ge m \ge N$, then $\left|\sum_{i=m}^{n} b_i\right| = \sum_{i=m}^{n} b_i < \varepsilon$. But then,

$$\left|\sum_{i=m}^{n} a_i\right| \le \sum_{i=m}^{n} |a_i| \le \sum_{i=m}^{n} b_i < \varepsilon$$

whenever $n \ge m \ge N$. Applying the Cauchy criterion in the other direction, this means $\sum_{i=1}^{\infty} a_i$ converges. Furthermore, $\sum_{i=1}^{n} a_i \le \sum_{i=1}^{n} |a_i| \le \sum_{i=1}^{n} b_i$ for all n, and therefore

$$\sum_{i=1}^{\infty} a_i = \lim_{n\to\infty} \sum_{i=1}^{n} a_i \le \lim_{n\to\infty} \sum_{i=1}^{n} b_i = \sum_{i=1}^{\infty} b_i$$

by Theorem 3.29. □

Example 3.39. Let $a_i = (-1)^i$, and consider the infinite series

$$\sum_{i=0}^{\infty} a_i = 1 - 1 + 1 - 1 + 1 - 1 + 1 - 1 + \ldots.$$

On the one hand, this series diverges by oscillation because its sequence of finite partial sums is $\langle 1, 0, 1, 0, 1, 0, 1, 0, \ldots \rangle$. On the other hand, we can obtain a convergent series from this one simply by regrouping the terms:

$$\sum_{i=0}^{\infty} (a_{2i} + a_{2i+1}) = (1-1) + (1-1) + (1-1) + (1-1) + \ldots = 0.$$ ☕

This example shows that, unlike finite sums, infinite series are not associative. Apparently, simply adding in parentheses can change a divergent series into a convergent series. More on the associativity of infinite series is explored in the exercises. Next we turn our attention to commutativity. The following example shows that infinite series need not be commutative, i.e., rearranging the order of the terms can affect the value of the series.

Example 3.40. Consider the infinite series

$$1 - 1 + \tfrac{1}{2} - \tfrac{1}{2} + \tfrac{1}{2} - \tfrac{1}{2} + \underbrace{\tfrac{1}{4} - \tfrac{1}{4} + \dots + \tfrac{1}{4} - \tfrac{1}{4}}_{\text{4 pairs of terms}} + \underbrace{\tfrac{1}{8} - \tfrac{1}{8} + \dots + \tfrac{1}{8} - \tfrac{1}{8}}_{\text{8 pairs of terms}} + \dots$$

This series converges to 0. Rearranging the order of the terms in this series, we get the new series:

$$1 - 1 + \tfrac{1}{2} + \tfrac{1}{2} - \tfrac{1}{2} - \tfrac{1}{2} + \underbrace{\tfrac{1}{4} + \dots + \tfrac{1}{4}}_{\text{4 terms}} - \underbrace{\tfrac{1}{4} - \dots - \tfrac{1}{4}}_{\text{4 terms}} + \underbrace{\tfrac{1}{8} + \dots + \tfrac{1}{8}}_{\text{8 terms}} + \dots$$

This latter series has infinitely many finite partial sums equal to 1 and infinitely many finite partial sums equal to 0, so it diverges by oscillation.

Next, let us rearrange the order of the terms in a different way as follows:

$$\begin{aligned} 1 + \tfrac{1}{2} + \tfrac{1}{2} - 1 + \underbrace{\tfrac{1}{4} + \dots + \tfrac{1}{4}}_{\text{4 terms}} + \underbrace{\tfrac{1}{8} + \dots + \tfrac{1}{8}}_{\text{8 terms}} - \tfrac{1}{2} - \tfrac{1}{2} \\ + \underbrace{\tfrac{1}{16} + \dots + \tfrac{1}{16}}_{\text{16 terms}} + \underbrace{\tfrac{1}{32} + \dots + \tfrac{1}{32}}_{\text{32 terms}} - \underbrace{\tfrac{1}{4} - \dots - \tfrac{1}{4}}_{\text{4 terms}} + \underbrace{\tfrac{1}{64} + \dots + \tfrac{1}{64}}_{\text{64 terms}} + \dots \end{aligned}$$

This series diverges to ∞.

Alternatively, the terms of this series can be reordered so that the new series still converges, but to a different value than the original: The series

$$\begin{aligned} 1 + \tfrac{1}{2} + \tfrac{1}{2} - 1 + \underbrace{\tfrac{1}{4} + \dots + \tfrac{1}{4}}_{\text{4 terms}} - \tfrac{1}{2} + \underbrace{\tfrac{1}{8} + \dots + \tfrac{1}{8}}_{\text{4 terms}} - \tfrac{1}{2} \\ + \underbrace{\tfrac{1}{8} + \dots + \tfrac{1}{8}}_{\text{4 terms}} - \tfrac{1}{4} + \underbrace{\tfrac{1}{16} + \dots + \tfrac{1}{16}}_{\text{4 terms}} - \tfrac{1}{4} + \underbrace{\tfrac{1}{16} + \dots + \tfrac{1}{16}}_{\text{4 terms}} - \dots \end{aligned}$$

converges to 2.

This example shows that rearranging the order of the terms in an infinite series can lead to a variety of behaviors: convergence to a different value, divergence to infinity, divergence by oscillation... seemingly anything at all. On the other hand, some series are not affected by rearrangement at all: Clearly, the infinite sum $0 + 0 + 0 + 0 + \dots$ converges to 0, and rearranging the terms has no affect whatsoever. In the remainder of this section, we will discover precisely when the outcome of a convergent series can be changed by rearranging its terms and precisely what effects can be achieved by rearrangement.

Definition 3.41. The *positive part* of an infinite series $\sum_{n=1}^{\infty} a_n$ is the extended real number $\sum_{n=1}^{\infty} \frac{1}{2}(a_n + |a_n|)$. Observe that

$$\tfrac{1}{2}(a_n + |a_n|) = \begin{cases} a_n & \text{if } a_n \geq 0, \\ 0 & \text{if } a_n < 0. \end{cases}$$

Thus, an alternative description of the positive part of $\sum_{n=1}^{\infty} a_n$ is that it is the sum of all the nonnegative terms of $\sum_{n=1}^{\infty} a_n$, ignoring the negative terms completely. The *negative part* of $\sum_{n=1}^{\infty} a_n$ is $\sum_{n=1}^{\infty} \frac{1}{2}(a_n - |a_n|)$, which is the sum of all the negative terms of $\sum_{n=1}^{\infty} a_n$.

Because every term of the series $\sum_{n=1}^{\infty} \frac{1}{2}(a_n + |a_n|)$ is nonnegative, the sequence of finite partial sums of this series is nondecreasing, so it converges in the extended reals by Theorem 3.31. Similarly, the series $\sum_{n=1}^{\infty} \frac{1}{2}(a_n - |a_n|)$ converges because all its terms are nonpositive, which means its sequence of finite partial sums is nonincreasing. Thus, the positive and negative parts of an infinite series are well-defined extended real numbers.

Proposition 3.42. *Let $\sum_{n=1}^{\infty} a_n$ be an infinite series. The series $\sum_{n=1}^{\infty} |a_n|$ converges (to a real number) if and only if the positive and negative parts of $\sum_{n=1}^{\infty} a_n$ are both finite.*

Proof. For the "if" direction, suppose the positive and negative parts of $\sum_{n=1}^{\infty} a_n$ are both finite. By Theorem 3.37, this implies

$$\sum_{n=1}^{\infty} |a_n| = \sum_{n=1}^{\infty} \tfrac{1}{2}(a_n + |a_n|) + \sum_{n=1}^{\infty} \tfrac{1}{2}(a_n - |a_n|)$$

also converges. For the "only if" direction, suppose $\sum_{i=1}^{\infty} |a_n|$ converges. Because $\frac{1}{2}(a_n + |a_n|) \leq |a_n|$ and $\frac{1}{2}(a_n - |a_n|) \leq |a_n|$ for all n, the comparison test implies that $\sum_{n=1}^{\infty} \frac{1}{2}(a_n + |a_n|)$ and $\sum_{n=1}^{\infty} \frac{1}{2}(a_n - |a_n|)$ both converge (to at most $\sum_{n=1}^{\infty} |a_n|$). □

Definition 3.43. An infinite series $\sum_{n=1}^{\infty} a_n$ is *absolutely convergent* if the series $\sum_{n=1}^{\infty} |a_n|$ converges. A series that converges to a real number, but is not absolutely convergent, is called *conditionally convergent*.

Example 3.44. The infinite series from the beginning of Example 3.40 is convergent, but not absolutely convergent. Thus, it is an example of a conditionally convergent series.

Definition 3.45. If $\sum_{n=1}^{\infty} a_n$ is an infinite series and $f : \mathbb{N} \to \mathbb{N}$ is a bijection, the series $\sum_{n=1}^{\infty} a_{f(n)}$ is called a *rearrangement* of $\sum_{n=1}^{\infty} a_n$.

Put together, the next two theorems reveal the dividing line between commutative and noncommutative series: Absolutely convergent series are not affected by rearrangement, but conditionally convergent series are, and furthermore, conditionally convergent series can be rearranged to have any behavior whatsoever. These theorems were first proved by the German mathematician Bernhard Riemann in 1868.

Theorem 3.46. *If $\sum_{i=1}^{\infty} a_i$ is an absolutely convergent series, then every rearrangement of the series converges to the same value as the original series.*

Proof. Suppose $\sum_{i=1}^{\infty} a_i$ is an absolutely convergent infinite series, and suppose $f : \mathbb{N} \to \mathbb{N}$ is a bijection. Let $r = \sum_{i=1}^{\infty} a_i$, and let $\varepsilon > 0$.

There is some $N_0 \in \mathbb{N}$ such that if $n \geq N_0$, then $\left|r - \sum_{i=1}^{n} a_i\right| < \varepsilon/2$. Applying the Cauchy criterion to the convergent series $\sum_{i=1}^{\infty} |a_i|$ converges, fix $N_1 \in \mathbb{N}$ such that if $n \geq m \geq N_1$, then $\left|\sum_{i=m}^{n} |a_i|\right| = \sum_{i=m}^{n} |a_i| < \varepsilon/2$. Let $N = \max\{N_0, N_1\}$.

Let $M = \max\{f^{-1}(1), f^{-1}(2), \ldots, f^{-1}(N)\}$, and let $m \geq M$. This choice of M implies that the finite partial sum $\sum_{i=1}^{m} a_{f(i)}$ includes the terms $a_1, a_2, \ldots, a_N$, though perhaps in a different order and with other terms added in as well. Consequently, the terms $a_1, a_2, \ldots, a_N$ all cancel in the difference $\sum_{i=1}^{m} a_{f(i)} - \sum_{i=1}^{N} a_i$, and applying the general rule that $|a + b| \leq |a| + |b|$ (i.e., the triangle inequality in $\mathbb{R}$) repeatedly, we get

$$\left|\sum_{i=1}^{m} a_{f(i)} - \sum_{i=1}^{N} a_i\right| \leq \sum_{i=1}^{m} |a_{f(i)}| - \sum_{i=1}^{N} |a_i|.$$

Let $\ell = \max\{f(1), f(2), \ldots, f(m)\}$. Then,

$$\left|\sum_{i=1}^{m} a_{f(i)} - \sum_{i=1}^{N} a_i\right| \leq \sum_{i=1}^{m} |a_{f(i)}| - \sum_{i=1}^{N} |a_i| \leq \sum_{i=N+1}^{\ell} |a_i| < \varepsilon/2,$$

where the last inequality follows from our choice of N_0. Consequently,

$$\begin{aligned}\left|\sum_{i=1}^{m} a_{f(i)} - r\right| &= \left|\sum_{i=1}^{m} a_{f(i)} - \sum_{i=1}^{N} a_i + \sum_{i=1}^{N} a_i - r\right| \\ &\leq \left|\sum_{i=1}^{m} a_{f(i)} - \sum_{i=1}^{N} a_i\right| + \left|\sum_{i=1}^{N} a_i - r\right| < \varepsilon/2 + \varepsilon/2 = \varepsilon.\end{aligned}$$

Thus, $\sum_{i=1}^{\infty} a_{f(i)} = \lim_{m\to\infty} \sum_{i=1}^{m} a_{f(i)} = r$. □

Lemma 3.47. *Every conditionally convergent series has an infinite positive part and an infinite negative part.*

Proof. If an infinite series has a finite positive part and a finite negative part, then it is absolutely convergent by Proposition 3.42. If the positive or negative part is infinite (but not both), then $\sum_{i=1}^{\infty} \frac{1}{2}(a_i + |a_i|) + \sum_{i=1}^{\infty} \frac{1}{2}(a_i - |a_i|)$ is a well-defined sum in the extended reals, and therefore

$$\begin{aligned}\sum_{i=1}^{\infty} a_i &= \sum_{i=1}^{\infty} \left(\tfrac{1}{2}(a_i + |a_i|) + \tfrac{1}{2}(a_i - |a_i|)\right) \\ &= \sum_{i=1}^{\infty} \tfrac{1}{2}(a_i + |a_i|) + \sum_{i=1}^{\infty} \tfrac{1}{2}(a_i - |a_i|)\end{aligned}$$

by Theorem 3.37(3). But this sum is either ∞ or $-\infty$, so in this case $\sum_{i=1}^{\infty} a_i$ does not converge, but diverges to ∞ or to $-\infty$. □

Theorem 3.48 (Riemann rearrangement theorem). *Suppose $\sum_{i=1}^{\infty} a_i$ is a conditionally convergent series. For every $r \in \mathbb{R}$, there is a rearrangement of $\sum_{i=1}^{\infty} a_i$ that converges to r. There are also rearrangements of $\sum_{i=1}^{\infty} a_i$ that diverge to ∞, rearrangements that diverge to $-\infty$, and rearrangements that diverge by oscillation.*

Proof. Rather than dealing with each case separately, we prove all the assertions in the theorem at once. To do this, first define a sequence $\langle r_n : n \in \mathbb{N} \rangle$ of real numbers as follows, depending on which case we wish to consider:

- To obtain a rearrangement of $\sum_{i=1}^{\infty} a_i$ converging to some real number r, let $r_n = r$ for all n.
- To obtain a rearrangement diverging to ∞, let $r_n = n$ for all n.
- To obtain a rearrangement diverging to $-\infty$, let $r_n = -n$ for all n.
- To get a rearrangement diverging to $-\infty$, let $r_n = (-1)^{n+1}$ for all n.

By Lemma 3.47, the positive and negative parts of $\sum_{i=1}^{\infty} a_i$ are both infinite. Among other things, this implies there are infinitely many positive terms and infinitely many negative terms. Let $pos(n)$ denote the nth member of $\{i \in \mathbb{N} : a_i \geq 0\}$, and let $neg(n)$ denote the nth member of $\{i \in \mathbb{N} : a_i < 0\}$. Because the positive and negative parts of $\sum_{i=1}^{\infty} a_i$ are both infinite,

$$\sum_{i=1}^{\infty} a_{pos(i)} = \infty \qquad \text{and} \qquad \sum_{i=1}^{\infty} a_{neg(i)} = -\infty.$$

We now describe how to get a bijection $f : \mathbb{N} \to \mathbb{N}$, built via recursion from the sequence $\langle r_n : n \in \mathbb{N} \rangle$, such that the corresponding rearrangement of $\sum_{i=1}^{\infty} a_i$ has the desired behavior. Roughly, the idea is to alternate blocks of positive terms with blocks of negative terms, so that, by the end of the nth block of positive terms, our rearrangement has a finite partial sum just above r_{2n-1}, and by the nth block of negative terms, the finite partial sum is just below r_{2n}.

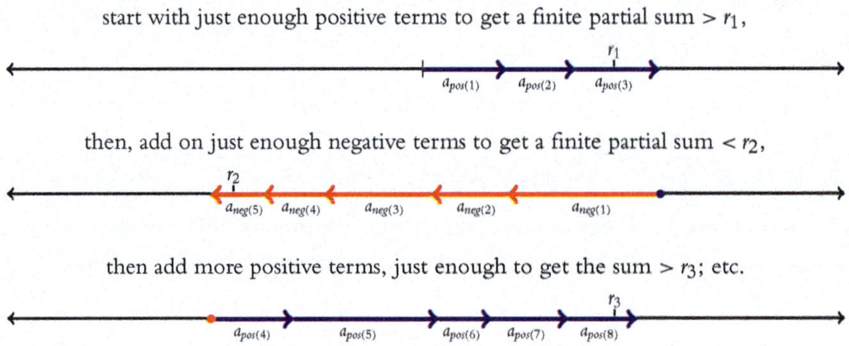

For the first step of the recursion, let p_1 be the least element of $\mathbb{N}$ with $\sum_{i=1}^{p_1} a_{pos(i)} > r_1$, and then let n_1 be the least element of $\mathbb{N}$ with $\sum_{i=1}^{p_1} a_{pos(i)} + \sum_{i=1}^{p_1} a_{neg(i)} < r_2$. Define $f(i) = pos(i)$ if $i \leq p_1$, and $f(i) = neg(i - p_1)$ if $p_1 < i \leq p_1 + n_1$. This defines $f(i)$ for all $i \leq p_1 + n_1$. Note that $\sum_{i=1}^{p_1} a_{f(i)} = \sum_{i=1}^{p_1} a_{pos(i)}$ and $\sum_{i=1}^{p_1+n_1} a_{f(i)} = \sum_{i=1}^{p_1} a_{pos(i)} + \sum_{i=1}^{n_1} a_{neg(i)}$.

In general, for the kth step of the recursion, suppose we have already found some natural number p_{k-1} and n_{k-1} and have already defined $f(i)$ for $i \leq p_{k-1} + n_{k-1}$ in such a way that

$$\{f(i) : i \leq p_{k-1} + n_{k-1}\} = \{pos(i) : i \leq p_{k-1}\} \cup \{neg(i) : i \leq n_{k-1}\}.$$

In other words, f maps $\{1, 2, \ldots, p_{k-1} + n_{k-1}\}$ bijectively to the first p_{k-1} indices of positive terms plus the first n_{k-1} indices of negative terms of our infinite series. Let p_k be the least element of $\mathbb{N}$ such that $p_k > p_{k-1}$ and $\sum_{i=1}^{p_k} a_{pos(i)} + \sum_{i=1}^{n_{k-1}} a_{neg(i)} > r_{2k-1}$, and then let n_k be the least element of $\mathbb{N}$ such that $n_k > n_{k-1}$ and $\sum_{i=1}^{p_k} a_{pos(i)} + \sum_{i=1}^{n_k} a_{neg(i)} < r_{2k}$. Define

$$f(i) = \begin{cases} pos(i - n_{k-1}) & \text{if } p_{k-1} + n_{k-1} < i \leq p_k + n_{k-1}, \\ neg(i - p_k) & \text{if } p_k + n_{k-1} < i \leq p_k + n_k. \end{cases}$$

Because $f(i)$ was already defined for $i \leq p_{k-1} + n_{k-1}$, this defines $f(i)$ for all $i \leq p_k + n_k$. Note that $\sum_{i=1}^{p_k+n_{k-1}} a_{f(i)} = \sum_{i=1}^{p_k} a_{pos(i)} + \sum_{i=1}^{n_{k-1}} a_{neg(i)}$, and $\sum_{i=1}^{p_k+n_k} a_{f(i)} = \sum_{i=1}^{p_k} a_{pos(i)} + \sum_{i=1}^{n_k} a_{neg(i)}$.

This completes the recursion. It is clear from the construction that f is a bijection $\mathbb{N} \to \mathbb{N}$, and we claim that the rearrangement $\sum_{i=1}^{\infty} a_{f(i)}$ has the desired properties. Let us analyze the first case, where $r_n = r$ for all $n \in \mathbb{N}$.

For each $k \in \mathbb{N}$, both $\sum_{i=1}^{p_k} a_{pos(i)} + \sum_{i=1}^{n_{k-1}} a_{neg(i)} > r_{2k-1} = r$ and $\sum_{i=1}^{p_k} a_{pos(i)} + \sum_{i=1}^{n_k} a_{neg(i)} < r_{2k} = r$ are finite partial sums of the rearrangement (the $(p_k + n_{k-1})^{\text{th}}$ and $(p_k + n_k)^{\text{th}}$ finite partial sums, respectively). Thus $\sum_{i=1}^{\infty} a_{f(i)}$ has infinitely many finite partial sums $> r$ and infinitely many $< r$; so, if it converges at all, then it converges to r.

To see that $\sum_{i=1}^{\infty} a_{f(i)}$ converges, let $\varepsilon > 0$. Because the original series converges, there is some $N_0 \in \mathbb{N}$ such that if $i \geq N_0$, then $|a_i| < \varepsilon$. There is some M large enough that if $m \geq M$, then $f(m) \geq N$. Let $m \geq M$, and suppose k is large enough that $m \leq p_k$. Then,

$$\sum_{i=1}^{p_k+n_{k-1}-1} a_{f(i)} < r \qquad \text{and} \qquad \sum_{i=1}^{p_k+n_{k-1}} a_{f(i)} > r,$$

which means $\left|r - \sum_{i=1}^{p_k+n_{k-1}} a_{f(i)}\right| < |a_{f(p_k+n_{k-1})}| < \varepsilon$. Similarly,

$$\sum_{i=1}^{p_k+n_k-1} a_{f(i)} > r \qquad \text{and} \qquad \sum_{i=1}^{p_k+n_k} a_{f(i)} < r,$$

which means $\left|r - \sum_{i=1}^{p_k+n_k} a_{f(i)}\right| < |a_{f(p_k+n_k)}| < \varepsilon$. Furthermore, the finite partial sums of the rearrangement are nonincreasing from $\sum_{i=1}^{p_k+n_{k-1}} a_{f(i)}$ to $\sum_{i=1}^{p_k+n_k} a_{f(i)}$ because only negative terms are being added in. In other words, if ℓ is between $p_k + n_{k-1}$ and $\leq p_k + n_k$, then $\sum_{i=1}^{\ell} a_{f(i)}$ is between $\sum_{i=1}^{p_k+n_{k-1}} a_{f(i)}$ and $\sum_{i=1}^{p_k+n_k} a_{f(i)}$. Thus, $r - \varepsilon < \sum_{i=1}^{\ell} a_{f(i)} < r + \varepsilon$ whenever $p_k + n_{k-1} \leq \ell \leq p_k + n_k$. Similarly, $r + \varepsilon > \sum_{i=1}^{\ell} a_{f(i)} > r + \varepsilon$ whenever $p_k + n_k \leq \ell \leq p_{k+1} + n_k$. This is true for all sufficiently large values of k, and therefore $\left|r - \sum_{i=1}^{\ell} a_{f(i)}\right| < \varepsilon$ for all sufficiently large values of ℓ. In other words, $\sum_{i=1}^{\infty} a_{f(i)} = \lim_{\ell\to\infty} \sum_{i=1}^{\ell} a_{f(i)} = r$.

For the case where $r_n = n$ for all n, a similar argument shows that for any $\varepsilon > 0$, if k is large enough, then $\sum_{i=1}^{\ell} a_{f(i)} > k - \varepsilon$ whenever $\ell \leq p_k + n_k$. This shows that $\sum_{i=1}^{\infty} a_{f(i)} = \lim_{\ell\to\infty} \sum_{i=1}^{\ell} a_{f(i)} = \infty$. In the case $r_n = -n$ for all n, we get $\sum_{i=1}^{\infty} a_{f(i)} = \lim_{\ell\to\infty} \sum_{i=1}^{\ell} a_{f(i)} = -\infty$ instead.

In the final case, where $r_n = (-1)^{n+1}$ for all n, we get $\sum_{i=1}^{p_k} a_{pos(i)} + \sum_{i=1}^{n_{k-1}} a_{neg(i)} > r_{2k-1} = 1$ and $\sum_{i=1}^{p_k} a_{pos(i)} + \sum_{i=1}^{n_k} a_{neg(i)} < r_{2k} = -1$ for all k. Thus, $\sum_{i=1}^{\infty} a_{f(i)}$ has infinitely many finite partial sums > 1 and infinitely many < -1, and therefore $\sum_{i=1}^{\infty} a_{f(i)}$ diverges by oscillation. □

Observe that this proof works not just for conditionally convergent series, but for any series $\sum_{i=1}^{\infty} a_i$ such that

- the positive and negative parts are both infinite, and
- $\lim_{i\to\infty} |a_i| = 0$.

Any such series can be rearranged to exhibit any of the behaviors described in the Riemann rearrangement theorem. Series satisfying the first condition but not the second can be rearranged to diverge to ∞ or $-\infty$, or to diverge by oscillation, like the series $1-1+1-1+1-1+1-\ldots$, but such series cannot be rearranged to converge to any $r \in \mathbb{R}$. This requires $\lim_{i\to\infty} |a_i| = 0$, by Proposition 3.35, and this limit cannot be changed to 0 by rearrangement.

The Baire Category Theorem

Recall from Chap. 1 that $D \subseteq \mathbb{R}$ is dense if every non-empty open interval in $\mathbb{R}$ contains a point of D. The following definition generalizes this idea from $\mathbb{R}$ to arbitrary metric spaces.

Definition 3.49. Let X be a metric space. A set $A \subseteq X$ is *dense* in X if $A \cap U \neq \emptyset$ for every non-empty open $U \subseteq X$. ☕

Proposition 3.50. *If X is a metric space and $A \subseteq X$, then A is dense in X if and only if $\overline{A} = X$.*

Proof. For the "if" direction, let us prove the contrapositive. Suppose A is not dense in X, and let $U \subseteq X$ be a non-empty open set with $U \cap A = \emptyset$. Then, $X \setminus U$ is a closed set containing A, so $\overline{A} \subseteq X \setminus U \neq X$.

For the "only if" direction, suppose A is dense in X. Let $x \in X$. For every $r > 0$, $B_r(x)$ is a non-empty open subset of X, so $A \cap B_r(x) \neq \emptyset$. As r was arbitrary, this means $x \in \overline{A}$. As x was arbitrary, this means $\overline{A} = X$. □

The following theorem, first proved by René-Louis Baire in 1899, is known as the *Baire Category Theorem*, or the *BCT* for short.

Theorem 3.51 (Baire Category Theorem). *Suppose X is a complete metric space, and $\mathcal{G}$ is a countable collection of dense open subsets of X. Then, $\bigcap \mathcal{G}$ is a dense subset of X.*

Example 3.52. Because $\mathbb{R}$ is a complete metric space, the BCT applies with $X = \mathbb{R}$. For each $r \in \mathbb{R}$, the set $\mathbb{R} \setminus \{r\}$ is a dense open subset of $\mathbb{R}$. If $A \subseteq \mathbb{R}$, then $\mathbb{R} \setminus A = \bigcap_{a\in A} (\mathbb{R} \setminus \{a\})$. If A is a countable subset of $\mathbb{R}$, then taking $\mathcal{G} = \{\mathbb{R} \setminus \{a\} : a \in A\}$, the Baire Category Theorem tells us that $\mathbb{R} \setminus A = \bigcap_{a\in A} (\mathbb{R} \setminus \{a\}) = \bigcap \mathcal{G}$ is dense in $\mathbb{R}$. ☕

Lemma 3.53. *Suppose X is a metric space and $x \in X$.*

(1) $\overline{B_r(x)} \subseteq \{y \in X : d(x, y) \leq r\}$ *for any $r > 0$, and*

(2) *For any open set $U \ni x$, there is some $r > 0$ such that $\overline{B_r(x)} \subseteq U$.*

Proof. Fix $r > 0$. To prove (1), let $y \in X$, and suppose $d(x, y) > r$. Let $\delta = d(x, y) - r$, and observe that if $z \in B_\delta(y)$, then

$$d(x, z) \geq d(x, y) - d(y, z) > d(x, y) - \delta = r,$$

which means $z \notin B_r(x)$. Hence, $B_\delta(y) \cap B_r(x) = \emptyset$. This shows $y \notin \overline{B_r(x)}$, which completes the proof of (1). For (2), note that if $x \in U$, then there is some r_0 such that $B_{r_0}(x) \subseteq U$ (because U is open). Setting $r = r_0/2$, the first part of the lemma implies that $\overline{B_r(x)} \subseteq \{y \in X \colon d(x, y) \leq r\} \subseteq \{y \in X \colon d(x, y) < r_0\} = B_{r_0}(x) \subseteq U$. □

Proof of Theorem 3.51. Suppose $\mathcal{G}$ is a countable collection of dense open subsets of X. If $\mathcal{G}$ is countably infinite, we may list the elements of $\mathcal{G}$ as $U_1, U_2, U_3, \ldots$. If $\mathcal{G}$ is finite, we may list the elements of $\mathcal{G}$ as $U_1, U_2, \ldots, U_k$, and then we may define $U_n = X$ for all $n > k$. Either way, we have a sequence $U_1, U_2, U_3, \ldots$ of dense open subsets of X with $\bigcap \mathcal{G} = \bigcap_{n=1}^{\infty} U_n$.

Let U be a non-empty open subset of X. Our goal is to show that $U \cap \bigcap_{n=1}^{\infty} U_n \neq \emptyset$.

To do this, we construct, via recursion, a sequence $\langle x_0, x_1, x_2, x_3, \ldots \rangle$ of points in X, and a sequence $\langle r_0, r_1, r_2, r_3, \ldots \rangle$ of positive real numbers, as follows. For the base step, let $x_0 \in U$, and fix some $r_0 > 0$ such that $\overline{B_{r_0}(x_0)} \subseteq U$. (Note that some such r_0 exists by Lemma 3.53.)

Now, suppose $n > 0$, and assume we already have some $x_{n-1} \in X$ and $r_{n-1} > 0$. Note that $U_n \cap B_{r_{n-1}}(x_{n-1}) \neq \emptyset$ because U_n is dense in X. Thus we may choose some

- $x_n \in U_n \cap B_{r_{n-1}}(x_{n-1})$,

and then choose some $r_n > 0$ such that

- $r_n \leq 1/n$, and
- $\overline{B_{r_n}(x_n)} \subseteq U_n \cap B_{r_{n-1}}(x_{n-1})$.

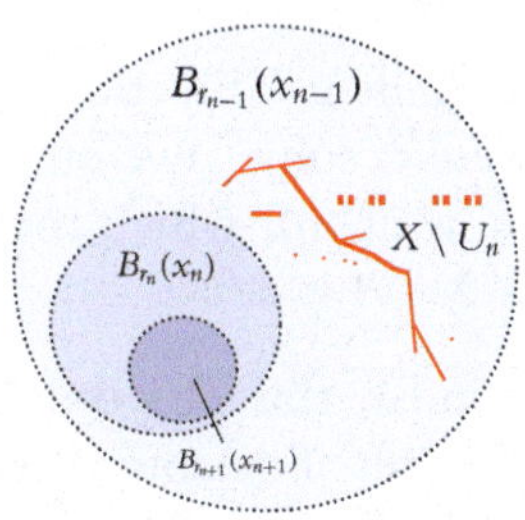

The fact that there is some number r_n satisfying these two properties follows from Lemma 3.53.

This completes the recursive construction. Observe that if $m < n$, then

$$x_n \in B_{r_n}(x_n) \subseteq \overline{B_{r_n}(x_n)} \subseteq B_{r_{n-1}}(x_{n-1}) \subseteq \cdots \subseteq \overline{B_{r_{m+1}}(x_{m+1})} \subseteq B_{r_m}(x_m).$$

We claim that the sequence $\langle x_n \colon n \in \mathbb{N} \rangle$ is Cauchy. To see this, let $\varepsilon > 0$, and fix some $N \in \mathbb{N}$ such that $2/N < \varepsilon$. If $m, n \geq N$, then $x_m, x_n \in B_{r_N}(x_N)$ (by the previous paragraph), and thus

$$d(x_m, x_n) \leq d(x_m, x_N) + d(x_N, x_n) < r_N + r_N \leq 1/N + 1/N < \varepsilon.$$

Because X is complete, every Cauchy sequence in X converges: So, there is some $x \in X$ such that $x = \lim_{n\to\infty} x_n$. Now fix some whole number $n \geq 0$. By construction, $\langle x_n, x_{n+1}, x_{n+2}, \ldots \rangle$ is a sequence of points in $B_{r_n}(x_n)$. This implies that the limit of this sequence is in $\overline{B_{r_n}(x_n)}$. But the limit of this sequence is x (because it is a subsequence of $\langle x_m : n \in \mathbb{N}\rangle$). Thus, $x \in \overline{B_{r_n}(x_n)}$ for every $n \geq 0$. Now, $x \in \overline{B_{r_0}(x_0)}$ implies $x \in U$, and if $n > 0$, then $x \in \overline{B_{r_{n-1}}(x_{n-1})}$ implies $x \in U_n$. Therefore, $x \in U \cap \bigcap_{n=1}^{\infty} U_n$, and in particular, $U \cap \bigcap_{n=1}^{\infty} U_n \neq \emptyset$. □

Definition 3.54. Let X be a metric space, and let $A \subseteq X$. We say A is *nowhere dense* in X if, for every non-empty open $U \subseteq X$, there is a non-empty open $V \subseteq U$ such that $V \cap A = \emptyset$. ☕

Example 3.55. The Cantor set K is a nowhere dense subset of $\mathbb{R}$.

.. ..

Recall from Chap. 2 that K contains no non-empty open intervals. To see that K is nowhere dense, suppose U is a non-empty open subset of $\mathbb{R}$. There is some non-empty interval $(a, b) \subseteq U$, and $K \not\supseteq (a, b)$. But then, because K is closed, $V = (a, b) \setminus K$ is a non-empty open set contained in U and disjoint from K. Hence, K is nowhere dense. ☕

Proposition 3.56. *If X is a metric space and $A \subseteq X$, then A is nowhere dense in X if and only if $X \setminus \overline{A}$ is dense.*

Proof. For the "if" direction, suppose $X \setminus \overline{A}$ is dense in X. Let U be a non-empty open subset of X. Because $X \setminus \overline{A}$ is dense, $U \cap (X \setminus \overline{A}) \neq \emptyset$. But this is an open set. Setting $V = U \cap (X \setminus \overline{A})$, we have $V \neq \emptyset$, $V \subseteq U$, and $V \cap \overline{A} = \emptyset$, which imply $V \cap A = \emptyset$.

For the "only if" direction, suppose A is nowhere dense. Given any non-empty open $U \subseteq X$, there is some non-empty open $V \subseteq U$ with $V \cap A = \emptyset$. But then, $V \cap \overline{A} = \emptyset$ (because V is open), which means $U \not\subseteq \overline{A}$. Thus, $U \cap (X \setminus \overline{A}) \neq \emptyset$. As U was arbitrary, this means $X \setminus \overline{A}$ is dense. □

In particular, this proposition implies that if $A \subseteq X$ is closed, then A is nowhere dense if and only if its complement is a dense open set. This observation leads us to the following "dual" version of the BCT:

Theorem 3.57 (Baire Category Theorem, version 2). *Suppose X is a complete metric space and $\mathcal{F}$ is a countable collection of nowhere dense closed subsets of X. Then, $\bigcup \mathcal{F} \neq X$. In fact, $X \setminus \bigcup \mathcal{F}$ is dense in X.*

Proof. Let $\mathcal{F}$ be a countable collection of nowhere dense closed subsets of X. Define

$$\mathcal{G} = \{X \setminus F : F \in \mathcal{F}\}.$$

Then, $\mathcal{G}$ is countable (it is clearly in bijection with $\mathcal{F}$), and by Proposition 3.56, every member of $\mathcal{G}$ is a dense open subset of X. By the first version of the BCT (Theorem 3.51), $\bigcap \mathcal{G}$ is dense in X. But, $\bigcap \mathcal{G} = X \setminus \bigcup \mathcal{F}$ by De Morgan's Laws, so this completes the proof. □

Meager and comeager sets

Definition 3.58. Let X be a metric space. A subset A of X is *meager* if there is a countable collection $\mathcal{F}$ of nowhere dense closed subsets of X that covers A (meaning that $\bigcup \mathcal{F} \supseteq A$).

Example 3.59. Every countable subset of $\mathbb{R}$ is meager. To see this, observe that for every $r \in \mathbb{R}$, the singleton $\{r\}$ is a nowhere dense subset of $\mathbb{R}$. Therefore, if $A \subseteq \mathbb{R}$ is countable, then $\mathcal{F} = \{\{a\} : a \in A\}$ is a countable collection of closed nowhere dense sets that covers A.

Every nowhere dense subset of $\mathbb{R}$ is also meager because, if $A \subseteq \mathbb{R}$ is nowhere dense, then so is $\overline{A}$ (by Proposition 3.56), and so $\{\overline{A}\}$ is a countable collection of closed nowhere dense sets covering A. This means, for example, that the Cantor set C is an uncountable meager subset of $\mathbb{R}$.

Example 3.60. For an example of a *non*meager subset of $\mathbb{R}$, consider $\mathbb{R}$ itself. If there were some countable collection $\mathcal{F}$ of nowhere dense closed subsets of $\mathbb{R}$ that covers $\mathbb{R}$, we would have $\bigcup \mathcal{F} = \mathbb{R}$, contradicting Theorem 3.57.

Similarly, $[a, b]$ is a nonmeager subset of $\mathbb{R}$ whenever $a < b$. To see this, suppose $\mathcal{F}$ is a countable collection of nowhere dense subsets of $\mathbb{R}$. Theorem 3.57 states not only that $\bigcup \mathcal{F} \neq \mathbb{R}$, but that $\mathbb{R} \setminus \bigcup \mathcal{F}$ is dense in $\mathbb{R}$. In particular, $\mathcal{F}$ does not cover the interval $[a, b]$.

Two important facts about countability, proved in Chap. 1, are: (1) a subset of a countable set is countable and (2) a countable union of countable sets is countable. The following theorem shows that the same things are true for meager sets.

Theorem 3.61. *Let X be a metric space.*

1. *If A is a meager subset of X and $B \subseteq A$, then B is also meager.*
2. *If $\mathcal{A}$ is a countable set of meager subsets of X, then $\bigcup \mathcal{A}$ is meager.*

Proof. For (1), suppose $A \subseteq X$ is meager. This means there is a countable collection $\mathcal{F}$ of nowhere dense closed subsets of X with $\bigcup \mathcal{F} \supseteq A$. But if $B \subseteq A$, then $\bigcup \mathcal{F} \supseteq B$ as well, so B is also meager.

For (2), suppose $\mathcal{A}$ is a countable collection of meager subsets of X. For each $A \in \mathcal{A}$, there is a countable collection $\mathcal{F}_A$ of nowhere dense closed subsets of X that covers A. Because a countable union of countable sets is countable, the union of all the $\mathcal{F}_A$'s, namely, $\bigcup \{\mathcal{F}_A : A \in \mathcal{A}\}$, is a countable collection of nowhere dense closed subsets of X. As $\bigcup \{\mathcal{F}_A : A \in \mathcal{A}\}$ covers $\bigcup \mathcal{A}$, this shows $\bigcup \mathcal{A}$ is meager. □

As the name suggests, meagerness is a kind of "smallness" property for subsets of $\mathbb{R}$. It takes uncountably many meager sets to union up to all of $\mathbb{R}$, or even to an interval (a, b). In fact, meagerness is one of three important smallness properties for subsets of

$\mathbb{R}$. The other two are countability, which we studied in Chap. 1, and being "Lebesgue null" or "measure 0," which we study later in Chaps. 6–7. If a set is countable, then it must be both meager and measure 0, but being meager does not imply being measure 0, and being measure 0 does not imply being meager.

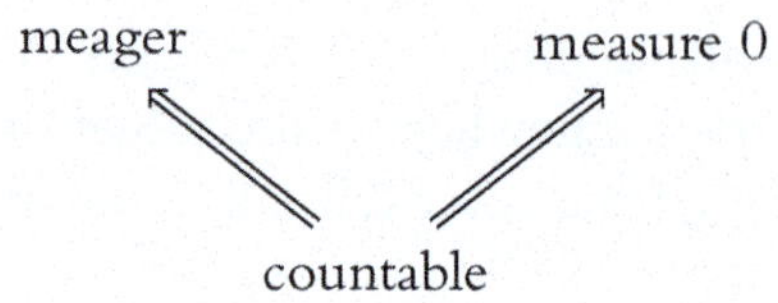

Definition 3.62. Let X be a metric space. A subset A of X is *comeager* if there is a countable set $\mathcal{G}$ of dense open subsets of X, such that $\bigcap \mathcal{G} \subseteq A$.

Theorem 3.63. *A subset of a metric space is comeager if and only if its complement is meager.*

Proof. Let X be a metric space and $A \subseteq X$.

Suppose A is comeager. There is a countable collection $\mathcal{G}$ of dense open subsets of X with $\bigcap \mathcal{G} \subseteq A$. Let $\mathcal{F} = \{X \setminus G \colon G \in \mathcal{G}\}$. Clearly, $|\mathcal{F}| = |\mathcal{G}|$, so $\mathcal{F}$ is countable. Furthermore, if G is a dense open set, then its complement $X \setminus G$ is a nowhere dense closed set. (Generally, the complement of a dense set need not be nowhere dense. For example, $\mathbb{Q}$ is dense in $\mathbb{R}$, as if $\mathbb{R} \setminus \mathbb{Q}$. But if G is dense and open, then, for any non-empty open $U \subseteq X$, we have $G \cap U \neq \emptyset$, so $V = G \cap U$ is a non-empty open subset of U with $V \cap (X \setminus G) = \emptyset$.) Thus, $\mathcal{F}$ is a countable collection of nowhere dense closed sets, and

$$\bigcup \mathcal{F} = \bigcup \{X \setminus G \colon G \in \mathcal{G}\} = X \setminus \bigcap \mathcal{G} \supseteq X \setminus A.$$

Hence, $X \setminus A$ is meager.

Conversely, suppose A is meager. Then, there is a countable collection $\mathcal{F}$ of nowhere dense closed subsets of X such that $\bigcup \mathcal{F} \supseteq A$. Let $\mathcal{G} = \{X \setminus F \colon F \in \mathcal{F}\}$. Clearly, $|\mathcal{G}| = |\mathcal{F}|$, so $\mathcal{G}$ is countable. Moreover, the complement of a nowhere dense closed set is a dense open set. Thus, $\mathcal{G}$ is a countable collection of dense open sets, and

$$\bigcap \mathcal{G} = \bigcup \{X \setminus F \colon F \in \mathcal{F}\} = X \setminus \bigcup \mathcal{F} \subseteq X \setminus A.$$

Hence, $X \setminus A$ is comeager. □

Corollary 3.64. *Let X be a metric space. If $\mathcal{A}$ is a countable collection of comeager subsets of X, then $\bigcap \mathcal{A}$ is comeager.*

Proof. This follows from Theorem 3.61 and De Morgan's laws. □

Theorem 3.65. *If X is a complete metric space, then every comeager subset of X is dense.*

Proof. This follows immediately from the Baire Category Theorem (the first version, Theorem 3.51). □

If we are to think of meager sets as being "very small" subsets of a complete metric space X, then Theorem 3.63 provides us with some intuition for how to think of comeager sets: They have very small complements, and each one is therefore "nearly all" of a metric space. To end this section, let us point out that these ideas only apply to *complete* metric spaces. While the definitions of "meager" and "comeager" still make sense in arbitrary metric spaces, they no longer carry the same intuitive meaning of being nearly none or nearly all of the space.

Example 3.66. The reasoning from Example 3.59, although stated there just for $\mathbb{R}$, actually shows that in any metric space X with no isolated points, every countable $A \subseteq X$ is meager. It follows that if $X = \mathbb{Q}$, then X itself is a meager subset of X. Every subset of X is also comeager (by Theorem 3.63). Note that $\mathbb{Q}$ is not a complete metric space, so the previous theorem does not apply.

Banach–Mazur games

In this last section, we introduce the concept of infinite games. The games we will look at, called Banach–Mazur games, turn out to be intimately related to the Baire Category Theorem and to meager and comeager sets.

Given a metric space X and some $A \subseteq X$, the *Banach–Mazur game* associated with A and X, denoted $\mathsf{BM}(X, A)$, is a two-player game played in infinitely many rounds. The players, called Hit and Miss, take turns constructing an infinite decreasing sequence of open subsets of X. In the end, Hit wins if the intersection of these sets contains a point of A, and otherwise Miss wins.

$$
\begin{array}{lccccccccccccc}
\text{Hit} & U_1 & & & U_2 & & & U_3 & & & U_n & & & U_{n+1} \\
 & & \supseteq & & \subseteq & \supseteq & & \subseteq & \cdots & \subseteq & & \supseteq & & \subseteq \quad \cdots \\
\text{Miss} & & & V_1 & & & V_2 & & & V_{n-1} & & & V_n &
\end{array}
$$

More precisely, in round 1 of the game, Hit plays first by selecting a non-empty open set $U_1 \subseteq X$. Miss plays next by choosing a non-empty open set $V_1 \subseteq U_1$. After the sets $U_1, V_1, U_2, V_2, \ldots, U_{n-1}, V_{n-1}$ have already been played, round n of the game begins with Hit selecting a non-empty open $U_n \subseteq V_{n-1}$, and then Miss chooses some non-empty open $V_n \subseteq U_n$.

Definition 3.67. Given a metric space X, let $\mathcal{O}^+(X)$ denote the set of all non-empty open subsets of X. Given $A \subseteq X$, a *play* of the game $\mathsf{BM}(X, A)$ is a sequence $\langle U_1, V_1, U_2, V_2, U_3, V_3, \ldots \rangle$ in $\mathcal{O}^+(X)$ with $U_1 \supseteq V_1 \supseteq U_2 \supseteq \ldots$.

A *tactic* for MISS in $\mathsf{BM}(X, A)$ is a function $\tau : \mathcal{O}^+(X) \to \mathcal{O}^+(X)$, such that $\tau(U) \subseteq U$ for every $U \in \mathcal{O}^+(X)$. It is a *winning tactic* for MISS if, for every play $\langle U_1, V_1, U_2, V_2, U_3, V_3, \dots \rangle$ of $\mathsf{BM}(X, A)$ for which $V_n = \tau(U_n)$ for every n, the intersection $\bigcap_{n=1}^{\infty} U_n = \bigcap_{n=1}^{\infty} V_n$ contains no points of A.

In other words, a tactic for MISS in $\mathsf{BM}(X, A)$ is a rule τ that can tell MISS what to play in any given round of the game, depending on what was just played by HIT. It is a winning tactic if playing according to that rule always results in MISS winning the game, no matter what moves HIT makes.

$$\begin{array}{lccccccccccccc}
\text{HIT} & U_1 & & & & U_2 & & & & U_3 & & U_n & & & & U_{n+1} \\
& & \supseteq & & \supseteq & & \supseteq & & \supseteq & \cdots & \supseteq & & \supseteq & & \supseteq & \cdots \\
\text{MISS} & & & \tau(U_1) & & & & \tau(U_2) & & & \tau(U_{n-1}) & & & \tau(U_n) & &
\end{array}$$

A winning tactic for HIT is defined similarly, but with one difference: Because HIT plays first in $\mathsf{BM}(X, A)$, a tactic for HIT must include an instruction for what to play in round 1 of the game. To streamline our notation, we just let $\tau(X)$ be the first move of the game.

$$\begin{array}{lccccccccccccc}
\text{HIT} & \tau(X) & & & & \tau(V_1) & & & & \tau(V_2) & & \tau(V_{n-1}) & & & & \tau(U_n) \\
& & \supseteq & & \supseteq & & \supseteq & & \supseteq & \cdots & \supseteq & & \supseteq & & \supseteq & \cdots \\
\text{MISS} & & & V_1 & & & & V_2 & & & U_{n-1} & & & V_n & &
\end{array}$$

Definition 3.68. A *tactic* for HIT in the Banach–Mazur game $\mathsf{BM}(X, A)$ is a function $\tau : \mathcal{O}^+(X) \to \mathcal{O}^+(X)$, such that $\tau(U) \subseteq U$ for every $U \in \mathcal{O}^+(X)$. It is a *winning tactic* if for every play $\langle U_1, V_1, U_2, V_2, U_3, V_3, \dots \rangle$ of $\mathsf{BM}(X, A)$ beginning with $U_1 = \tau(X)$ and having $U_n = \tau(V_{n-1})$ for every $n > 1$, the intersection $\bigcap_{n=1}^{\infty} U_n = \bigcap_{n=1}^{\infty} V_n$ contains at least one point of A.

Example 3.69. If A is a countable subset of $\mathbb{R}$, then MISS has a winning tactic in the game $\mathsf{BM}(\mathbb{R}, A)$.

If A is finite, then the tactic defined by taking $\tau(U) = U \setminus A$ for all $U \in \mathcal{O}^+(\mathbb{R})$ is a winning tactic for MISS.

If A is countably infinite, this simple idea may no longer work because $U \setminus A$ may not be open (e.g., if $A = \mathbb{Q}$). Let $n \mapsto a_n$ denote a bijection $\mathbb{N} \to A$, so that $A = \{a_n : n \in \mathbb{N}\}$. For every $U \in \mathcal{O}^+(X)$, define

$$\tau(U) = \begin{cases} U & \text{if } A \cap U = \emptyset, \\ U \setminus \{a_n\} & \text{if } A \cap U \neq \emptyset \text{ and } n \text{ is the least natural number} \\ & \quad \text{with the property that } a_n \in A \cap U. \end{cases}$$

This defines a function $\mathcal{O}^+(\mathbb{R}) \to \mathcal{O}^+(\mathbb{R})$, and we claim it is a winning tactic for Miss. To see this, suppose $\langle U_1, V_1, U_2, V_2, U_3, V_3, \ldots \rangle$ is a play of $\mathsf{BM}(\mathbb{R}, A)$ with $V_n = \tau(U_n)$ for all n. Using induction, it is not difficult to show that $a_1, a_2, \ldots, a_n \notin V_n$, for every $n \in \mathbb{N}$. But then, $\bigcap_{n\in\mathbb{N}} A_n$ does not contain any points of A, which means that Miss wins this play of $\mathsf{BM}(\mathbb{R}, A)$. Since this was an arbitrary play of the game with Miss playing according to τ, this shows τ is a winning tactic for Miss. ☕

Intuitively, the smaller A is, the easier it is for Miss to win $\mathsf{BM}(X, A)$. This fits with the previous example: Countability is a notion of smallness for subsets of $\mathbb{R}$, and Miss wins $\mathsf{BM}(\mathbb{R}, A)$ when A is countable. The next theorem uses a similar idea to show something stronger along these lines.

Theorem 3.70. *If A is a meager subset of a metric space X, then* Miss *has a winning tactic in* $\mathsf{BM}(X, A)$.

Proof. Let $A \subseteq X$ be meager. Fix a sequence $F_1, F_2, F_3, \ldots$ of nowhere dense closed subsets of X such that $A \subseteq \bigcup_{n=1}^{\infty} F_n$.

Given $U \in \mathcal{O}^+(X)$, note that, because each F_n is closed and nowhere dense, $U \setminus F_n \in \mathcal{O}^+(X)$ as well. For each $U \in \mathcal{O}^+(X)$, define

$$\tau(U) = \begin{cases} U & \text{if } A \cap U = \emptyset, \\ U \setminus F_n & \text{if } A \cap U \neq \emptyset \text{ and } n \text{ is the least natural number} \\ & \quad \text{with the property that } F_n \cap U \neq \emptyset. \end{cases}$$

Observe that if $U \cap A \neq \emptyset$, then $U \cap \bigcup_{n=1}^{\infty} F_n \neq \emptyset$ (because $A \subseteq \bigcup_{n=1}^{\infty} F_n$), which means that $U \cap F_n \neq \emptyset$ for some n, and therefore this function is well-defined. Thus, τ is a function $\mathcal{O}^+(X) \to \mathcal{O}^+(X)$, and we claim it is a winning tactic for Miss. To this end, suppose $\langle U_1, V_1, U_2, V_2, U_3, V_3, \ldots \rangle$ is a play of $\mathsf{BM}(X, A)$ with $V_n = \tau(U_n)$ for all n.

Let us prove by induction that $F_n \cap V_n = \emptyset$ for all $n \in \mathbb{N}$. For the base case, the definition of τ implies that either $U_1 \cap F_1 = \emptyset$ and $V_1 \subseteq U_1$, or else $V_1 = \tau(U_1) = U_1 \setminus F_1$; either way, $F_1 \cap V_1 = \emptyset$. For the inductive step, fix n and suppose $F_i \cap V_i = \emptyset$ for all $i < n$. Because $V_1 \supseteq V_2 \supseteq \cdots \supseteq V_{n-1} \supseteq U_n$, we have $F_i \cap U_n = \emptyset$ for all $i < n$. But then, the definition of τ means that either $U_n \cap F_n = \emptyset$ already, and $V_n \subseteq U_n$, or else $V_n = \tau(U_n) = U_n \setminus F_n$; either way, $F_n \cap V_n = \emptyset$.

Thus, $F_n \cap V_n = \emptyset$ for all $n \in \mathbb{N}$. It follows that $F_n \cap \bigcap_{i=1}^{\infty} V_i = \emptyset$ for all n, which means $\left(\bigcup_{n=1}^{\infty} F_n\right) \cap \left(\bigcap_{i=1}^{\infty} V_i\right) = \emptyset$, hence $A \cap \left(\bigcap_{i=1}^{\infty} V_i\right) = \emptyset$. So Miss wins this play of $\mathsf{BM}(X, A)$. As this was an arbitrary play of $\mathsf{BM}(X, A)$ with Miss playing according to τ, τ is a winning tactic for Miss. □

The converse of this theorem is also true: For any metric space X, Miss has a winning tactic in $\mathsf{BM}(X, A)$ if and only if A is a meager subset of X. We next prove a special case of the converse, the case where X is separable.

Definition 3.71. A metric space X is *separable* if there is a countable dense subset of X.

Example 3.72. The real line $\mathbb{R}$ is separable because $\mathbb{Q}$ is dense in $\mathbb{R}$. More generally, the Euclidean space $\mathbb{R}^n$ is separable because $\mathbb{Q}^n$ is a countable dense subset of $\mathbb{R}^n$.

Lemma 3.73. *Suppose X is a separable metric space, and τ is a function $\mathcal{O}^+(X) \to \mathcal{O}^+(X)$. For every non-empty open $W \subseteq X$, there is a collection $\mathcal{V}$ of disjoint open subsets of W such that (a) $\overline{\bigcup \mathcal{V}} = \overline{W}$, and (b) for each $V \in \mathcal{V}$, there is some $U \subseteq W$ with $V = \tau(U)$.*

In order to see the point of this lemma, imagine τ is a tactic for Miss in a Banach–Mazur game $\mathsf{BM}(X, A)$. If W is a move for Miss in some round of the game, then $\mathcal{V}$ represents a collection of *possible future moves* for Miss in the next round of the game. This is essentially what (*b*) says: Hit could have played some U that results in Miss playing $V = \tau(U)$. The lemma states that there is a pairwise disjoint collection $\mathcal{V}$ of possible future moves that "fills up" W, in the sense that $\overline{\bigcup \mathcal{V}} = \overline{W}$.

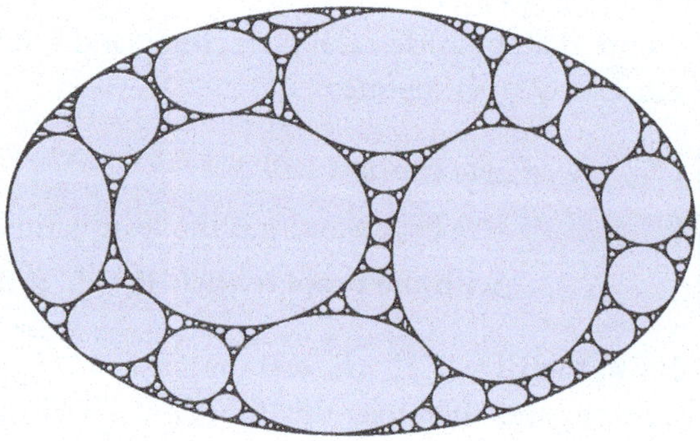

Proof of Lemma 3.73. Let $W \in \mathcal{O}^+(X)$. Let D be a countable dense subset of X, and let $\mathcal{B}_0 = \{B_{1/n}(d) : d \in D, n \in \mathbb{N}\}$. Note that $\mathcal{B}_0$ is countable (because $\mathbb{N} \times D$ is countable, and the map $(n, d) \to B_{1/n}(d)$ is a surjection $\mathbb{N} \times D \to \mathcal{B}_0$). Therefore, $\mathcal{B} = \{U \in \mathcal{B}_0 : U \subseteq W\}$ is countable as well. Let $n \mapsto U_n$ be a bijection $\mathbb{N} \to \mathcal{B}$, so that $\mathcal{B} = \{U_n : n \in \mathbb{N}\}$.

We construct the set $\mathcal{V}$ via recursion as follows. For the base step, let $V_1 = \tau(U_1)$. At step n, assuming we have already chosen some disjoint open sets $V_1, V_2, \ldots, V_{n-1}$, let

$$V_n = \begin{cases} \emptyset & \text{if } U_n \cap \bigcup_{j<n} V_j \neq \emptyset, \\ \tau(U_n) & \text{if } U_n \cap \bigcup_{j<n} V_j = \emptyset. \end{cases}$$

In the end, we have a set $\{V_n : n \in \mathbb{N}\}$ of open subsets of X, some of which may be $\emptyset$. Let $\mathcal{V} = \{V_n : n \in \mathbb{N}\} \setminus \{\emptyset\}$.

It is clear from the construction that the members of $\mathcal{V}$ are pairwise disjoint, non-empty open subsets of W, and for each $V \in \mathcal{V}$ there is some $U \subseteq W$ with $\tau(U) = V$. To finish the proof, we need to show $\overline{\bigcup \mathcal{V}} = \overline{W}$.

Let $x \in \overline{W}$, and let $\varepsilon > 0$. Then $B_\varepsilon(x) \cap W$ is a non-empty open set. Because D is dense, there is some $d \in D \cap B_\varepsilon(x) \cap W$. Because $B_\varepsilon(x) \cap W$ is open and contains d, there is some $m \in \mathbb{N}$ such that $B_{1/m}(d) \subseteq B_\varepsilon(x) \cap W$. In other words, there is some $n \in \mathbb{N}$ such that $U_n \subseteq B_\varepsilon(x) \cap W$. At stage n of our recursion, either (1) there was already some $j < n$ such that $V_j \cap U_n \neq \emptyset$, or (2) $V_n = \tau(U_n) \subseteq U_n$. Either way, there is some $j \leq n$ such that $V_j \cap U_n \neq \emptyset$. Thus, $B_\varepsilon(x) \cap \mathcal{V} \neq \emptyset$. As ε was arbitrary, this means $x \in \overline{\bigcup \mathcal{V}}$. As $x \in \overline{W}$ was arbitrary, this shows $\overline{W} \subseteq \overline{\bigcup \mathcal{V}}$. The reverse inclusion is easy: Each $V \in \mathcal{V}$ is a subset of W, which implies $\bigcup \mathcal{V} \subseteq W$, which implies $\overline{\bigcup \mathcal{V}} \subseteq \overline{W}$. □

Theorem 3.74. *Let X be a separable metric space, and let $A \subseteq X$.* Miss *has a winning tactic in* BM(X, A) *if and only if A is a meager subset of X.*

Proof. The "if" direction was already proved as Theorem 3.70. For the "only if" direction, fix $A \subseteq X$, and let τ be a winning tactic for Miss in BM(X, A).

We construct a sequence $G_1, G_2, G_3, \ldots$ of dense open subsets of X by recursion. Setting $W = X$ in Lemma 3.73, there is a collection $\mathcal{V}_1$ of pairwise disjoint open subsets of X such that $\overline{\bigcup \mathcal{V}_1} = X$, and for each $V \in \mathcal{V}_1$ there is some $U \subseteq X$ such that $V = \tau(U)$. Intuitively, we think of $\mathcal{V}_1$ as a set of possible round-1 moves for Miss in BM(X, A). Let $G_1 = \bigcup \mathcal{V}_1$. Observe that G_1 is open (the union of a collection of open sets) and dense (because $\overline{G_1} = X$). This completes the first step of the recursion.

For the second step, for each $W \in \mathcal{V}_1$, apply Lemma 3.73 to obtain a collection $\mathcal{V}_2^W$ of pairwise disjoint open subsets of W such that $\overline{\bigcup \mathcal{V}_2^W} = \overline{W}$ and, for each $V \in \mathcal{V}_2^W$, there is some $U \subseteq W$ with $V = \tau(U)$. Intuitively, we think of $\mathcal{V}_2^W$ as a set of possible round-2 moves for Miss, assuming that Miss played W in round 1. Let $\mathcal{V}_2 = \bigcup_{W \in \mathcal{V}_1} \mathcal{V}_2^W$, and let $G_2 = \bigcup \mathcal{V}_2$. Then,

$$\overline{\bigcup \mathcal{V}_2} \supseteq \bigcup_{W \in \mathcal{V}_1} \overline{\bigcup \mathcal{V}_2^W} = \bigcup W \in \mathcal{V}_1 \overline{W} \supseteq \bigcup \mathcal{V}_1 = G_1.$$

This implies $\overline{\bigcup \mathcal{V}_2} \supseteq \overline{G_1} = X$, and therefore $\bigcup \mathcal{V}_2$ is dense in X (and open). This completes step 2 of the recursion.

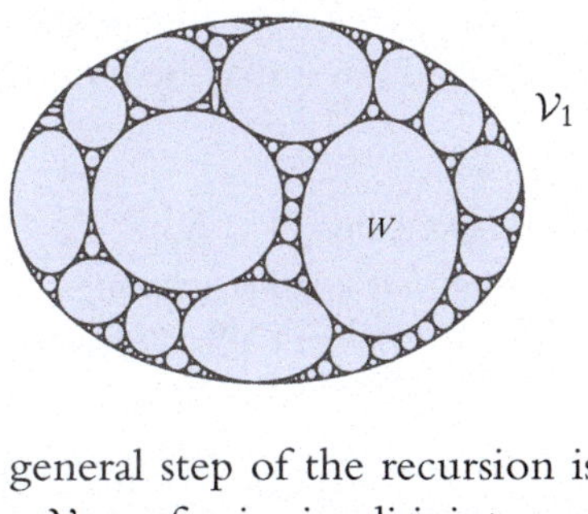

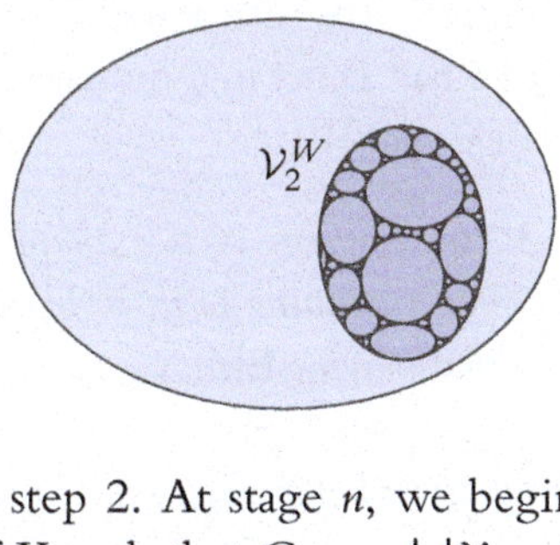

The general step of the recursion is similar to step 2. At stage n, we begin with a collection $\mathcal{V}_{n-1}$ of pairwise disjoint open subsets of X such that $G_{n-1} = \bigcup \mathcal{V}_{n-1}$ is dense. Applying Lemma 3.73, for each $W \in \mathcal{V}_{n-1}$, there is a collection $\mathcal{V}_n^W$ of pairwise disjoint open subsets of W such that $\overline{\bigcup \mathcal{V}_n^W} = \overline{W}$ and, for each $V \in \mathcal{V}_n^W$, there is some $U \subseteq W$ with $V = \tau(U)$. Let $\mathcal{V}_n = \bigcup_{W \in \mathcal{V}_{n-1}} \mathcal{V}_n^W$ and $G_n = \bigcup \mathcal{V}_n$. Observe that

$$\overline{\bigcup \mathcal{V}_n} \supseteq \bigcup_{W \in \mathcal{V}_{n-1}} \overline{\bigcup \mathcal{V}_n^W} = \bigcup W \in \mathcal{V}_{n-1} \overline{W} \supseteq \bigcup \mathcal{V}_{n-1} = G_{n-1},$$

which implies $\overline{\bigcup \mathcal{V}_n} \supseteq \overline{G_{n-1}} = X$, and therefore G_n is dense (and open).

This recursion produces a sequence $G_1, G_2, G_3, \ldots$ of dense open subsets of X. Thus, the set $H = \bigcap_{n=1}^{\infty} G_n$ is comeager. To finish the proof, we will show $A \cap H = \emptyset$. (Because H is comeager, this will imply A is meager.)

Let $x \in H$. This means $x \in G_n$ for every n, which means that, for every n, there is some $V_n \in \mathcal{V}_n$ with $x \in V_n$.

By our choice of $\mathcal{V}_1$, there is some $U_1 \in \mathcal{O}^+(X)$ with $\tau(U_1) = V_1$. For each $n > 1$, observe that $V_n \cap V_{n-1} \neq \emptyset$ (because both contain x). This implies $V_n \in \mathcal{V}_n^{V_{n-1}}$ because, if $V \in \mathcal{V}_n \setminus \mathcal{V}_n^{V_{n-1}}$, then $V \in \mathcal{V}_{n-1}^W$ for some $W \neq V_{n-1}$, but then $W \cap V_{n-1} = \emptyset$, which implies $W \cap V = \emptyset$. Because $V_n \in \mathcal{V}_n^{V_{n-1}}$, there is some $U_n \in \mathcal{O}^+(X)$ with $V_{n-1} \subseteq U_n$ and $\tau(U_n) = V_n$.

Thus, there is a sequence $U_1, U_2, U_3, \ldots$ of open subsets of X such that $\langle U_1, V_1, U_2, V_2, U_3, V_3, \ldots \rangle$ is a play of $\mathsf{BM}(X, A)$ in which Miss plays according to the tactic τ. Because τ is a winning tactic for Miss, this implies $\bigcap_{n=1}^{\infty} V_n$ contains no points of A. But of course, $x \in \bigcap_{n=1}^{\infty} V_n$, so $x \notin A$.

Because x was an arbitrary point of H, this shows $H \cap A = \emptyset$. Because H is comeager, this means A is meager. □

Given Theorem 3.74, of course one wonders whether there might be a similar characterization of those $A \subseteq X$ for which Hit has a winning strategy in $\mathsf{BM}(X, A)$. We end this chapter by proving such a characterization.

Definition 3.75. Given a metric space X, the *diameter* of a subset A of X is $\operatorname{diam}(A) = \sup\{d(x, y) : x, y \in A\}$.

For example, the diameter of an interval (a, b) in $\mathbb{R}$ is $b - a$, and in general, the diameter of a ball $B_r(x)$ in a metric space is at most $2r$ (though it may be less, depending on the metric space).

Lemma 3.76. *Suppose X is a complete separable metric space, and $A \subseteq X$.*

If Miss *has a winning tactic σ in* $\mathsf{BM}(X, A)$*, then* Miss *has another winning tactic τ with the property that playing* $\mathsf{BM}(X, A)$ *according to τ ensures that the set $\bigcap_{n=1}^{\infty} V_n$ contains exactly one point (a point not in A).*

Similarly, if Hit *has a winning tactic σ in* $\mathsf{BM}(X, A)$*, then* Hit *has another winning tactic τ with the property that playing* $\mathsf{BM}(X, A)$ *according to τ ensures that the set $\bigcap_{n=1}^{\infty} V_n$ contains exactly one point (a point in A).*

Proof. Fix $A \subseteq X$, and suppose σ is a winning tactic for MISS in $\mathsf{BM}(X, A)$. Let D be a countable dense subset of X, and let $\mathcal{B} = \{B_{1/n}(d) \colon d \in D, n \in \mathbb{N}\}$ (like in the proof of Lemma 3.73). Observe that $\mathcal{B}$ is countable, and let $n \mapsto W_n$ be a bijection $\mathbb{N} \to \mathcal{B}$, so that $\mathcal{B} = \{W_n \colon n \in \mathbb{N}\}$.

Suppose $\varepsilon > 0$ and U is a non-empty open subset of X. There is some $d \in D \cap U$, and (because U is open) there is some $n \in \mathbb{N}$ with $B_{1/n}(d) \subseteq U$. Then, there is some $N \geq n$ such that $2/N < \varepsilon$ and $\overline{B_{1/N}(d)} \subseteq B_{1/n}(d) \subseteq U$. This shows that for any non-empty open U and any $\varepsilon > 0$, there is some $W_m \in \mathcal{B}$ such that $\overline{W_m} \subseteq U$ and $\mathrm{diam}(W_m) < \varepsilon$.

To each $U \in \mathcal{O}^+(X)$, we associate some $W(U) \in \mathcal{B}$ as follows. If U is unbounded, then let $m = \min\{n \in \mathbb{N} \colon \overline{W_n} \subseteq U\}$, and let $W(U) = W_n$. If U is bounded, then let $m = \min\{n \in \mathbb{N} \colon \overline{W_n} \subseteq U \text{ and } \mathrm{diam}(W_n) \leq \frac{1}{2}\mathrm{diam}(U)\}$, and let $W(U) = W_n$. Observe that for any $U \in \mathcal{O}^+(X)$,

- $W(U) \in \mathcal{B}$ and $\overline{W(U)} \subseteq U$, and
- $W(U)$ is bounded, and if U is bounded, then $\mathrm{diam}(W_n) \leq \frac{1}{2}\mathrm{diam}(U)$.

Define $\tau : \mathcal{O}^+(X) \to \mathcal{O}^+(X)$ by setting

$$\tau(U) = \sigma\big(W(U)\big).$$

Note that this is a tactic because $\sigma\big(W(U)\big)$ is open and $\emptyset \neq \sigma\big(W(U)\big) \subseteq W(U) \subseteq U$ for every $U \in \mathcal{O}^+(X)$.

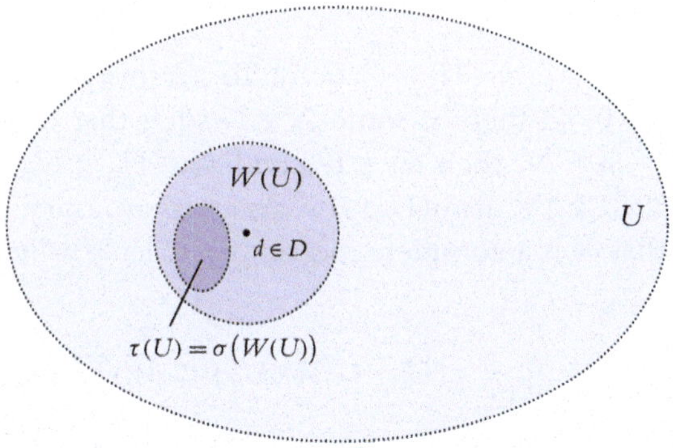

We aim to show that τ is a winning tactic for MISS and that if MISS plays according to τ, then $\bigcap_{n=1}^{\infty} U_n$ contains exactly one point. Suppose $\langle U_1, V_1, U_2, V_2, U_3, V_3, \ldots \rangle$ is a play of $\mathsf{BM}(X, A)$ in which $V_n = \tau(U_n) \supseteq U_{n+1}$ for all n, i.e., where MISS plays according to τ.

$$\begin{array}{lccccccccc}
\text{HIT} & U_1 & & & & U_2 & & & & U_3 \;\supseteq\; \cdots \\
 & \supseteq & & & \supseteq & \supseteq & & & \supseteq & \\
 & & W(U_1) & & & & W(U_2) & & & \\
 & & & \supseteq & & & & \supseteq & & \\
\text{MISS} & & & \tau(U_1) = V_1 & & & & \tau(U_2) = V_2 & &
\end{array}$$

To see that τ is a winning tactic for Miss, note that for every n,

$$U_n \supseteq W(U_n) \supseteq \sigma\big(W(U_n)\big) = \tau(U_n) = V_n \supseteq U_{n+1}.$$

In particular, $W(U_1) \supseteq V_1 \supseteq W(U_2) \supseteq V_2 \supseteq W(U_3) \supseteq V_3 \supseteq \ldots$. This means that $\langle W(U_1), V_1, W(U_2), V_2, W(U_3), V_3, \ldots \rangle$ is a play of $\mathsf{BM}(X, A)$, but, in this play of the game, Miss plays according to σ rather than τ because $V_n = \tau(U_n) = \sigma\big(W(U_n)\big)$ for all n. Because σ is a winning tactic for Miss, this means $\bigcap_{n=1}^{\infty} V_n$ contains no points of A. But then, this means that Miss wins the original play $\langle U_1, V_1, U_2, V_2, U_3, V_3, \ldots \rangle$ of $\mathsf{BM}(X, A)$. Because this was an arbitrary legal gameplay of $\mathsf{BM}(X, A)$ in which Miss played according to τ, this shows that τ is a winning tactic for Miss.

It remains to show that $\bigcap_{n=1}^{\infty} V_n$ contains exactly one point. First, notice that $W(U_1)$ is bounded, which means $V_1 \subseteq W(U_1)$ is bounded as well. Then, notice that for all $n \in \mathbb{N}$, $U_{n+1} \subseteq W(U_n) \subseteq U_n$, which implies

$$\operatorname{diam}(U_{n+1}) \leq \operatorname{diam}\big(W(U_n)\big) \leq \tfrac{1}{2}\operatorname{diam}(U_n).$$

Hence $\lim_{n\to\infty} \operatorname{diam}(U_n) = 0$. If $x, y \in \bigcap_{n=1}^{\infty} V_n = \bigcap_{n=1}^{\infty} U_n$, then $x, y \in U_n$ for all n. Because $\lim_{n\to\infty} \operatorname{diam}(U_n) = 0$, this implies $x = y$. Therefore, $\bigcap_{n=1}^{\infty} V_n$ contains at most one point.

The only thing left is to show $\bigcap_{n=1}^{\infty} V_n = \bigcap_{n=1}^{\infty} U_n \neq \emptyset$. To do this, first pick some point $x_n \in U_n$ for every n. (For example, x_n could be the point $d \in D$ at the center of the ball $W(U_n)$.)

We claim the sequence $\langle x_n : n \in \mathbb{N} \rangle$ is Cauchy. To see this, fix $\varepsilon > 0$. We know already that $\lim_{n\to\infty} \operatorname{diam}(U_n) = 0$, so there is some $N \in \mathbb{N}$ such that $\operatorname{diam}(U_n) < \varepsilon$ for all $n \geq N$. In particular, if $n \geq m \geq N$, then $x_m \in U_m$ and $x_n \in U_n \subseteq U_m$. By the definition of diameter, this means $d(x_m, x_n) \leq \operatorname{diam}(U_m) < \varepsilon$. As ε was arbitrary, this shows $\langle x_n : n \in \mathbb{N} \rangle$ is Cauchy. Recalling that X is a complete metric space, let $x = \lim_{n\to\infty} x_n$.

If $m < n$, then

$$x_n \in U_n \subseteq V_m = \tau(U_m) \subseteq W(U_m) \subseteq \overline{W(U_m)} \subseteq U_m.$$

The fact that this is true for all $n > m$ implies $x = \lim_{n\to\infty} x_n \in \overline{W(U_m)}$. Because this is true for all m, $x \in \bigcap_{m=1}^{\infty} \overline{W(U_m)}$. But, $V_m \supseteq \overline{W(U_{m+1})} \supseteq V_{m+1}$ for all m, and this implies that $\bigcap_{m=1}^{\infty} \overline{W(U_m)} = \bigcap_{m=1}^{\infty} V_m$. Hence, $x \in \bigcap_{m=1}^{\infty} V_m$, and in particular $\bigcap_{m=1}^{\infty} V_m \neq \emptyset$.

This concludes the proof that if Miss has a winning tactic, then Miss has a winning tactic that ensures $\bigcap_{n=1}^{\infty} V_n$ contains exactly one point. The second assertion of the theorem, which says the same thing but for Hit instead of Miss, is proved in exactly the same way. □

We are now ready to prove the final theorem of the chapter, which characterizes those $A \subseteq X$ for which Hit has a winning tactic in $\mathsf{BM}(X, A)$. The proof is an example of what is known as a "strategy stealing argument."

Theorem 3.77. *Let X be a complete separable metric space, and let $A \subseteq X$. HIT has a winning tactic in $\mathsf{BM}(X, A)$ if and only if there is some non-empty open $U \subseteq X$ such that $U \setminus A$ is meager.*

Proof. To prove the "if" direction of the theorem, fix $A \subseteq X$, and suppose there is some $U \in \mathcal{O}^+(X)$ such that $U \setminus A$ is meager. By Theorem 3.70, MISS has a winning tactic τ in the game $\mathsf{BM}(X, U \setminus A)$. Furthermore, by Lemma 3.76, we may (and do) assume that when MISS plays $\mathsf{BM}(X, U \setminus A)$, using the tactic τ, $\bigcap_{n=1}^{\infty} V_n$ contains exactly one point (a point not in $U \setminus A$). Define $\bar{\tau} : \mathcal{O}^+(X) \to \mathcal{O}^+(X)$ by setting

$$\bar{\tau}(V) = \begin{cases} U & \text{if } V = X, \\ \tau(V) & \text{if } V \neq X. \end{cases}$$

In other words, HIT opens the game by playing $U_1 = \bar{\tau}(X) = U$ in round 1, and afterward plays using the tactic τ. Roughly, the idea is that after opening with $U_1 = U$, HIT pretends to be the second player (normally the role of MISS) in the game $\mathsf{BM}(X, U \setminus A)$, and using a tactic known to be winning for the second player of that game.

MISS			V_1		V_2			V_{n-1}		V_n		
		$\supseteq$	$\supseteq$	$\supseteq$	$\supseteq$		$\cdots$	$\supseteq$	$\supseteq$	$\supseteq$		$\cdots$
HIT	U_1		U_2		U_3				U_n		U_{n+1}	

The point is that if $\langle U_1, V_1, U_2, V_2, U_3, V_3, \ldots \rangle$ is a play of the game $\mathsf{BM}(X, A)$ in which the first player plays according to the tactic $\bar{\tau}$, then $\langle V_1, U_2, V_2, U_3, V_3, U_4, \ldots \rangle$ is a play of the game $\mathsf{BM}(X, U \setminus A)$ with the second player playing according to the tactic τ. Because τ is a winning tactic for the second player in $\mathsf{BM}(X, U \setminus A)$, the unique point in the set $\bigcap_{n=1}^{\infty} V_n$ is not in $U \setminus A$. But, $V_1 \subseteq U_1 = U$, so this point is in U; thus, the unique point of $\bigcap_{n=1}^{\infty} V_n$ is in A. Hence, $A \cap \bigcap_{n=1}^{\infty} V_n \neq \emptyset$, which means HIT wins the play $\langle U_1, V_1, U_2, V_2, U_3, V_3, \ldots \rangle$ of the game $\mathsf{BM}(X, A)$. This shows that $\bar{\tau}$ is a winning strategy for HIT in $\mathsf{BM}(X, A)$.

For the "only if" direction of the theorem, suppose HIT has a winning tactic τ in the game $\mathsf{BM}(X, A)$. By Lemma 3.76, we may (and do) assume that when HIT plays $\mathsf{BM}(X, A)$ using τ, $\bigcap_{n=1}^{\infty} V_n$ contains exactly one point (a point of A). Let $W = \tau(X)$ and define $\bar{\tau} : \mathcal{O}^+(X) \to \mathcal{O}^+(X)$ by

$$\bar{\tau}(U) = \begin{cases} U & \text{if } U \not\subseteq W, \\ \tau(U \cap W) & \text{if } U \subseteq W. \end{cases}$$

We claim $\bar{\tau}$ is a winning tactic for MISS in $\mathsf{BM}(X, U \setminus A)$. Roughly, this is true because MISS is taking the role of the first player (normally HIT) in the game $\mathsf{BM}(X, U \setminus A)$, and using a tactic known to be winning for the first player of that game.

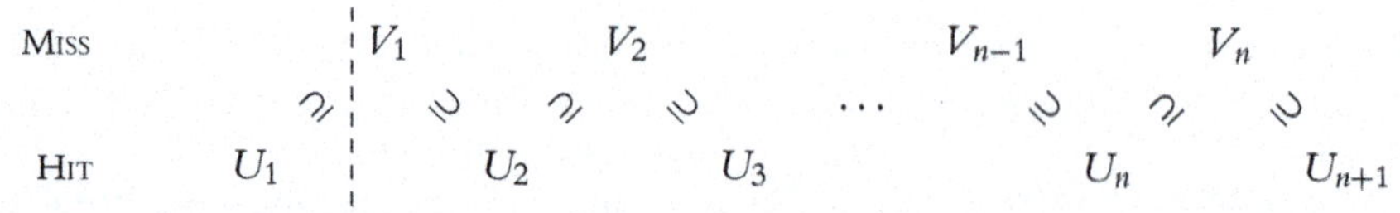

More precisely, if $\langle U_1, V_1, U_2, V_2, U_3, V_3, \ldots \rangle$ is a valid play of the game $\mathsf{BM}(X, A)$ in which the first player plays according to the tactic τ, then $\langle V_1, U_2, V_2, U_3, V_3, U_4, \ldots \rangle$ is a play of the game $\mathsf{BM}(X, U \setminus A)$ with $V_1 \subseteq U_1 = U$ and the second player playing according to the tactic $\bar{\tau}$. Because τ is a winning tactic for the first player in $\mathsf{BM}(X, A)$, the unique point in the set $\bigcap_{n=1}^{\infty} V_n$ is in A. Hence, $(U \setminus A) \cap \bigcap_{n=1}^{\infty} V_n = \emptyset$, which means the second player wins the play $\langle U_1, V_1, U_2, V_2, U_3, V_3, \ldots \rangle$ of the game $\mathsf{BM}(X, U \setminus A)$. This shows that $\bar{\tau}$ is a winning strategy for Miss in $\mathsf{BM}(X, U \setminus A)$. By Theorem 3.74, this means $U \setminus A$ is meager. □

Theorems 3.74 and 3.77 combined raise an interesting question: Is there a metric space X and a subset A of X such that neither player has a winning tactic in $\mathsf{BM}(X, A)$? Or, using Theorems 3.74 and 3.77, an equivalent rephrasing of this question is: Is there a metric space X and a subset A of X such that A is not meager and $U \setminus A$ is not meager for any non-empty open $U \subseteq X$? It turns out that this question is more subtle than it first appears. We will have to wait until Chap. 7 to find out the answer.

Exercises

1. An easy way to misstate the definition of a Cauchy sequence goes like this: *a Cauchy sequence is a sequence of points where the distance between each term of the sequence and the next approaches* 0. Provide an example of a sequence of real numbers that satisfies this property but fails to be Cauchy: in other words, a non-Cauchy sequence $\langle r_n : n \in \mathbb{N} \rangle$ in $\mathbb{R}$ with $\lim_{n \to \infty} |r_{n+1} - r_n| = 0$.
2. Given a metric space X, is it true that if $x \in X$ and $r > 0$, then $\overline{B_r(x)} \neq B_r(x)$? If it is true then prove it, and if not, then find a counterexample.
3. Prove that, if X is a metric space and $\langle x_n : n \in \mathbb{N} \rangle$ is a sequence of points in X, then the set of all subsequential limits of $\langle x_n : n \in \mathbb{N} \rangle$ is a closed subset of X.
4. Conversely, show that if $C \subseteq \mathbb{R}$ is closed, then there is a sequence $\langle r_n : n \in \mathbb{N} \rangle$ of real numbers such that

$$C = \{x \in \mathbb{R} : x \text{ is a subsequential limit of } \langle r_n : n \in \mathbb{N} \rangle\}.$$

5. Let $A \subseteq \mathbb{R}$. Prove that if A is not compact, then there is an infinite sequence of points in A with no subsequential limit in A. Combining this with Lemma 3.19, conclude that $A \subseteq \mathbb{R}$ is compact if and only if every sequence of points in A has a subsequential limit in A.

6. Prove that if X is a complete metric space, then a subspace $A \subseteq X$ is complete if and only if A is closed.
7. Show that completeness is preserved by isometries, but not by homeomorphisms. In other words, (1) prove that if a metric space is complete, then so is every metric space isometric to it, but (2) find an example of two homeomorphic metric spaces where one is complete and the other is not.
8. In light of the previous problem, let us say that a metric space is *completely metrizable* if it is homeomorphic to a complete metric space. Prove that open subspaces of complete metric spaces are completely metrizable. (Contrast this with Problem 6.)
9. Prove that if a completely metrizable space does not have any isolated points, then it is uncountable. Deduce from this that the space $\mathbb{Q}$ of rational numbers is not completely metrizable.
10. Prove that if $-1 < r < 1$, then $\sum_{n=0}^{\infty} r^n = \frac{1}{1-r}$.
11. Prove that the series $\sum_{n=0}^{\infty} \frac{1}{n!}$ converges. Define $e = \sum_{n=0}^{\infty} \frac{1}{n!}$, and prove that e is irrational.
12. Compute $\sum_{n=1}^{\infty} \frac{1}{n^2+n}$. Using the fact that this series converges and the comparison test, show that $\sum_{n=1}^{\infty} \frac{1}{n^2}$ converges also.
13. Let $\sum_{n=1}^{\infty} a_n$ be an infinite series, and let $\langle n_k \colon k \in \mathbb{N} \rangle$ be an increasing sequence of natural numbers with $n_1 = 1$. For each $k \in \mathbb{N}$, define $b_k = \sum_{i=n_k}^{n_{k+1}} a_i$. Prove the following:
   (a) If $\sum_{n=1}^{\infty} a_n$ converges, then $\sum_{k=1}^{\infty} b_k$ converges also, and furthermore $\sum_{n=1}^{\infty} a_n = \sum_{k=1}^{\infty} b_k$.
   (b) If $a_n \geq 0$ for all n, then the converse is also true: If $\sum_{k=1}^{\infty} b_k$ converges, then $\sum_{n=1}^{\infty} a_n$ converges also, and $\sum_{n=1}^{\infty} a_n = \sum_{k=1}^{\infty} b_k$.
   (c) However, it is possible to have a divergent series $\sum_{n=1}^{\infty} a_n$ and a sequence $\langle n_k \colon k \in \mathbb{N} \rangle$ with $n_1 = 1$ such that $\sum_{k=1}^{\infty} b_k$ converges.
14. Let a_n denote the nth natural number that does not contain the digit 9 in its decimal representation. (So, $a_1 = 1$, $a_2 = 2$, $\ldots$, $a_8 = 8$, $a_9 = 10$, $a_{10} = 11$, $\ldots$, $a_{18} = 20$, $\ldots$.) Prove $\sum_{n=1}^{\infty} \frac{1}{a_n}$ converges.
   (*Hint:* Group the terms $\frac{1}{a_n}$ into blocks, according to how many digits a_n has. Within each block, figure out how many nonzero terms there are, and then overestimate the sum over each block as the number of terms in the block times the biggest value of $\frac{1}{a_n}$ in the block. Show this overestimation converges, then use the comparison test.)
   (*Hint for how to use the previous hint:* The previous exercise tells you what happens when you group terms into blocks like this.)
15. Suppose $\langle a_n \colon n \in \mathbb{N} \rangle$ is an infinite sequence in $\mathbb{R}$ such that (*a*) the sequence $\langle |a_n| \colon n \in \mathbb{N} \rangle$ is nonincreasing, (*b*) $\lim_{n\to\infty} |a_n| = 0$, and (*c*) $a_n \geq 0$ if and only if n is odd. Prove that $\sum_{n=1}^{\infty} a_n$ converges.

16. Give an example of an infinite series $\sum_{n=1}^{\infty} a_n$ that converges, but $\sum_{n=1}^{\infty}(a_n)^2$ diverges, and give an example of an infinite series $\sum_{n=1}^{\infty} b_n$ that diverges, but $\sum_{n=1}^{\infty}(b_n)^2$ converges.

17. Suppose X is a complete metric space, and $f : X \to X$ is a *contraction mapping*, which means that there is some $c < 1$ such that $d(f(x), f(y)) \leq cd(x, y)$ for all $x, y \in X$. Prove that there is some $x \in X$ with $f(x) = x$.

18. Let X be a separable metric space. Prove that if $\mathcal{U}$ is a collection of pairwise disjoint open subsets of X, then $\mathcal{U}$ is countable.

19. Prove $\mathbb{R} \setminus \mathbb{Q}$ is not equal to a countable union of closed subsets of $\mathbb{R}$.

20. Suppose X is a metric space and the set of isolated points of X is dense in X. Prove that HIT has a winning tactic in $\mathsf{BM}(X, A)$ if and only if A contains an isolated point.

21. Our proof of Lemma 3.76 uses the fact that X is complete. Show that this assumption on X is necessary by finding an example of a noncomplete metric space X, together with a subset $A \subseteq X$, such that MISS has a winning tactic in $\mathsf{BM}(X, A)$, but does not have a winning tactic that always results in $\bigcap_{n=1}^{\infty} V_n \neq \emptyset$.

22. This exercise guides you through a proof of the following:

Theorem. *Every metric space is isometric to a dense subspace of an essentially unique complete metric space, called its* Cauchy completion.

Let (X, d) be a metric space, and let $\mathcal{C}$ denote the set of all Cauchy sequences in X.

(a) Prove that the relation

$$\langle x_n : n \in \mathbb{N}\rangle \sim \langle y_n : n \in \mathbb{N}\rangle \qquad \Leftrightarrow \qquad \lim_{n\to\infty} d(x_n, y_n) = 0$$

is an equivalence relation on $\mathcal{C}$.

(b) Let $\langle x_n^1\rangle, \langle x_n^2\rangle, \langle y_n^1\rangle, \langle y_n^2\rangle \in \mathcal{C}$, and suppose that $\langle x_n^1\rangle \sim \langle x_n^2\rangle$ and $\langle y_n^1\rangle \sim \langle y_n^2\rangle$. Prove that

$$\lim_{n\to\infty} d(x_n^1, y_n^1) = \lim_{n\to\infty} d(x_n^2, y_n^2).$$

Conclude from this that the function

$$\delta([\langle x_n\rangle], [\langle y_n\rangle]) = \lim_{n\to\infty} d(x_n, y_n)$$

is a well-defined map from pairs of $\sim$-equivalence classes to $\mathbb{R}$.

(c) Let $\mathcal{C}/\sim$ denote the set of all $\sim$-equivalence classes. Prove that the defined function δ is a complete metric on $\mathcal{C}/\sim$.

(d) Show that (X, d) is isometric to a dense subspace of $(\mathcal{C}/\sim, \delta)$.

(e) Suppose that (Y, e) is a complete metric space and that (X, d) is isometric to a dense subspace of (Y, e). Prove that (Y, e) is isometric to $(\mathcal{C}/\sim, \delta)$. (This is what is meant by the phrase "essentially unique" in the statement of the theorem.)

CHAPTER 4

Continuity

We saw the definition of a continuous function already in Chap. 2. We proved that continuous functions preserve connectedness and compactness, leading the Intermediate Value Theorem and the Extreme Value Theorem. We also proved a theorem that gives an alternative way to think about continuity: A function $f : X \to Y$ is continuous if and only if $f^{-1}[U]$ is open in X for every open $U \subseteq Y$.

In this chapter, we take a closer look at continuous functions. In addition to the one already mentioned, we will provide several other equivalent ways to think about continuity. We will classify some of the ways in which a function can fail to be continuous and to measure discontinuities with a number called oscillation. At the end of the chapter we will encounter an especially bizarre kind of function known as a space-filling curve: a continuous function that maps the unit interval $[0, 1]$ surjectively onto the unit square $[0, 1] \times [0, 1]$.

Continuity via limits

Definition 4.1. Let X and Y be metric spaces, $D \subseteq X$, and $f : D \to Y$. If x is a limit point of D, then we write

$$\lim_{t \to x} f(t) = y$$

to mean that for every $\varepsilon > 0$ there is some $\delta > 0$ such that if $t \in D$, $t \neq x$, and $d_X(t, x) < \delta$, then $d_Y(f(t), y) < \varepsilon$. If there is no $y \in Y$ with this property, then we say that $\lim_{t \to x} f(t)$ diverges.

For example, the piecewise linear function

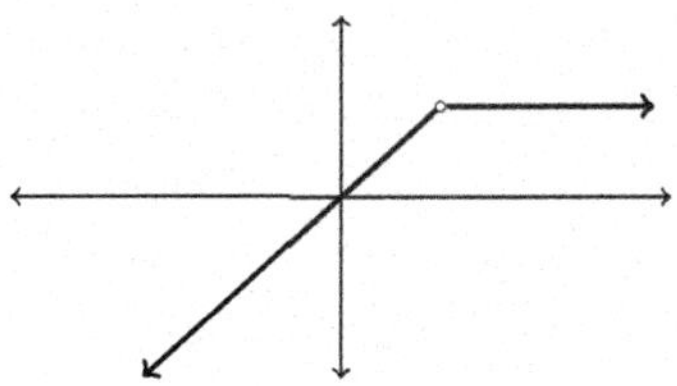

$$f(x) = \begin{cases} x & \text{if } x < 1, \\ 1 & \text{if } x > 1 \end{cases}$$

has domain $D = \mathbb{R} \setminus \{1\}$. Even though f is not defined at $x = 1$, we nevertheless have $\lim_{t \to 1} f(t) = 1$.

Foundations of Real Analysis
https://doi.org/10.1016/B978-0-44-348527-5.00008-1

Theorem 4.2. *Let X and Y be metric spaces, $D \subseteq X$, and $f : D \to Y$. If $x \in D$ is a limit point of D, then f is continuous at x if and only if* $\lim_{t\to x} f(t) = f(x)$.

Proof. This is essentially just a rephrasing of the original ε-δ definition of continuity. More precisely, if $f : D \to Y$ is continuous at x, then this means that for every $\varepsilon > 0$ there is some $\delta > 0$ such that $d_X(x, z) < \delta$ and $z \in D$ implies $d_Y(f(x), f(z)) < \varepsilon$; but this is precisely what $\lim_{t\to x} f(t) = f(x)$ means, according to Definition 4.1. Conversely, assuming $\lim_{t\to x} f(t) = f(x)$ and plugging in $f(x)$ for y in the definition, we get the definition of "f is continuous at x" from Chap. 2. □

Definition 4.3. Let X be a metric space, let $D \subseteq X$, and let $f : D \to \mathbb{R}$. If x is a limit point of D, then we write

$$\lim_{t\to x} f(t) = \infty$$

to mean that, for every $r \in \mathbb{R}$, there is some $\delta > 0$ such that if $t \in D$, $t \neq x$, and $d(t, x) < \delta$, then $f(x) > r$. Similarly, we write

$$\lim_{t\to x} f(t) = -\infty$$

to mean that for every $r \in \mathbb{R}$ there is some $\delta > 0$ such that if $t \in D$, $t \neq x$, and $d(t, x) < \delta$, then $f(x) < r$. If neither of these things happens and $\lim_{t\to x} f(t)$ diverges, then we say $\lim_{t\to x} f(t)$ *does not exist.*

Lemma 4.4. *Let X and Y be metric spaces, let $D \subseteq X$, and let $f : D \to Y$. If $x \in X$ is a limit point of D, then*

1. $\lim_{t\to x} f(t) = y$ *if and only if, for every sequence $\langle t_n : n \in \mathbb{N}\rangle$ of points in $D \setminus \{x\}$, if* $\lim_{n\to\infty} t_n = x$, *then* $\lim_{n\to\infty} f(t_n) = y$.
2. *If $Y = \mathbb{R}$, then $\lim_{t\to x} f(t) = \infty$ (or, respectively, $-\infty$) if and only if for every sequence $\langle t_n : n \in \mathbb{N}\rangle$ of points in $D \setminus \{x\}$, if $\lim_{n\to\infty} t_n = x$, then $\lim_{n\to\infty} f(t_n) = \infty$ (respectively, $-\infty$).*

Proof. To prove the "only if" part of (1), suppose that $\lim_{t\to x} f(x) = y$. Let $\langle t_n : n \in \mathbb{N}\rangle$ be a sequence in $D \setminus \{x\}$ converging to x, and let $\varepsilon > 0$. Because $\lim_{t\to x} f(x) = y$, there is some $\delta > 0$ such that if $t \in D$, $t \neq x$, and $d_X(t, x) < \delta$, then $d_Y(f(t), y) < \varepsilon$. Because $\langle t_n : n \in \mathbb{N}\rangle$ converges to x, there is some $N \in \mathbb{N}$ such that if $n \geq N$, then $d_X(t_n, x) < \delta$. If $n \geq N$, then $d_X(t_n, x) < \delta$ and, consequently, $d_Y(f(t_n), y) < \varepsilon$. As ε was arbitrary, this shows $\lim_{n\to\infty} f(t_n) = y$.

To prove the "if" part of (1), we show the contrapositive. So assume $\lim_{t\to x} f(t) \neq y$, which means there is some fixed $\varepsilon > 0$ such that for any $\delta > 0$, there is some $t \in D$ (exactly which value of t may depend on δ) such that $t \neq x$, $d_X(t, x) < \delta$ and $d_Y(f(t), y) \geq \varepsilon$. In particular, for every $n \in \mathbb{N}$ (setting $\delta = 1/n$ in the previous sentence), there is some $t_n \in D$ such that $t_n \neq x$, $d_X(t_n, x) < \delta$, and $d_Y(f(t_n), y) \geq \varepsilon$. Then, $\langle t_n : n \in \mathbb{N}\rangle$ is a sequence of points in $D \setminus \{x\}$ such that $\lim_{n\to\infty} t_n = x$, but $\lim_{n\to\infty} f(t_n) \neq y$ because $d_Y(f(t_n), y) \geq \varepsilon$ for every n.

The proof of (2) uses essentially the same kind of argument, and the details are left as a homework exercise. □

The following theorem gives us a new characterization of continuous functions via converging sequences.

Theorem 4.5. *Let X and Y be metric spaces, $D \subseteq X$, and $f : D \to Y$. Then, f is continuous at $x \in X$ if and only if for every sequence $\langle t_n : n \in \mathbb{N}\rangle$ of points in D, if $\lim_{n\to\infty} t_n = x$, then $\lim_{n\to\infty} f(t_n) = f(x)$.*

Proof. If x is an isolated point of D, then every function $D \to Y$ is continuous at x. (To see this, fix $\delta > 0$ such that $D \cap B_\delta(x) = \{x\}$, and note that this δ works for any $\varepsilon > 0$ in our original definition of continuity.) And if x is isolated, then the only sequences $\langle t_n : n \in \mathbb{N}\rangle$ of points in D that converge to x are eventually constant (i.e., $t_n = x$ for all but finitely many n), in which case clearly $\lim_{n\to\infty} f(t_n) = f(x)$. In other words, both sides of our "if and only if" assertion are true when x is isolated, which makes the assertion true.

Now, suppose $x \in D$ is a limit point of D. In this case, the theorem follows immediately from Theorem 4.2 and Lemma 4.4, plus the observation that a sequence $\langle t_n : n \in \mathbb{N}\rangle$ of point in D converges to x if and only if the subsequence consisting of all the terms not equal to x also converges to x (or is finite). □

In the following theorem, and from here on out, if f is a function $\mathbb{R} \to \mathbb{R}$ and $c \in \mathbb{R}$, then cf denotes the function that maps t to $c \cdot f(t)$. Similarly, if f and g both are functions $\mathbb{R} \to \mathbb{R}$, then $f + g$ denotes the function $t \mapsto f(t) + g(t)$, fg denotes the function $t \mapsto f(t)g(t)$, and f/g denotes the function $t \mapsto f(t)/g(t)$, which is well-defined assuming $g(t) \neq 0$.

Theorem 4.6. *Suppose f and g are functions $\mathbb{R} \to \mathbb{R}$, and both are continuous at some point x. Then,*

1. *For any $c \in \mathbb{R}$, the function cf is continuous at x.*
2. *The function $f + g$ is continuous at x.*
3. *The function fg is continuous at x.*
4. *If $f(x) \neq 0$, then the function $1/f$ is continuous at x.*

Proof. Recall that part (1) of the Algebraic Limit Theorem states that if $\lim_{n\to\infty} r_n = y$ then $\lim_{n\to\infty} cr_n = cy$. Combining this fact with Theorem 4.5 proves part (1) of the present theorem. Similarly, parts (2), (3), and (4) of the present theorem follow from combining Theorem 4.5 with parts (2), (3), and (4), respectively, of the Algebraic Limit Theorem. □

Example 4.7. Theorem 4.6 allows us to build new continuous functions out of old ones, as follows.

First observe that the identity function $\mathrm{id} : \mathbb{R} \to \mathbb{R}$, defined by setting $\mathrm{id}(x) = x$ for all x, is continuous. (This follows easily from any of our characterizations of continuity: in the original definition, just take $\delta = \varepsilon$; in the open sets characterization we have $\mathrm{id}^{-1}[U] = U$ for any open U, etc.)

Starting with the identity function $\mathrm{id}(x) = x$, we can deduce that many other functions are continuous via Theorem 4.6. For example, the function $x \mapsto x^2$ is just $\mathrm{id} \cdot \mathrm{id}$, so it is continuous by Theorem 4.6(3). Applying Theorem 4.6 again with $f = \mathrm{id}$ and $g(x) = x^2$, the function $x \mapsto x^3$ is continuous as well. Continuing this way, a simple induction shows that the function $x \mapsto x^n$ is continuous for all $n \in \mathbb{N}$.

Once we know this, Theorem 4.6(1) tells us that for any $c \in \mathbb{R}$ and $n \in \mathbb{N}$, the function $x \mapsto cx^n$ is continuous. Then applying Theorem 4.6(2) a few times tells us that every polynomial function $x \mapsto c_n x^n + \cdots + c_1 x + c_0$ is continuous as well.

Combining this with Theorem 4.6(4), we conclude that if f and g are polynomials, then f/g is continuous wherever $g(x) \neq 0$. In other words, every rational function is continuous on its domain. ☕

Uniform continuity

According to our original definition, a function f is continuous if for every $\varepsilon > 0$ and every $x \in \mathrm{dom}(f)$, there is some $\delta > 0$ such that points within δ of x map within ε of $f(x)$. In general the number δ depends on both ε and x; changing ε or x may require a different value of δ to make the definition work. The idea of a "uniformly" continuous function is that δ can be chosen independently of x (uniformly across the whole domain of f), depending only on the value of ε.

Definition 4.8. Let X and Y be metric spaces. A function $f : X \to Y$ is *uniformly continuous* if for every $\varepsilon > 0$ there is some $\delta > 0$ such that if $x, z \in X$ and $d_X(x, z) < \delta$, then $d_Y(f(x), f(z)) < \varepsilon$. ☕

$$\text{continuous}: \ \forall \varepsilon, x \ \exists \delta \ \forall z \quad d(x, z) < \delta \Rightarrow d(f(x), f(z)) < \varepsilon.$$
$$\text{uniformly continuous}: \ \forall \varepsilon \ \exists \delta \ \forall x, z \quad d(x, z) < \delta \Rightarrow d(f(x), f(z)) < \varepsilon.$$

Proposition 4.9. *Every uniformly continuous function is continuous. In other words, uniform continuity is a strengthening of continuity.*

Example 4.10. Let $f : \mathbb{R} \to \mathbb{R}$ be the polynomial function $f(x) = x^2$. This function is continuous (see Example 4.7), but not uniformly continuous. To see that f is not uniformly continuous, let $\varepsilon = 1$. If f were uniformly continuous, there would be some

$\delta > 0$ such that $|x^2 - z^2| < 1$ whenever $|x - z| < \delta$. But this is not the case. No matter how small δ is, we may take $x = 1/\delta$ and $z = x + \delta/2$. Then, $|x - z| = \delta/2 < \delta$, but $|x^2 - z^2| = (1/\delta + \delta/2)^2 - (1/\delta)^2 = 1 + \delta^2/4 > \varepsilon$.

Differentiable functions, which will be the topic of Chap. 5, are closely approximated by linear functions on the microscopic scale. If δ is very small, then the difference between $f(x)$ and $f(x+\delta)$ is approximately $f'(x) \cdot \delta$, just like a line with slope $f'(x)$. For a fixed value of ε, this means that if $f'(x)$ gets very big, then δ will need to be very small in order to ensure $|f(x) - f(x+\delta)| < \varepsilon$. Very roughly, this is why the function $f(x) = x^2$ fails to be uniformly continuous: its derivative, $f'(x) = 2x$, takes on arbitrarily large values, so that the slope of the graph of $f(x)$ gets steeper and steeper for larger and larger x, which means that for a given ε, smaller and smaller values of δ are needed to witness continuity at x.

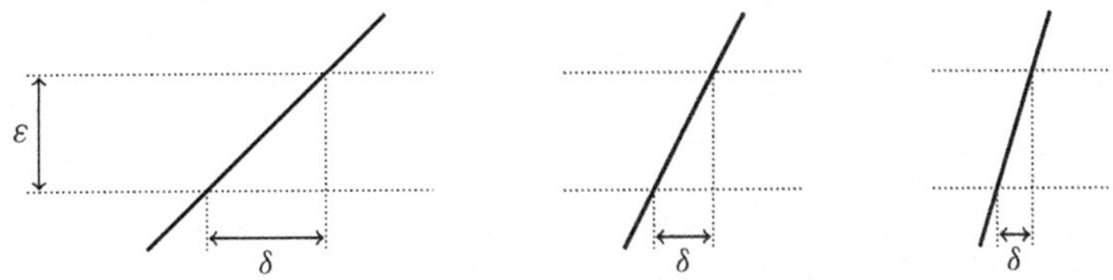

Example 4.11. Let $f : (0, 1] \to \mathbb{R}$ be the function $f(x) = 1/x$. This function is continuous (again by Example 4.7), but not uniformly continuous. To see that f is not uniformly continuous, fix any $\varepsilon > 0$. If f were uniformly continuous, there would be some $\delta > 0$ such that $|1/x - 1/z| < \varepsilon$ whenever $|x - z| < \delta$. But this is not the case. In fact, the image of the interval $(0, \delta)$ is unbounded, so one may find $x, z \in (0, \delta)$ such that $|1/x - 1/z| > \varepsilon$, or indeed so that $1/x$ and $1/z$ are as far apart as one wants.

The previous two examples show that, unlike continuity, uniform continuity is not preserved by taking products or quotients of uniformly continuous functions, i.e., parts (3) and (4) of Theorem 4.6 do not work for uniform continuity. However, uniform continuity is preserved under composition, addition, and multiplication by constants. This is proved in one of the end-of-chapter exercises.

Theorem 4.12. *Suppose X and Y are metric spaces and $f : X \to Y$. If f is continuous and X is compact, then f is uniformly continuous.*

Proof. Suppose X is compact and $f : X \to Y$ is continuous. Let $\varepsilon > 0$. Because $f : X \to Y$ is continuous, for every $x \in X$, there is some $\delta_x > 0$ such that if $z \in X$ and $d_X(x, z) < \delta_x$ then $d_Y(f(x), f(z)) < \varepsilon/2$.

Let $\mathcal{U} = \left\{ B_{\frac{1}{2}\delta_x}(x) : x \in X \right\}$. This is an open cover for X, and therefore, because X is compact, it has a finite subcover. In other words, there are some finitely many points $x_1, x_2, \ldots, x_n \in X$ such that $X = \bigcup_{i \le n} B_{\frac{1}{2}\delta_{x_i}}(x_i)$. Let $\delta = \min\{\frac{1}{2}\delta_{x_1}, \frac{1}{2}\delta_{x_2}, \ldots, \frac{1}{2}\delta_{x_n}\}$.

To see that this value of δ works, let $x, z \in X$ with $d_X(x, z) < \delta$. Because $X = \bigcup_{i \leq n} B_{\frac{1}{2}\delta_{x_i}}(x_i)$, there is some $i \leq n$ such that $x \in B_{\frac{1}{2}\delta_{x_i}}(x_i)$. This implies that

$$d_X(x_i, z) \leq d_X(x_i, x) + d_X(x, z) \leq \tfrac{1}{2}\delta_{x_i} + \delta \leq \delta_{x_i}.$$

Also, $d_X(x_i, x) < \frac{1}{2}\delta_{x_i} < \delta_{x_i}$. By our choice of the number δ_{x_i}, this means $d_Y(f(x_i), f(z)) < \varepsilon/2$ and $d_Y(f(x_i), f(x)) < \varepsilon/2$, which implies

$$d_Y(f(x), f(z)) \leq d_Y(f(x), f(x_i)) + d_Y(f(x_i), f(z)) < \varepsilon/2 + \varepsilon/2 = \varepsilon.$$

As ε was arbitrary, this shows f is uniformly continuous. □

Some functions that are not continuous

In this section, we classify various ways in which functions $\mathbb{R} \to \mathbb{R}$ can be discontinuous. In order to describe the top level of our classification, we first need one-sided versions of Definitions 4.1 and 4.3.

Definition 4.13. Let f be a function $\mathbb{R} \to \mathbb{R}$, and let $x, y \in \mathbb{R}$. We write

$$\lim_{t \to x^-} f(t) = y$$

to mean that for every $\varepsilon > 0$, there is some $\delta > 0$ such that if $x - \delta < t < x$, then $|y - f(t)| < \varepsilon$. Similarly, we write

$$\lim_{t \to x^-} f(t) = \infty \qquad \text{or} \qquad \lim_{t \to x^-} f(t) = -\infty$$

to mean that for every $r \in \mathbb{R}$, there is some $\delta > 0$ such that if $x - \delta < t < x$, then $f(t) > r$ (for divergence to ∞) or $f(t) < r$ (for divergence to $-\infty$). If none of these things happens, then we say that $\lim_{t \to x^-} f(t)$ *does not exist.*

In the same way, we write $\lim_{t \to x^+} f(t) = y$ to mean that for every $\varepsilon > 0$, there is some $\delta > 0$ such that if $x < t < x + \delta$, then $|y - f(t)| < \varepsilon$. Similarly, we write $\lim_{t \to x^+} f(t) = \infty$ or $\lim_{t \to x^+} f(t) = -\infty$ to mean that for every $r \in \mathbb{R}$, there is some $\delta > 0$ such that if $x < t < x + \delta$, then $f(t) > r$ (for divergence to ∞) or $f(t) < r$ (for divergence to $-\infty$). If none of these things happens, then we say that $\lim_{t \to x^+} f(t)$ *does not exist.*

Observation. For any $f : \mathbb{R} \to \mathbb{R}$ and any $x \in \mathbb{R}$ and $y \in \mathbb{R} \cup \{\pm\infty\}$, $\lim_{t \to x} f(t) = y$ if and only if both $\lim_{t \to x^-} f(t) = y$ and $\lim_{t \to x^+} f(t) = y$.

This observation leads to the top level of our taxonomy of discontinuous functions: Either the one-sided limits $\lim_{t \to x^-} f(t)$ and $\lim_{t \to x^+} f(t)$ both exist, or they do not.

Definition 4.14. Let f be a function $\mathbb{R} \to \mathbb{R}$ and let $x \in \mathbb{R}$. If $\lim_{t \to x^-} f(t)$ and $\lim_{t \to x^+} f(t)$ both exist, but still $\lim_{t \to x} f(t) \neq f(x)$, then f has a *discontinuity of the first kind* at x. If either $\lim_{t \to x^-} f(t)$ or $\lim_{t \to x^+} f(t)$ does not exist, then f has a *discontinuity of the second kind* at x. In either case, we say that f is *discontinuous* at x, and x is a *point of discontinuity* for f.

Example 4.15. Recall that the *Dirichlet function* is defined as

$$f(x) = \begin{cases} 1 & \text{if } x \in \mathbb{Q}, \\ 0 & \text{if } x \notin \mathbb{Q}. \end{cases}$$

This function has a discontinuity of the second kind at every point.

Example 4.16. The function

$$f(x) = \begin{cases} \sin(1/x) & \text{if } x < 0, \\ x & \text{if } x \geq 0, \end{cases}$$

is a bounded function $\mathbb{R} \to \mathbb{R}$ with only one point of discontinuity (at $x = 0$), and it is a discontinuity of the second kind.

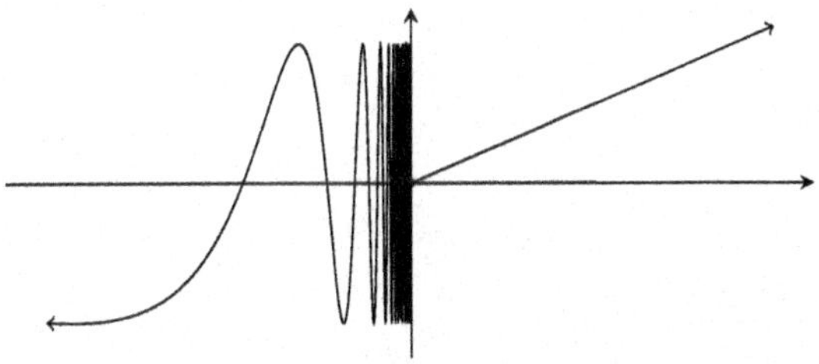

Example 4.17. Given a continuous function $g : \mathbb{R} \to \mathbb{R}$, there is a function $f : \mathbb{R} \to \mathbb{R}$ with the property that f is continuous at x if and only if $x = 0$, and f has a discontinuity of the second kind at every other point. To see this, define f to be the product of g with the following Dirichlet function:

$$f(x) = \begin{cases} g(x) & \text{if } x \in \mathbb{Q}, \\ 0 & \text{if } x \notin \mathbb{Q}. \end{cases}$$

To see that this function is continuous at x wherever $g(x) = 0$, use the original ε-δ definition of continuity as follows. Suppose $g(x) = 0$, and let $\varepsilon > 0$. There is some $\delta > 0$ such that if $|x - z| < \delta$, then $|g(x) - g(z)| = |g(z)| < \varepsilon$. But $f(z)$ is always either equal to 0 or to $g(z)$, so $|f(z)| \leq |g(z)|$ for all z. Hence, if $|x - z| < \delta$, then $|f(x) - f(z)| = |f(z)| \leq |g(z)| < \varepsilon$.

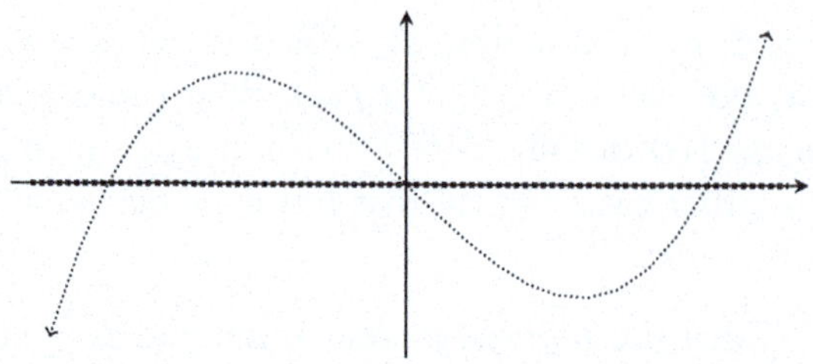

To see that f has a discontinuity of the second kind wherever g is nonzero, fix x with $g(x) \neq 0$, and, aiming for a contradiction, suppose $\lim_{t \to x^-} f(t) = y$ for some y. Let $\varepsilon = |g(x)|/3$, and fix δ such that if $x - \delta < t < x$, then $|f(t) - y| < \varepsilon$. There is some (irrational) z with $x - \delta < z < x$ and $f(z) = 0$, and there is some (rational) q with $x - \delta < q < x$ and $|g(x) - f(q)| = |g(x) - g(q)| < \varepsilon$ (because g is continuous). By the triangle inequality,

$$|g(x)| = |f(z) - g(x)| \leq |f(z) - y| + |y - f(q)| + |f(q) - g(x)| < \varepsilon + \varepsilon + \varepsilon = |g(x)|,$$

a contradiction. Hence, $\lim_{t \to x^-} f(t)$ does not exist.

As for discontinuities of the first kind, we distinguish three basic types:

Definition 4.18. Let f be a function $\mathbb{R} \to \mathbb{R}$ and let $x \in \mathbb{R}$.

If $\lim_{t \to x^-} f(t)$ and $\lim_{t \to x^+} f(t)$ both converge, but to different values, then we say that f has a *jump discontinuity* at x.

If $\lim_{t \to x^-} f(t)$ and $\lim_{t \to x^+} f(t)$ both converge to the same value, but not to $f(x)$, then we say that f has a *removable discontinuity* at x.

If $\lim_{t \to x^-} f(t)$ and $\lim_{t \to x^+} f(t)$ both exist, but one or both of them is $\pm\infty$, then we say that f has a *vertical asymptote* at x.

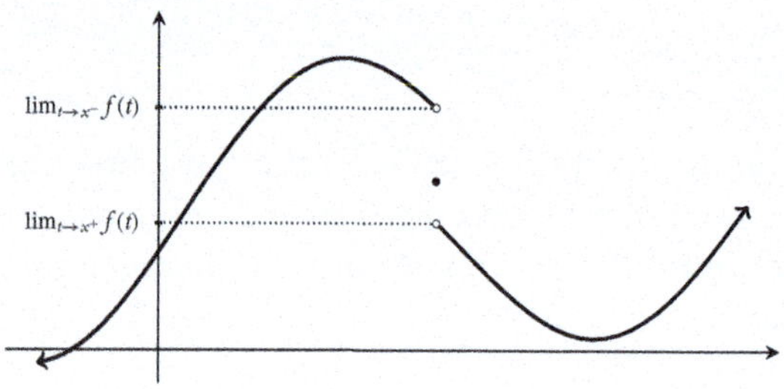

jump discontinuity

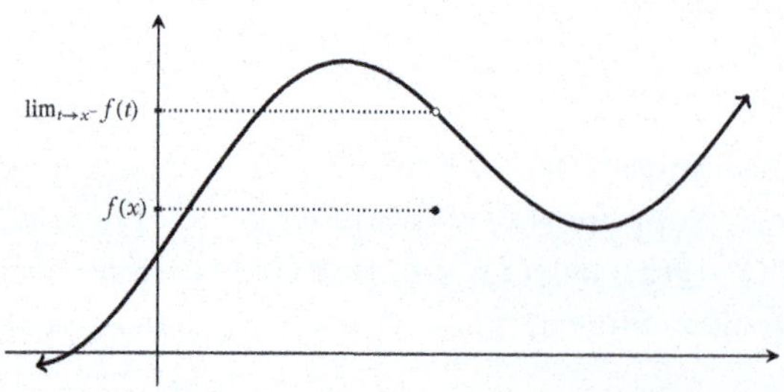

removable discontinuity

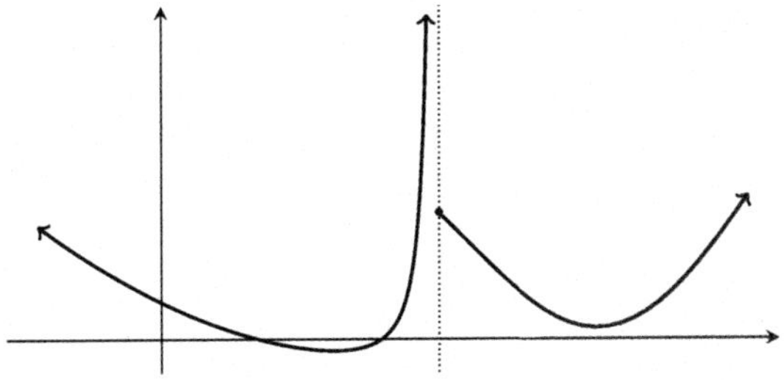

We end this section with a result known as *Froda's Theorem*, which shows that an important class of functions, the monotone functions, cannot be too badly discontinuous: They have only countably many discontinuities, and all are of the first kind.

Definition 4.19. Given $D \subseteq \mathbb{R}$, a function $f : D \to \mathbb{R}$ is *nondecreasing* if $f(a) \leq f(b)$ whenever $a \leq b$. Similarly, f is *nonincreasing* if $f(a) \geq f(b)$ whenever $a \leq b$. A function that is either nondecreasing or nonincreasing is called *monotone*. ☕

Theorem 4.20 (Froda's Theorem). *Let f be a monotone function $\mathbb{R} \to \mathbb{R}$. Then, f has only countably many points of discontinuity, and they are all jump discontinuities.*

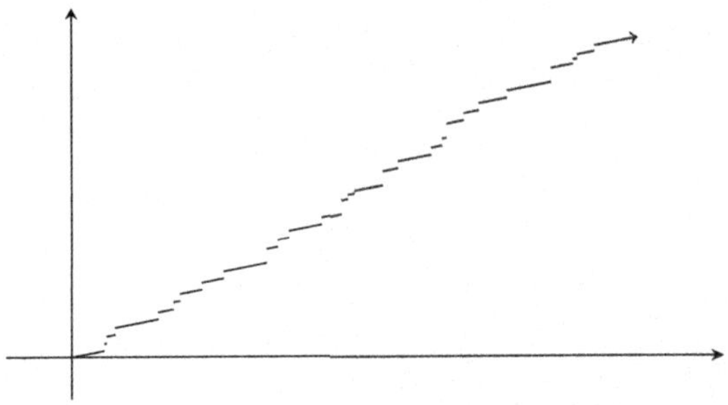

Our proof of the following lemma is written assuming that f is nondecreasing. The proof for nonincreasing functions is essentially the same, but interchanging $<$ and $\leq$ with $>$ and $\geq$, interchanging inf with sup, etc.

Lemma 4.21. *If $f : \mathbb{R} \to \mathbb{R}$ is a nondecreasing function, then the limits $\lim_{t \to x^-} f(t)$ and $\lim_{t \to x^+} f(t)$ both exist for every $x \in (a, b)$. Furthermore,*

$$\begin{aligned} \sup \{f(t) : t < x\} = \lim_{t \to x^-} f(t) &\leq f(x) \\ &\leq \lim_{t \to x^+} f(t) = \inf \{f(t) : t > x\}. \end{aligned}$$

Proof. Because f is nondecreasing, $f(t) \leq f(x)$ whenever $t < x$. Hence $\{f(t) : t < x\}$ is bounded above (and it is non-empty), so $\sup \{f(t) : t < x\}$ exists. Let $\alpha = \sup \{f(t) : t < x\}$. Note that $\alpha \leq f(x)$, because $f(x)$ is an upper bound for $\sup \{f(t) : t < x\}$.

Let $\varepsilon > 0$. Then $\alpha - \varepsilon$ is not an upper bound for $\{f(t) : t < x\}$, so there is some $t_0 < x$ with $\alpha - \varepsilon < f(t_0) \leq \alpha$. Because f is nondecreasing, if $t_0 < t < x$, then $\alpha - \varepsilon <$

$f(t_0) \le f(t) \le \alpha$. As ε was arbitrary, this shows $\lim_{t\to x^-} f(t) = \alpha = \sup\{f(t) : t < x\} \le f(x)$, as required.

Essentially the same argument is used to prove that $\inf\{f(t) : t > x\}$ exists, and $\lim_{t\to x^+} f(t) = \inf\{f(t) : t > x\} \ge f(x)$. □

Proof of Theorem 4.20. Our lemma says that $\lim_{t\to x^-} f(t)$ and $\lim_{t\to x^+} f(t)$ both exist for every x and that $\lim_{t\to x^-} f(t) \le f(x) \le \lim_{t\to x^+} f(t)$. This implies that if $\lim_{t\to x^-} f(t) = \lim_{t\to x^+} f(t)$, then f is continuous at x. Hence, f has a discontinuity at x only if $\lim_{t\to x^-} f(t) < \lim_{t\to x^+} f(t)$. Therefore, f has only jump discontinuities, and it remains to show that there are only countably many of them.

Let $D = \{x \in \mathbb{R} : f \text{ is not continuous at } x\}$. By the previous paragraph, if $x \in D$, then $\lim_{t\to x^-} f(t) < \lim_{t\to x^+} f(t)$. In particular, for every $x \in D$ there is a rational number q_x such that

$$\lim_{t\to x^-} f(t) < q_x < \lim_{t\to x^+} f(t).$$

Now, suppose $x, y \in D$ and $x < y$. Fix some number t_0 with $x < t_0 < y$. By the above lemma,

$$\begin{aligned} q_x &< \lim_{t\to x^+} f(t) = \inf\{f(t) : t > x\} \\ &\le f(t_0) \le \sup\{f(t) : t < y\} = \lim_{t\to y^-} f(t) < q_y. \end{aligned}$$

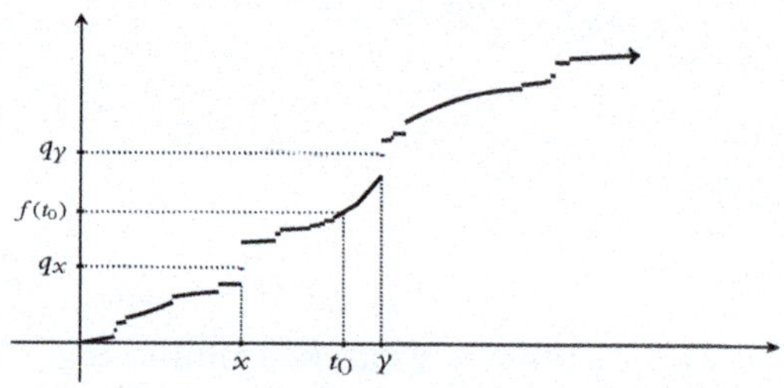

In particular, if $x, y \in D$ and $x \ne y$, then $q_x \ne q_y$. In other words, the map $x \mapsto q_x$ is an injective function $D \to \mathbb{Q}$. Because $\mathbb{Q}$ is countable, the existence of an injection $D \to \mathbb{Q}$ shows that D is countable. □

Oscillation

Given a function $f : \mathbb{R} \to \mathbb{R}$, consider (as in our proof of Froda's Theorem) the set of its discontinuity points $D_f = \{x \in \mathbb{R} : f \text{ is not continuous at } x\}$. What can this set look like? It is easy to construct examples of functions discontinuous on a given finite set, so D_f can be equal to any finite $F \subseteq \mathbb{R}$. On the other extreme, in Example 4.17 we saw that for any continuous $g : \mathbb{R} \to \mathbb{R}$, there is a function $f : \mathbb{R} \to \mathbb{R}$ such that $D_f = \{x \in \mathbb{R} : g(x) \ne 0\}$. Using this fact, it is not hard to see that D_f can be equal to the complement of any finite

$F \subseteq \mathbb{R}$. Other sets are possible as well: In the exercises for Chap. 2, we saw a function f, Thomae's function, for which $D_f = \mathbb{Q}$. Could there be a function f with $D_f = \mathbb{R} \setminus \mathbb{Q}$? Are there any restrictions at all?

In this section, we introduce the notion of oscillation, which measures numerically how discontinuous a function is at a given point. This idea plays an important part later in our study of integration. And as we shall soon see, it is also one of the key ideas involved in answering the question of the previous paragraph.

Definition 4.22. Let X be a metric space, and let $f : X \to \mathbb{R}$. If $U \subseteq X$ and $U \neq \emptyset$, then the *oscillation of f on U* is

$$\omega_f(U) = \sup\left\{f(t) - f(s) : s, t \in U\right\}.$$

Note that this supremum is taken in the extended reals and may be equal to ∞ (e.g., if f has a vertical asymptote at x). Given some $x \in X$, the *oscillation of f at x* is

$$\omega_f(x) = \lim_{r \to 0} \omega_f(B_r(x)).$$

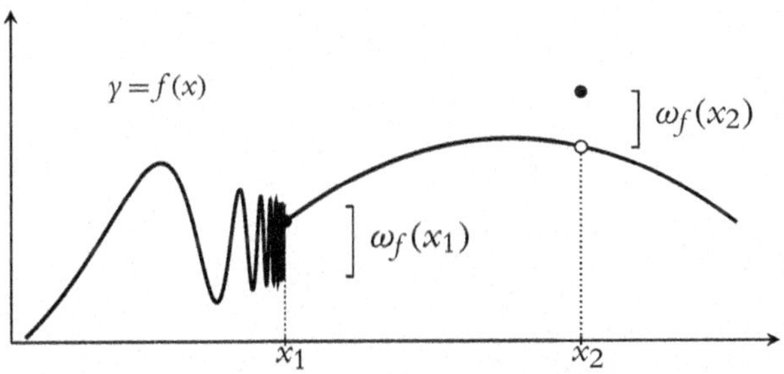

If f has a jump discontinuity at x, then $\omega_f(x)$ measures the size of the jump. If f has a removable discontinuity at x, then $\omega_f(x)$ is the difference between $f(x)$ and $\lim_{t \to x} f(t)$. As we shall see in Theorem 4.25, f is continuous at x if and only if $\omega_f(x) = 0$.

The supremum in the definition of $\omega_f(U)$ cannot be negative: If $U \neq \emptyset$, then there is some $x \in U$, and $0 = f(x) - f(x)$ is in the set whose supremum we are taking. The following lemma shows, among other things, that $\omega_f(x)$ is well-defined for any x and any f.

Lemma 4.23. *Let X be a metric space and let $f : X \to \mathbb{R}$.*

1. *For any $U, V \subseteq X$, if $U \subseteq V$ then $\omega_f(U) \leq \omega_f(V)$.*
2. $\lim_{r \to 0} \omega_f(B_r(x)) = \inf\left\{\omega_f(B_r(x)) : r > 0\right\}$.
3. *The limit in the definition of $\omega_f(x)$ exists for any $x \in X$. In other words, $\omega_f(x)$ is well-defined.*
4. *If $\omega_f(B_r(x)) \leq c$ for any $r > 0$, then $\omega_f(x) \leq c$.*

Proof. For (1), simply observe that if $U \subseteq V$, then

$$\{f(t) - f(s) : s, t \in U\} \subseteq \{f(t) - f(s) : s, t \in V\},$$

which implies that

$$\sup\{f(t) - f(s) : s, t \in U\} \leq \sup\{f(t) - f(s) : s, t \in V\},$$

which means $\omega_f(U) \leq \omega_f(V)$.

(2) and (4) are immediate consequences of (1). For (3), note that the set $\{\omega_f(B_r(x)) : r > 0\}$ is bounded below by 0 because $\omega_f(U) \geq 0$ for any $U \neq \emptyset$. Thus, the infimum in (2) exists (and the limit does, too). □

Theorem 4.24. *Let X be a metric space, let $f : X \to \mathbb{R}$, and let $x \in X$. If $\omega_f(x) \leq c$ for some $c \geq 0$, then for all $\varepsilon > 0$, there is some $\delta > 0$ such that if $t \in X$ and $d_X(x, t) < \delta$, then $|f(x) - f(t)| < c + \varepsilon$.*

Proof. Let $x \in X$ and $c \geq 0$, and suppose that $\omega_f(x) \leq c$. Fix $\varepsilon > 0$. By Lemma 4.23(2), there is some $\delta > 0$ such that $\omega_f(B_\delta(x)) < c + \varepsilon$. Because $x \in B_\delta(x)$, we have $f(x) - f(t) \leq \omega_f(B_\delta(x)) < c + \varepsilon$ and $f(t) - f(x) \leq \omega_f(B_\delta(x)) < c + \varepsilon$ for any $t \in B_\delta(x)$. In other words, if $d_X(t, x) < \delta$, then $|f(x) - f(t)| < c + \varepsilon$. □

Notice that the statement implied by $\omega_f(x) \leq c$ in Theorem 4.24 is exactly like the ε-δ definition of continuity, except that at the end we require only $|f(x) - f(y)| < c + \varepsilon$ rather than $|f(x) - f(y)| < \varepsilon$. It is continuity with an "error" of c. It follows immediately that if $\omega_f(x) = 0$, then f is continuous at x. And it turns out that the converse is also true.

Theorem 4.25. *Let X be a metric space, and let $f : X \to \mathbb{R}$. For any $x \in X$, f is continuous at x if and only if $\omega_f(x) = 0$.*

Proof. If $\omega_f(x) = 0$, then setting $c = 0$ in Theorem 4.24 implies that f is continuous at x. To prove the converse, suppose f is continuous at x. Let $\varepsilon > 0$. There is some $\delta > 0$ such that if $t \in X$ and $d_X(x, t) < \delta$, then $|f(x) - f(t)| < \varepsilon/2$. By the triangle inequality, if $s, t \in B_\delta(x)$, then

$$|f(t) - f(s)| \leq |f(t) - f(x)| + |f(x) - f(s)| < \varepsilon/2 + \varepsilon/2 = \varepsilon,$$

which means $\omega_f(B_\delta(x)) \leq \varepsilon$. By Lemma 4.23, this implies $\omega_f(x) \leq \varepsilon$. Because ε was arbitrary, this shows $\omega_f(x) = 0$. □

As already mentioned, one may think of the condition at the end of Theorem 4.24 as being like continuity but with an error of c. Similarly, the conclusion of the following theorem is like uniform continuity with an error of c, and the theorem itself is a generalization of Theorem 4.12.

Theorem 4.26. *Let X be a metric space and $f : X \to \mathbb{R}$, and suppose that $\omega_f(x) \leq c$ for every $x \in X$. If X is compact, then for any $\varepsilon > 0$, there is some $\delta > 0$ such that if $x, z \in X$ and $d_X(x, z) < \delta$, then $|f(x) - f(z)| < c + \varepsilon$.*

Proof. Let $\varepsilon > 0$. Because $\omega_f(x) \leq c$ for every $x \in X$, by Lemma 4.23(2), there is for each $x \in X$ some $\delta_x > 0$ such that $\omega_f(B_{\delta_x}(x)) < c + \varepsilon$.

Let $\mathcal{U} = \left\{B_{\frac{1}{2}\delta_x}(x) : x \in X\right\}$. This is an open cover of X, so, because X is compact, it must have a finite subcover. That is, there are some finitely many points $x_1, x_2, \ldots, x_n \in X$ such that $X = \bigcup_{i \leq n} B_{\frac{1}{2}\delta_{x_i}}(x_i)$. Let $\delta = \min\left\{\frac{1}{2}\delta_{x_1}, \frac{1}{2}\delta_{x_2}, \ldots, \frac{1}{2}\delta_{x_n}\right\}$.

To see that this value of δ works, fix $x, z \in X$ with $d_X(x, z) < \delta$. There is some $i \leq n$ such that $x \in B_{\frac{1}{2}\delta_{x_i}}(x_i)$. This implies $d_X(x, x_i) < \delta_{x_i}$ and

$$d_X(z, x_i) \leq d_X(z, x) + d_X(x, x_i) < \delta + \tfrac{1}{2}\delta_{x_i} \leq \tfrac{1}{2}\delta_{x_i} + \tfrac{1}{2}\delta_{x_i} = \delta_{x_i}.$$

Consequently, $x, z \in B_{\delta_{x_i}}(x_i)$. Because $\omega_f(B_{\delta_x}(x)) < c + \varepsilon$, this implies $|f(x) - f(z)| < c + \varepsilon$. □

Definition 4.27. Let X be a metric space. A subset of X is called F_σ if it is a countable union of closed sets. A subset of X is called G_δ if it is a countable intersection of open sets.

For example, $\mathbb{Q}$ is an F_σ subset of $\mathbb{R}$ because $\mathbb{Q}$ is countable and therefore $\mathbb{Q} = \bigcup\left\{\{q\} : q \in \mathbb{Q}\right\}$ is a countable union of closed sets. Similarly, the set $\mathbb{R} \setminus \mathbb{Q}$ of irrationals is G_δ because $\mathbb{R} \setminus \mathbb{Q} = \bigcap\left\{\mathbb{R} \setminus \{q\} : q \in \mathbb{Q}\right\}$.

Lemma 4.28. *Let X be a metric space, and let $f : X \to \mathbb{R}$. For any given $c \geq 0$, $\left\{x \in X : \omega_f(x) \geq c\right\}$ is a closed subset of X.*

Proof. Let $c \geq 0$, and let $E = \left\{x \in [a, b] : \omega_f(x) \geq c\right\}$. To prove the theorem, we want to show $\overline{E} = E$.

Let $x_0 \in \overline{E}$, and let $\varepsilon > 0$. Because $x_0 \in \overline{E}$, there is some $x \in E \cap B_\varepsilon(x_0)$. Because $B_\varepsilon(x_0)$ is open, there is some $\delta > 0$ such that $B_\delta(x) \subseteq B_\varepsilon(x_0)$. By Lemma 4.23,

$$\omega_f(B_\varepsilon(x_0)) \geq \omega_f(B_\delta(x)) \geq \omega_f(x) \geq c.$$

Because ε was arbitrary, this implies $\omega_f(x_0) \geq c$, and therefore $x_0 \in E$. Hence, $\overline{E} \subseteq E$, which means $\overline{E} = E$. □

Theorem 4.29. *Let X be a metric space. For any function $f : \mathbb{R} \to \mathbb{R}$,*

$$D_f = \left\{x \in X : f \text{ is not continuous at } x\right\}$$

is an F_σ subset of X.

Proof. For each $c > 0$, let $D_{\geq c} = \{x \in X\colon \omega_f(x) \geq c\}$. By Theorem 4.25,

$$D_f = \{x \in X\colon \omega_f(x) > 0\} = \bigcup_{n\in\mathbb{N}} D_{\geq 1/n},$$

and by Lemma 4.28, $D_{\geq 1/n}$ is closed for each n. □

Corollary 4.30. *There is no function $\mathbb{R} \to \mathbb{R}$ that is continuous at x if and only if x is irrational.*

Proof. This follows from the previous theorem together with Exercise 19 from the previous chapter, which states that $\mathbb{R} \setminus \mathbb{Q}$ is not F_σ. □

To end this section, we prove a kind of converse to Theorem 4.29: If $A \subseteq \mathbb{R}$ is F_σ, then there is a function $f : \mathbb{R} \to \mathbb{R}$ with $D_f = A$. We give the simpler but similar argument for a closed set first.

Example 4.31. If $F \subseteq \mathbb{R}$ is closed, then there is a function $f : \mathbb{R} \to \mathbb{R}$ with $D_f = \{x\colon f \text{ is discontinuous at } x\} = F$. To see this, let $B = F \setminus \mathrm{Int}(F)$, and define

$$f(x) = \begin{cases} 1 & \text{if } x \in B \text{ or } x \in \mathrm{Int}(F) \cap \mathbb{Q}, \\ 0 & \text{otherwise.} \end{cases}$$

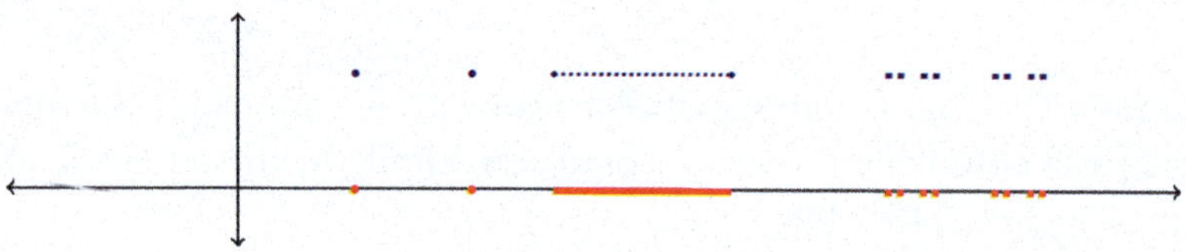

If $x \notin F$, then, because F is closed, there is some $\delta > 0$ such that $B_\delta(x) \cap F = \emptyset$. Then, $f(t) = 0$ for all $t \in B_\delta(x)$, so $\omega_f(x) = \omega_f(B_\delta(x)) = 0$, which means f is continuous at x.

If $x \in F \setminus \mathrm{Int}(F)$, then for every $\delta > 0$ there are points of $\mathbb{R} \setminus F$ in $B_\delta(x)$. In particular, $B_\delta(x)$ contains points $t \in \mathbb{R} \setminus F$ with $f(t) = 0$, and $B_\delta(x)$ contains a point x with $f(x) = 1$. Hence, $\omega_f(B_\delta(x)) = 1$. As this is true for every $\delta > 0$, this shows $\omega_f(x) = 1$, which means f is not continuous at x.

If $x \in \mathrm{Int}(F)$, then for every $\delta > 0$, there are both rational and irrational points of $\mathrm{Int}(F)$ in $B_\delta(x)$. From this and the definition of f, it follows that $\omega_f(B_\delta(x)) = 1$. As δ was arbitrary, this shows $\omega_f(x) = 1$, and as before, this means f is not continuous at x.

Theorem 4.32. *If $A \subseteq \mathbb{R}$ is F_σ, then there is a function $f : \mathbb{R} \to \mathbb{R}$ such that f is discontinuous at x if and only if $x \in A$.*

Proof. Suppose $A \subseteq \mathbb{R}$ is F_σ, and fix a sequence $F_1, F_2, F_3, \ldots$ of closed sets such that $A = \bigcup_{n\in\mathbb{N}} F_n$. For each $n \in \mathbb{N}$, let $B_n = F_n \setminus \mathrm{Int}(F_n)$. Define

$$f(x) = \begin{cases} \frac{1}{n} & \text{if } x \notin \bigcup_{i<n} F_i \text{ but either } x \in B_n \text{ or } x \in \mathrm{Int}(F_n) \cap \mathbb{Q}, \\ 0 & \text{otherwise.} \end{cases}$$

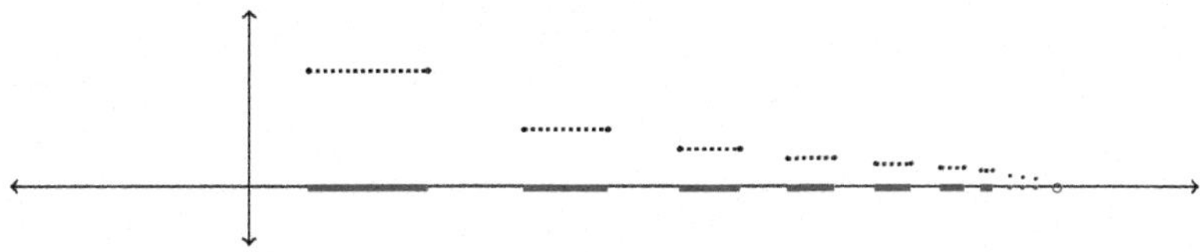

First, let us show that f is continuous at points not in A. Fix $x \notin A$. Let $\varepsilon > 0$, and fix $N \in \mathbb{N}$ such that $1/N < \varepsilon$. Because $\bigcup_{n \leq N} F_n$ is closed and $x \notin \bigcup_{n \leq N} F_n$, there is some $\delta > 0$ such that $B_\delta(x) \cap \bigcup_{n \leq N} F_n = \emptyset$. Then, $0 \leq f(t) \leq 1/N$ for all $t \in B_\delta(x)$, which implies $\omega_f(B_\delta(x)) \leq \varepsilon$, so $\omega_f(x) < \varepsilon$. As ε was arbitrary, this shows $\omega_f(x) = 0$, so f is continuous at x.

Now let us show f is discontinuous at points in A. Fix $x \in A$, and let N be the unique number with $x \in F_N \setminus \bigcup_{n<N} F_n$. Let $\varepsilon = \frac{1}{N} - \frac{1}{N+1}$. If $x \in B_N$, then for any $\delta > 0$ there is some $z \in B_\delta(x)$ with $z \notin F_N$. In this case, $f(x) = \frac{1}{N}$ and either $f(z) \geq \frac{1}{N-1}$ or $f(z) \leq \frac{1}{N+1}$. Either way, $|f(x) - f(z)| \geq \varepsilon$, so $\omega_f(B_\delta(x)) \geq \varepsilon$. As δ was arbitrary, this means $\omega_f(x) \geq \varepsilon$, and in particular $\omega_f(x) \neq 0$, so f is not continuous at x. If instead $x \in \text{Int}(F)$, then x is in the open set $\text{Int}(F) \setminus \bigcup_{n<N} F_n$, so there is some $\delta > 0$ such that $B_\delta(x) \subseteq \text{Int}(F) \setminus \bigcup_{n<N} F_n$. For any $r \leq \delta$, the rational points in $B_r(x)$ have $f(x) = 1/N$, and the irrational points in $B_r(x)$ have $f(x) = 0$. It follows that $\omega_f(x) = 1/N \neq 0$, so f is not continuous at x. □

Combining Theorems 4.29 and 4.32, we have answered the question posed at the beginning of this section: A set $A \subseteq \mathbb{R}$ can be equal to the set of discontinuity points of a function $f : \mathbb{R} \to \mathbb{R}$ if and only if A is F_σ.

Space-filling curves

For the remainder of this chapter, our goal is to prove there is a continuous surjection from the unit interval $[0, 1]$ onto the unit square $[0, 1]^2$.

Definition 4.33. A continuous surjection from the unit interval $[0, 1]$ onto the unit square $[0, 1]^2$ is called a *space-filling curve*. More generally, a continuous surjection from the unit interval $[0, 1]$ onto $[0, 1]^n$, for any $n \geq 2$, is also called a space-filling curve. ☕

The first example of a space-filling curve was defined by Peano in 1890. Roughly, Peano's idea was to define closer and closer approximations to a space-filling curve, with the actual space-filling curve emerging as the limit of these approximations. The version of this idea shown in the next picture was presented by David Hilbert.

In Chap. 8, we explore the idea of taking limits of sequences of functions. For now, we approach the construction of space-filling curves via a different path. Our approach brings us through two other topics that are important in their own right: extending continuous functions to have a larger domain and mapping the Cantor set onto the unit interval.

Extending continuous functions

Definition 4.34. Suppose X and Y are sets, $A \subseteq B \subseteq X$, and f is a function $A \to Y$. If g is a function $B \to Y$ with $g|A = f$, then g is called an *extension of f from A to B.*

Notice that the codomain Y of f is also the codomain of g in this definition. When we extend a function's domain, we keep the codomain fixed and do not allow it to expand, too.

Example 4.35. Let $Y = \{1, 2\}$, the 2-point subset of $\mathbb{R}$. Let $A = \{0, 2\}$, and let $B = [0, 2]$. If f is the map $A \to Y$ mapping 0 to 1 and 2 to 2, then

$$g(x) = \begin{cases} 1 & \text{if } x \leq 1 \\ 2 & \text{if } x > 1 \end{cases}$$

is an extension of f from A to B.

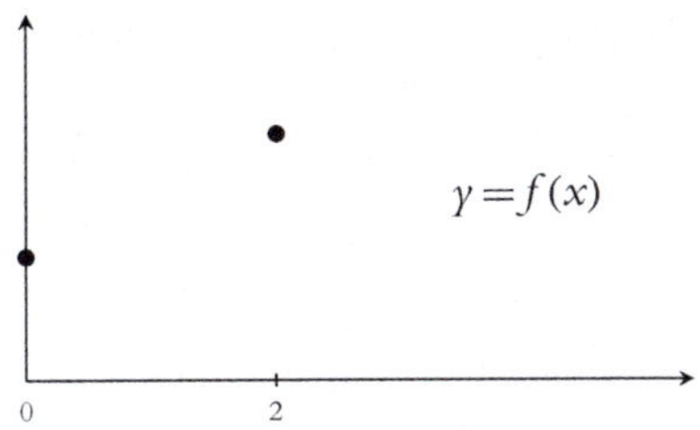

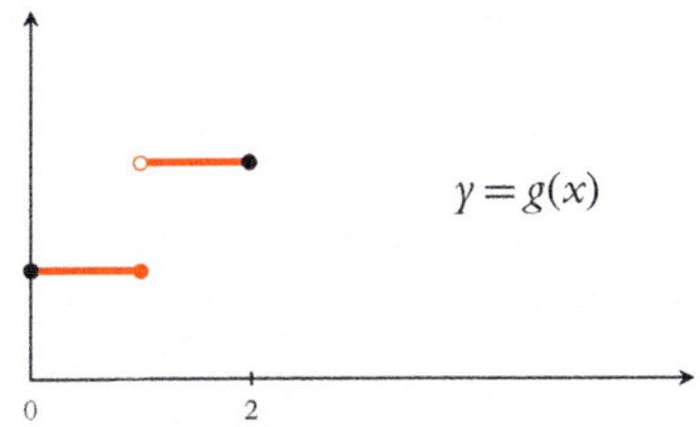

On the other hand, there is no *continuous* extension g of f from A to B. To see this, suppose, aiming for a contradiction, that $g : [0,2] \to \{0,1\}$ is such a function. Then, $g^{-1}(1)$ and $g^{-1}(2)$ are disjoint subsets of $[0,2]$, with $g^{-1}(1) \cup g^{-1}(2) = [0,2]$. Furthermore, if g is continuous, then $g^{-1}(1)$ is an open subset of $[0,2]$ because $\{1\}$ is an open subset of $\{1,2\}$, and similarly $g^{-1}(2)$ is an open subset of $[0,2]$. Therefore, $g^{-1}(1)$ and $g^{-1}(2)$ form a separation of $[0,2]$, showing that $[0,2]$ is not connected, a contradiction.

Example 4.36. Let $f : \mathbb{Q} \to \mathbb{R}$ denote the function

$$f(x) = \begin{cases} 0 & \text{if } x < \sqrt{2} \\ 1 & \text{if } x > \sqrt{2}. \end{cases}$$

Then, the function

$$g(x) = \begin{cases} 0 & \text{if } x \le \sqrt{2} \\ 1 & \text{if } x > \sqrt{2} \end{cases}$$

is an extension of f from $\mathbb{Q}$ to $\mathbb{R}$. However, like in the previous example, there is no continuous extension of f from $\mathbb{Q}$ to $\mathbb{R}$.

Any function can be extended to have a larger domain with the same codomain, provided only that the codomain is non-empty. Indeed, if $A \subseteq B$ and $f : A \to Y$, then let y be any point of Y and define

$$g(x) = \begin{cases} f(x) & \text{if } x \in A \\ y & \text{if } x \in B \setminus A. \end{cases}$$

Then, g is an extension of f from A to B.

But as the two examples suggest, we are sometimes interested in whether a function can be extended continuously from A to B. And as these examples show, this is not always possible. The following theorem provides a counterpoint to these examples: a large class of functions that can always be extended continuously.

Theorem 4.37. *Suppose A is a closed subset of $\mathbb{R}$, and f is a continuous function $A \to \mathbb{R}$. Then, there is a continuous extension of f from A to $\mathbb{R}$.*

Proof. Roughly, the idea is to extend f "linearly" to points of $\mathbb{R} \setminus A$, as in the following picture.

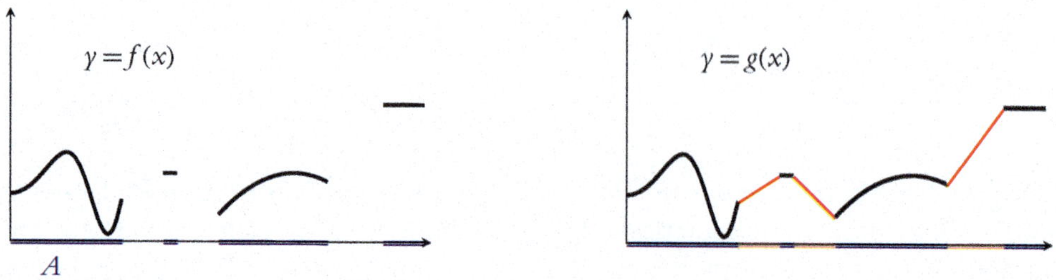

Let us give some further details of this idea. First, consider the case that A has no upper bound and no lower bound. We begin by defining two new functions $\mathbb{R} \to \mathbb{R}$, denoted ℓ and r, as follows:

$$\ell(x) = \sup\{t \in A : t \leq x\}, \text{ and}$$
$$r(x) = \inf\{t \in A : t \geq x\}.$$

Note that $\ell(x)$ is well-defined because $\{t \in A : t \leq x\}$ is non-empty (A has no lower bound) and bounded above (by x); similarly, $r(x)$ is well-defined. Because A is closed, $\ell(x) \in A$ and $r(x) \in A$ for every x. (Recall that the supremum of a subset of a closed set, when it exists, is in that closed set.) If $x \in A$, then $\ell(x) = r(x) = x$. If $x \notin A$, then $\ell(x)$ and $r(x)$ are the "endpoints" of A to the immediate left and right of x. For $x \in \mathbb{R} \setminus A$, define

$$g(x) = f(\ell(x)) + \tfrac{x-\ell(x)}{r(x)-\ell(x)}\big(f(r(x)) - f(\ell(x))\big).$$

This expression is well-defined when $x \notin A$ since $x \notin A$ implies $\ell(x) \neq r(x)$. It is the equation of the line passing through the points $\big(\ell(x), f(\ell(x))\big)$ and $\big(r(x), f(r(x))\big)$. If $x \in A$, then (of course) set $g(x) = f(x)$.

Clearly g is an extension of f to $\mathbb{R}$, and we need to prove it is continuous.

First, suppose $x \notin A$. Then, there is an open set containing x and disjoint from A, and on this open set g restricts to the equation of a line. So clearly g is continuous at x.

Next, we need to show g is continuous at points of A. Before doing this, let us point out that this is not an immediate consequence of the fact that $g\restriction A = f$ and f is continuous on A. For example, if we had been more careless in our definition of g and only required g to be continuous (rather than linear) between adjacent members of A, then g would not necessarily remain continuous on A:

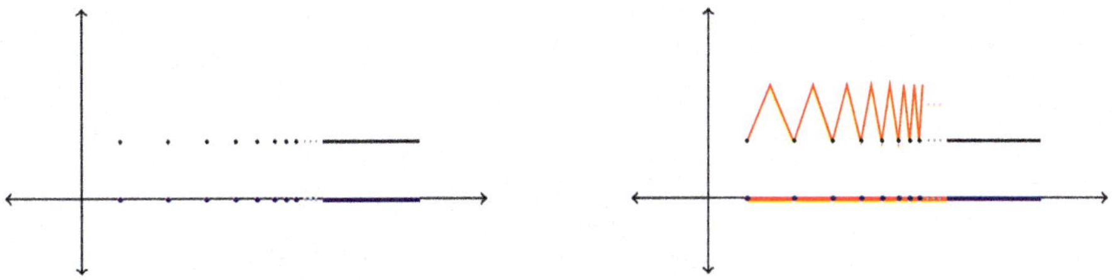

Let $x \in A$, and let $\varepsilon > 0$. First, suppose $x = r(t)$ for some $t < x$, which means the interval $[t, x)$ is disjoint from A. Then, $g{\restriction}[t, x]$ is just a line, which means there is some $\delta > 0$ such that $\omega_g([x - \delta, x]) < \varepsilon$. Next, suppose $x \neq r(t)$ for any t, which means there are points of A in the interval $(x - r, x)$ for every $r > 0$. Because f is continuous on A, there is some $r > 0$ such that $\omega_f(B_r(x) \cap A) < \varepsilon$. The interval $(x - r, x)$ contains some $t \in A$. Then, $[t, x] \subseteq B_r(x)$, so $\omega_f([t, x] \cap A) \leq \omega_f(B_r(x) \cap A) < \varepsilon$. Because $t, x \in A$ and because g extends linearly between adjacent members of A, $\omega_g([t, x]) = \omega_f([t, x]) < \varepsilon$. Setting $\delta = x - t$, there is some $\delta > 0$ such that $\omega_g([x - \delta, x]) < \varepsilon$.

In any case, the previous paragraph shows there is some $\delta > 0$ such that $\omega_g([x - \delta, x]) < \varepsilon$. An essentially identical argument shows there is some $\delta' > 0$ such that $\omega_g([x, x + \delta']) < \varepsilon$. Taking $r = \min\{\delta, \delta'\}$, we have $r > 0$, $\omega_g([x - r, x]) < \varepsilon$, and $\omega_g([x, x + r]) < \varepsilon$. It follows that if $s, t \in B_r(x)$, then $|g(x) - g(t)| < \varepsilon$ and $|g(x) - g(s)| < \varepsilon$, hence $|f(s) - f(t)| \leq |g(s) - g(x)| + |g(x) - g(t)| < \varepsilon + \varepsilon = 2\varepsilon$. Thus, $\omega_g([x - r, x + r]) < 2\varepsilon$, which means $\omega_g(x) \leq \omega_g(B_r(x)) \leq \omega_g([x - r, x + r]) < 2\varepsilon$. As ε was arbitrary, this shows $\omega_g(x) = 0$. Hence, g is continuous at x.

This completes the proof in the case that A has no upper bound and no lower bound. If A has a lower bound, then we simply extend the definition of g by setting $g(x) = f(\inf A)$ for all $x < \inf A$. (Note that $\inf A$ is in A because A is closed.) Similarly, if A has an upper bound, then we set $g(x) = f(\sup A)$ for all $x > \sup A$. □

Corollary 4.38. *Suppose A is a closed subset of $\mathbb{R}$, f is a continuous function $A \to \mathbb{R}$, and B is another subset of $\mathbb{R}$ with $A \subseteq B$. Then, there is a continuous extension of f from A to B.*

Proof. By Theorem 4.37, there is a continuous $g : \mathbb{R} \to \mathbb{R}$ that extends f. The function $g{\restriction}B$ satisfies the conclusion of the corollary. □

Before moving on, let us point out that Theorem 4.37 is really just a special case of a more general result known as the *Tietze extension theorem*. This more general theorem, which is also more difficult to prove, states: *If X is a normal topological space, A is a closed subset of X, and $f : A \to \mathbb{R}$ is continuous, then there is a continuous extension of f from A to X.* We have not defined what a "normal topological space" is, but every metric space is one, and $\mathbb{R}$ in particular is one, so that Theorem 4.37 is just the special case of the Tietze extension theorem obtained by setting $X = \mathbb{R}$.

Cantor's function

Recall that the Cantor set is defined as the decreasing intersection of a sequence of closed sets, starting with $[0, 1]$, where the nth set consists of 2^n closed intervals, and the $(n+1)^{\text{st}}$ set is obtained by removing the middle third from each of those closed intervals.

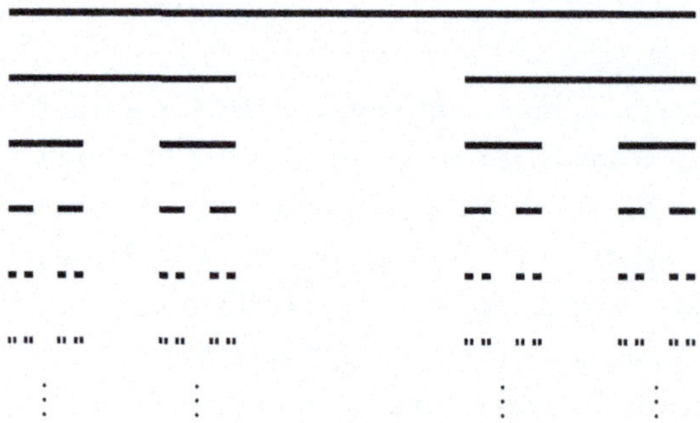

Recall also that the Cantor space $2^{\mathbb{N}}$ is the set of all functions $\mathbb{N} \to \{0, 1\}$, with the metric

$$d(f, g) = \begin{cases} 1/2^n & \text{if } f(i) = g(i) \text{ for all } i < n \text{ but } f(n) \neq g(n), \\ 0 & \text{if } f = g. \end{cases}$$

We proved near the end of Chap. 2 that the Cantor space $2^{\mathbb{N}}$ is homeomorphic to the Cantor set.

Theorem 4.39. *There is a continuous surjection from the Cantor set to* $[0, 1]$.

Proof. Because the Cantor space and the Cantor set are homeomorphic, it suffices to prove there is a continuous surjection from the Cantor space onto $[0, 1]$. This is somewhat similar to the proof in Chap. 2 that $2^{\mathbb{N}}$ is homeomorphic to the Cantor set.

First, we define, via recursion, a labeling of certain subintervals of $[0, 1]$ with finite sequences of 0's and 1's, as follows. To begin, set $I_{\emptyset} = [0, 1]$ (the interval $[0, 1]$ is labeled with the empty sequence). Next, let $I_{\langle 0 \rangle} = [0, \frac{1}{2}]$ and $I_{\langle 1 \rangle} = [\frac{1}{2}, 1]$. In general, given some function $s : \{1, 2, \ldots, n\} \to \{0, 1\}$ (a length-n sequence of 0's and 1's) and an interval I_s, define $I_{s^\frown 0}$ to be the left half of I_s and define $I_{s^\frown 1}$ to be the right half of I_s.

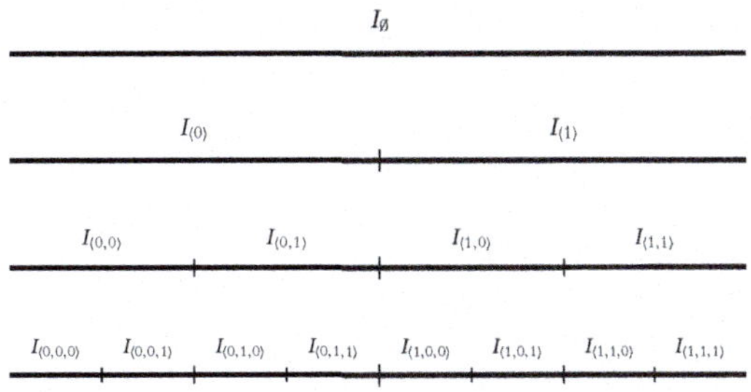

Having defined these intervals, define $\phi : 2^{\mathbb{N}} \to [0, 1]$ by taking $\phi(f)$ to be the unique point in the set $\bigcap_{n=1}^{\infty} I_{f\restriction\{1,2,\ldots,n\}}$.

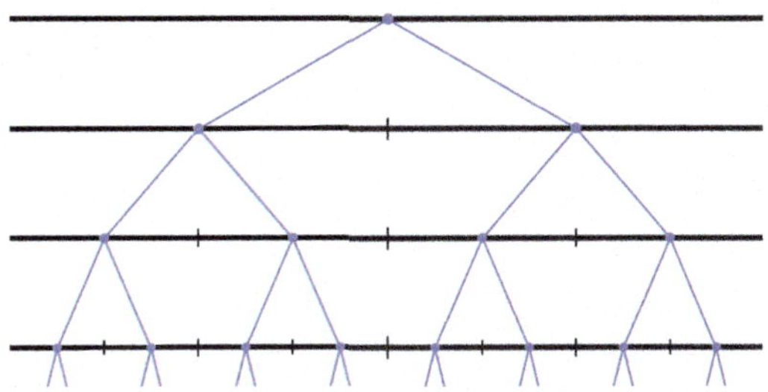

We must check that ϕ is well-defined, i.e., that the set $\bigcap_{n=1}^{\infty} I_{f\restriction\{1,2,\ldots,n\}}$ contains exactly one point for every $f \in 2^{\mathbb{N}}$. The fact that $\bigcap_{n=1}^{\infty} I_{f\restriction\{1,2,\ldots,n\}}$ is non-empty follows from the Nested Interval Property (Theorem 2.60 of Chap. 2) because $I_{f\restriction\{1\}} \supseteq I_{f\restriction\{1,2\}} \supseteq I_{f\restriction\{1,2,3\}} \supseteq \ldots$. Furthermore, $\bigcap_{n=1}^{\infty} I_{f\restriction\{1,2,\ldots,n\}}$ cannot contain two different points x and y because there is some N large enough that the length of $I_{f\restriction\{1,2,\ldots,N\}}$ is less than $|x-y|$, meaning that $I_{f\restriction\{1,2,\ldots,N\}}$ cannot contain both x and y, so neither does $\bigcap_{n=1}^{\infty} I_{f\restriction\{1,2,\ldots,n\}}$. Hence, $\bigcap_{n=1}^{\infty} I_{f\restriction\{1,2,\ldots,n\}}$ contains exactly one point, which we define to be $\phi(f)$.

To see that ϕ is surjective, fix $x \in [0, 1]$. Either $x \in I_{\langle 0\rangle}$ or $x \in I_{\langle 1\rangle}$ (or both, if $x = 1/2$) because $[0, 1] = I_{\langle 0\rangle} \cup I_{\langle 1\rangle}$. Generally, if $x \in I_s$, then either $x \in I_{s^\frown 0}$ or $x \in I_{s^\frown 1}$ (possibly both) because $I_s = I_{s^\frown 0} \cup I_{s^\frown 1}$. Using this, one may (via recursion) define a function $f : \mathbb{N} \to \{0, 1\}$ such that $x \in I_{f\restriction\{1,2,\ldots,n\}}$ for all $n \in \mathbb{N}$. This implies $\phi(f) = x$.

Finally let us check that ϕ is continuous. If $f, g \in 2^{\mathbb{N}}$ and $d(f, g) = 1/2^n$, then $f\restriction\{1, 2, \ldots, n\} = g\restriction\{1, 2, \ldots, n\}$. Let s denote this function, i.e., $s = f\restriction\{1, 2, \ldots, n\} = g\restriction\{1, 2, \ldots, n\}$. By our definition of ϕ, both $\phi(f)$ and $\phi(g)$ are in I_s, and I_s has length $1/2^n$. Thus, $|\phi(f) - \phi(g)| \leq 1/2^n = d(f, g)$. In other words, ϕ does not increase the distance between any two points. Consequently, if $\varepsilon > 0$, then taking $\delta = \varepsilon$ ensures that if $d(f, g) < \delta$, then $|\phi(f) - \phi(g)| < \varepsilon$. Thus, ϕ is continuous. □

Example 4.40. The function ϕ from the previous proof can also be described using the infinite series

$$\phi(f) = \textstyle\sum_{n=1}^{\infty} f(n)\left(\frac{1}{2}\right)^n.$$

The homeomorphism h from $2^{\mathbb{N}}$ to the Cantor set K, which is described in Chap. 2, admits the similar following description:

$$\phi(f) = \sum_{n=1}^{\infty} f(n)\left(\tfrac{2}{3}\right)^n.$$

The composition $\phi \circ h^{-1}$ is the following continuous surjection $K \to [0, 1]$:

$$\sum_{n=1}^{\infty} f(n)\left(\tfrac{2}{3}\right)^n \mapsto \sum_{n=1}^{\infty} f(n)\left(\tfrac{1}{2}\right)^n.$$

This function is order-preserving, and it maps adjacent endpoints of K to the same value (as in the picture after the statement of Theorem 4.39). If we follow the proof of Theorem 4.37 and extend $\phi \circ h^{-1}$ linearly from K to $[0, 1]$, we obtain a rather curious function $[0, 1] \to [0, 1]$, known as *Cantor's function*, also known as *Cantor's staircase function* or the *devil's staircase*.

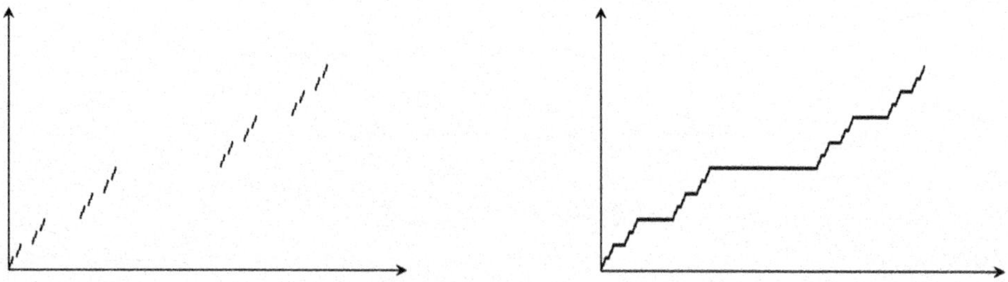

This function (shown on the right) is a nondecreasing map $[0, 1] \to [0, 1]$ that is *nowhere strictly increasing* (i.e., there is no interval (a, b) on which the function is strictly increasing), but not constant.

A space-filling curve

Lemma 4.41. *There is a continuous surjection from the Cantor set to* $[0, 1]^2$.

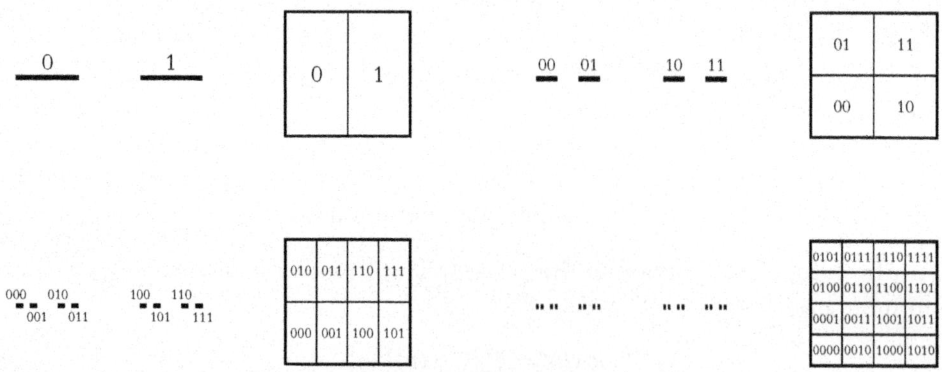

Proof. As in the proof of Theorem 4.39, it suffices to find a continuous surjection $2^{\mathbb{N}} \to [0, 1]^2$ because the Cantor set is homeomorphic to the Cantor space $2^{\mathbb{N}}$.

Given $f \in 2^{\mathbb{N}}$, define two new functions $f_{\text{odd}}, f_{\text{even}} \in 2^{\mathbb{N}}$ by setting

$$f_{\text{odd}}(n) = f(2n - 1) \qquad \text{and} \qquad f_{\text{even}}(n) = f(2n).$$

In other words, if we identify the members of $2^{\mathbb{N}}$ in the natural way with infinite sequences of 0's and 1's, then f_{odd} is the subsequence of f consisting of the terms with odd indices, and f_{even} is the subsequence of f consisting of the terms with even indices.

Let ϕ denote a continuous surjection $2^{\mathbb{N}} \to [0, 1]$, such as the one described in the proof of Theorem 4.39, and define $\Phi : 2^{\mathbb{N}} \to [0, 1]^2$ by setting

$$\Phi(f) = \big(\phi(f_{\text{odd}}), \phi(f_{\text{even}})\big).$$

To see that Φ is a surjection, fix some $(x, y) \in [0, 1]^2$. Because ϕ is surjective, there are some $g, h \in 2^{\mathbb{N}}$ such that $\phi(g) = x$ and $\phi(h) = y$. Define $f \in 2^{\mathbb{N}}$ by setting

$$f(n) = \begin{cases} g(\frac{n+1}{2}) & \text{if } n \text{ is odd,} \\ h(\frac{n}{2}) & \text{if } n \text{ is even.} \end{cases}$$

Then, $f_{\text{odd}} = g$ and $f_{\text{even}} = h$, so $\Phi(f) = \big(\phi(g), \phi(h)\big) = (x, y)$.

To see that Φ is continuous, first observe that if $d(f, g) < 2/2^{2n}$, then $d(f_{\text{odd}}, g_{\text{odd}}) < 1/2^n$ and $d(f_{\text{even}}, g_{\text{even}}) < 1/2^n$. Let $\varepsilon > 0$. Because ϕ is a continuous function, there is some $N \in \mathbb{N}$ such that if $d(f, g) < \frac{1}{2^N}$, then $|\phi(f) - \phi(g)| < \varepsilon/2$. Let $\delta = 1/2^{2N}$. If $d(f, g) < \delta$, then $d(f_{\text{odd}}, g_{\text{odd}}) < 1/2^N$ and $d(f_{\text{even}}, g_{\text{even}}) < 1/2^N$, which means

$$\begin{aligned} d\big(\Phi(f), \Phi(g)\big) &= \sqrt{|\phi(f_{\text{odd}}) - \phi(g_{\text{odd}})|^2 + |\phi(f_{\text{even}}) - \phi(g_{\text{even}})|^2} \\ &< \sqrt{(\varepsilon/2)^2 + (\varepsilon/2)^2} < \varepsilon. \end{aligned}$$

As ε was arbitrary, this shows Φ is continuous. □

We are finally ready to prove the main theorem of this section: the existence of a space-filling curve.

Theorem 4.42. *There is a continuous surjection* $F : [0, 1] \to [0, 1]^2$.

Proof. Let K denote the Cantor set, and let $\Phi : K \to [0, 1]^2$ be a continuous surjection, such as the one from the proof of the previous lemma.

Let $\pi_X : [0, 1]^2 \to [0, 1]$ denote the map $\pi_X(a, b) = a$. This is called the *projection mapping* onto the X-coordinate. Similarly, let $\pi_Y : [0, 1]^2 \to [0, 1]$ denote the map

$\pi_Y(a,b) = b$, the projection mapping onto the Y-coordinate. If $(a,b), (c,d) \in [0,1]^2$, then

$$|a-c| = \sqrt{|a-c|^2} \le \sqrt{|a-c|^2 + |b-d|^2} \quad \text{and}$$
$$|b-d| = \sqrt{|b-d|^2} \le \sqrt{|a-c|^2 + |b-d|^2}.$$

In other words, $|\pi_X(a,b) - \pi_X(c,d)|$ and $|\pi_Y(a,b) - \pi_Y(c,d)|$ are both less than the distance in $[0,1]^2$ from (a,b) to (c,d). By taking $\delta = \varepsilon$ in the ε-δ definition of continuity, it follows that π_X and π_Y are both continuous. Defining $\Phi_X = \pi_X \circ \Phi$ and $\Phi_Y = \pi_Y \circ \Phi$, it follows that Φ_X and Φ_Y are both continuous functions $K \to [0,1]$. Furthermore,

$$\Phi(z) = \big(\Phi_X(z), \Phi_Y(z)\big)$$

for all $z \in K$.

Because K is closed, there is by Theorem 4.37 a pair of continuous functions $F_X : [0,1] \to [0,1]$ and $F_Y : [0,1] \to [0,1]$ that extend Φ_X and Φ_Y, respectively, from K to $[0,1]$. Define

$$F(z) = \big(F_X(z), F_Y(z)\big)$$

for all $z \in [0,1]$.

Note that F is an extension of Φ from K to $[0,1]$. Because Φ is a surjection, its extension F is, too. To see that F is continuous, simply note that the continuity of F_X and F_Y implies

$$\lim_{t \to z} F(t) = \lim_{t \to z} \big(F_X(t), F_Y(t)\big) = \big(F_X(z), F_Y(z)\big) = F(z)$$

for all $z \in [0,1]$. □

The following pictures indicate the geometry of the space-filling curve described in this section.

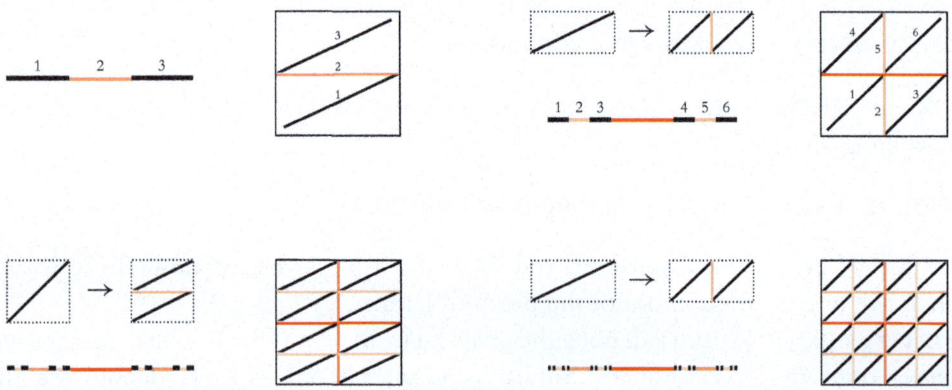

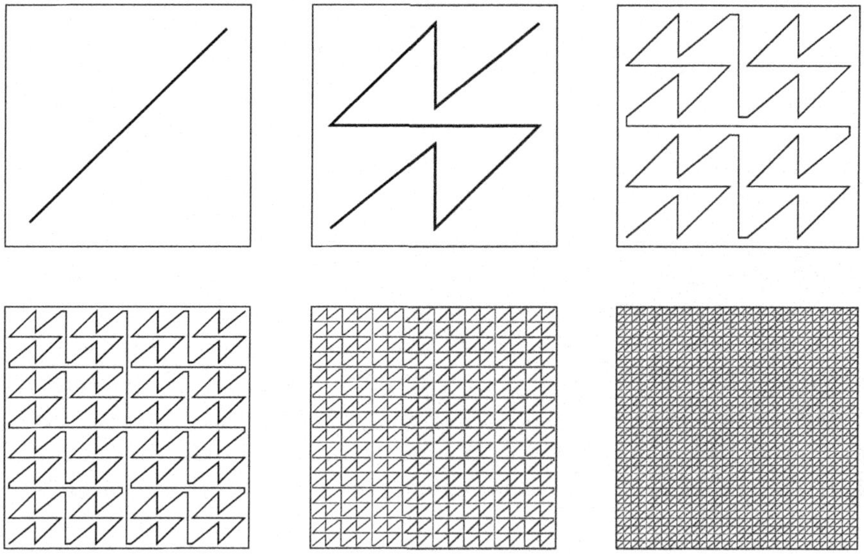

The arguments from this section can be modified to prove that there are higher-dimensional space-filling curves, that is, continuous surjections from $[0, 1]$ onto $[0, 1]^n$ for $n \geq 2$. Sorting out the exact details of this is left as an exercise.

Exercises

1. Prove that the function

$$f(x) = \begin{cases} x^p \sin(1/x) & \text{if } x \neq 0 \\ 0 & \text{if } x = 0 \end{cases}$$

is continuous at 0 if and only if $p > 0$.

2. Prove part (2) of Lemma 4.4.

3. Suppose $f, g, h : \mathbb{R} \to \mathbb{R}$, and $f(t) \leq g(t) \leq h(t)$ for all $t \in \mathbb{R}$. Prove that if f and h are both continuous at some point x and $f(x) = h(x)$, then g is also continuous at x.

4. Let X, Y, and Z be metric spaces, and suppose $f : X \to Y$ and $g : Y \to Z$ are uniformly continuous. Prove that $g \circ f$ is also uniformly continuous.

5. Prove that if f is a uniformly continuous function $\mathbb{R} \to \mathbb{R}$ and $c \in \mathbb{R}$, then cf is uniformly continuous. Also prove that if f and g are uniformly continuous functions $\mathbb{R} \to \mathbb{R}$, then $f + g$ is uniformly continuous.

6. Prove that if $A \subseteq \mathbb{R}$ is not closed, then there is a continuous function $A \to \mathbb{R}$ that is not uniformly continuous.

7. Prove that every function $\mathbb{N} \to \mathbb{R}$ is uniformly continuous. In particular, there are unbounded sets $A \subseteq \mathbb{R}$ such that every continuous function $A \to \mathbb{R}$ is uniformly continuous.

8. On the other hand, prove that if $A \subseteq \mathbb{R}$ and the set of limit points of A is unbounded, then there is a continuous function $A \to \mathbb{R}$ that is not uniformly continuous.

9. Prove that boundedness is not preserved by uniformly continuous functions. In other words, find an example of two metric spaces X and Y, a bounded subset $A \subseteq X$, and a uniformly continuous function $f : X \to Y$ such that $f[A]$ is not bounded in Y.

10. On the other hand, prove that boundedness is preserved by uniformly continuous functions on $\mathbb{R}$. In other words, prove that if Y is a metric space and $f : \mathbb{R} \to Y$ is uniformly continuous, then $f[A]$ is bounded in Y for every bounded $A \subseteq \mathbb{R}$.

11. Suppose $f : \mathbb{R} \to \mathbb{R}$ has a discontinuity of the first kind. Prove that there are some $a, b \in \mathbb{R}$, with $f(a) \neq f(b)$, and some λ between $f(a)$ and $f(b)$, such that there is no x between a and b with $f(x) = \lambda$. In other words, you are proving that if f has a discontinuity of the first kind, then f does not satisfy the conclusion of the Intermediate Value Theorem.

12. Froda's theorem states that monotone functions have only countably many points of discontinuity. Complementing this result, prove that for every countable $S \subseteq \mathbb{R}$, there is a monotone function $f : \mathbb{R} \to \mathbb{R}$ such that $S = D_f = \{x : f \text{ is not continuous at } x\}$.

13. Let X, Y be metric spaces and $f : X \to Y$. We say that f is a *bounded function* if $f[X]$ is a bounded subset of Y. Prove that if f is a bounded function $X \to \mathbb{R}$ and $U \subseteq X$, then

$$\omega_f(U) = \sup\{f(t) : t \in U\} - \inf\{f(t) : t \in U\}.$$

14. Let $A \subseteq \mathbb{R}$, let $x \in \overline{A} \setminus A$, and suppose $f : A \to \mathbb{R}$ is continuous. Prove that f has a continuous extension to $A \cup \{x\}$ if and only if $\omega_f(x) = 0$.

15. Prove that if $A \subseteq \mathbb{R}$ and A is not closed, then there is a continuous function $f :\to \mathbb{R}$ that has no continuous extension to $\mathbb{R}$.

16. Provide a counterexample to the following (false) assertion, which is a modification of Theorem 4.37: for every closed $A \subseteq \mathbb{R}$ and every uniformly continuous $f : A \to \mathbb{R}$, there is a uniformly continuous extension of f to $\mathbb{R}$.

17. Prove that a function $f : [0, 1] \to [0, 1]$ is continuous if and only if its graph $G_f = \{(x, y) \in [0, 1]^2 : f(x) = y\}$ is a closed subset of $[0, 1]^2$.

18. Suppose X is a compact metric space and $f : X \to Y$ is a continuous bijection. Prove that for every open $U \subseteq X$, $f[U]$ is open in Y.

19. Building on the previous exercise, prove there is no continuous bijection $[0, 1] \to [0, 1]^2$.

20. Modifying the arguments presented in the final section of this chapter, prove there is a continuous surjection $[0, 1] \to [0, 1]^n$ for all $n \geq 2$.

CHAPTER 5

Differentiation

This chapter is all about *derivatives*, one of the core concepts of analysis. The basic idea is that given a sufficiently well-behaved function $f(x)$, there is another function, called the derivative of $f(x)$ and denoted $f'(x)$, that tells us how sensitive the value of $f(x)$ is to making small changes in x.

Imagine looking at the graph of a (sufficiently well-behaved) function $f(x)$ under a microscope, using successively more powerful levels of magnification. As you zoom in on some particular point of this graph, things begin to smooth out. Ultimately, zooming in far enough makes the graph look completely smooth and straight—in other words, it looks like a line.

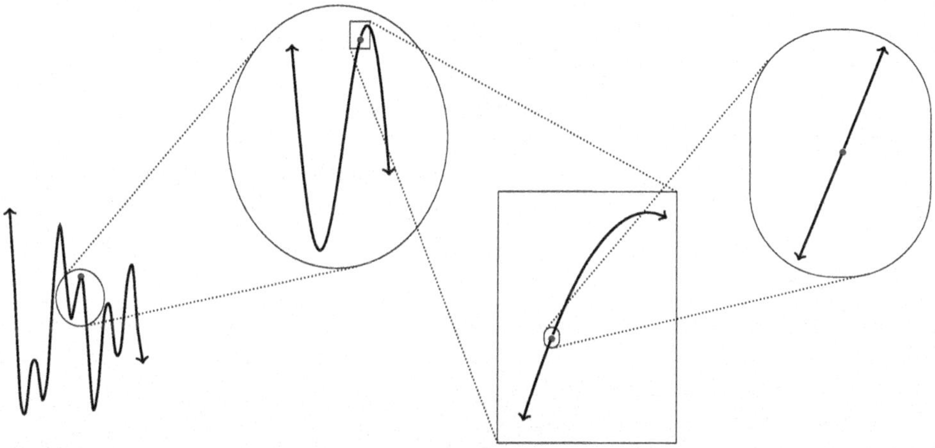

The derivative $f'(x)$ tells you the slope of that line, for a given value of x. If the value of x changes slightly, by some "microscopically small" number δ, then the value of $f(x)$ changes by $\delta \cdot f'(x)$ because the graph of the function $f(x)$, on a sufficiently small scale, looks like a line with slope $f'(x)$. Thus, $f'(x)$ measures the sensitivity of f to small changes in x: A small change in x results in a change $f'(x)$ times as large in $f(x)$.

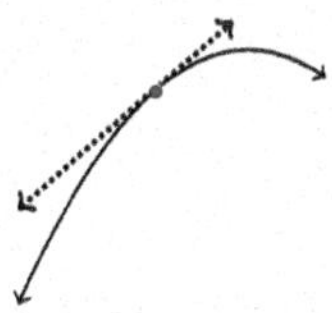

From a geometric point of view, this can be made precise using tangent lines. At a given point $(x, f(x))$ on the graph of a function, the line obtained by zooming in on the graph of f, as previously described, is the same as the line tangent to the graph at that point. The value of $f'(x)$ is therefore the slope of this tangent line.

Foundations of Real Analysis
https://doi.org/10.1016/B978-0-44-348527-5.00009-3

Definition and preliminaries

The slope of a tangent line is computed by taking a limit. Given a point $(x, f(x))$ on the graph of f, we imagine taking lines through this point and other points on the graph, of the form $(t, f(t))$, called secant lines. As t approaches x, the slopes of these secant lines approach the slope of the tangent line at $(x, f(x))$.

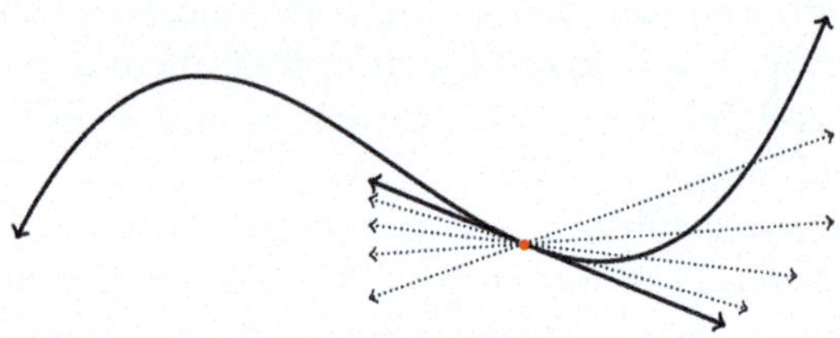

The slope of the line passing through $(x, f(x))$ and $(t, f(t))$ is $\frac{f(t)-f(x)}{t-x}$, which gives us the following definition.

Definition 5.1. Let $D \subseteq \mathbb{R}$, let $f : D \to \mathbb{R}$ be a function, and suppose $x \in \text{Int}(D)$. Then,

$$f'(x) = \lim_{t\to x} \frac{f(t) - f(x)}{t - x}$$

when this limit exists. If this limit does exist, we say that f is *differentiable* at x and $f'(x)$ is the *derivative* of f at x. If $A \subseteq D$ and f is differentiable at every $x \in A$, then we say that f is differentiable on A. If f is differentiable on its entire domain D, then f is called *differentiable.*

Example 5.2. A function of the form $f(x) = mx + b$ is differentiable. Indeed,

$$f'(x) = \lim_{t\to x} \frac{mt + b - (mx + b)}{t - x} = \lim_{t\to x} \frac{m(t - x)}{t - x} = m$$

for every x. This accords with our intuitive account of differentiation as addressed here: The graph of this function is a line of slope m (at any level of magnification), so it is differentiable with derivative m.

Theorem 5.3. *If a function is differentiable at some point, then it is also continuous at that point.*

Proof. Suppose a function f is differentiable at some point x. Then,

$$\begin{aligned}
\lim_{t\to x} f(t) &= f(x) + \lim_{t\to x} \big(f(t) - f(x)\big) \\
&= f(x) + \lim_{t\to x} (t - x) \frac{f(t) - f(x)}{t - x} \\
&= f(x) + \lim_{t\to x} (t - x) \cdot \lim_{t\to x} \frac{f(t) - f(x)}{t - x} = f(x) + 0 \cdot f'(x) = f(x),
\end{aligned}$$

and this implies f is continuous at x. $\square$

Although differentiability implies continuity, the converse is not generally true, as the next example shows.

Example 5.4. The absolute value function $f(x) = |x|$ is continuous, but it is not differentiable at $x = 0$. Indeed, $\lim_{t\to 0^+} \frac{f(t)-f(0)}{t-0} = \frac{t}{t} = 1$, while $\lim_{t\to 0^-} \frac{f(t)-f(0)}{t-0} = \frac{-t}{t} = -1$, and hence $\lim_{t\to 0} \frac{f(t)-f(0)}{t-0}$ does not exist. ☕

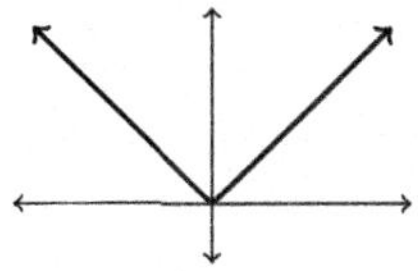

In the last section of this chapter, we will see a much more extreme example of the gap between continuity and differentiability: a function that is continuous at every point, but not differentiable at any point.

Differentiation rules

Theorem 5.5 (Sums rule). *Let $D \subseteq \mathbb{R}$, and suppose f, g are functions $D \to \mathbb{R}$ both differentiable at some point x. Then, the function $f + g$ is differentiable at x, and $(f+g)'(x) = f'(x) + g'(x)$.*

Proof. If f and g are both differentiable at x, then

$$\begin{aligned}
&\lim_{t\to x} \frac{(f(t)+g(t)) - (f(x)+g(x))}{t-x} \\
&= \lim_{t\to x} \left(\frac{f(t)-f(x)}{t-x} + \frac{g(t)-g(x)}{t-x} \right) \\
&= \lim_{t\to x} \frac{f(t)-f(x)}{t-x} + \lim_{t\to x} \frac{g(t)-g(x)}{t-x} = f'(x) + g'(x). \qquad \square
\end{aligned}$$

Theorem 5.6 (Product rule). *Let $D \subseteq \mathbb{R}$, and suppose f and g are functions $D \to \mathbb{R}$ both differentiable at some point x. Then, the function fg is differentiable at x, and $(fg)'(x) = f(x)g'(x) + f'(x)g(x)$.*

Proof. If f and g are both differentiable at x, then

$$\begin{aligned}
&\lim_{t\to x} \frac{(f(t)g(t)) - (f(x)g(x))}{t-x} = \lim_{t\to x} \frac{f(t)g(t) - f(x)g(x)}{t-x} \\
&= \lim_{t\to x} \frac{f(t)g(t) - f(t)g(x) + f(t)g(x) - f(x)g(x)}{t-x} \\
&= \lim_{t\to x} \frac{f(t)\big(g(t)-g(x)\big) + g(x)\big(f(t)-f(x)\big)}{t-x}
\end{aligned}$$

$$= \lim_{t\to x} f(t)\frac{g(t)-g(x)}{t-x} + \lim_{t\to x} g(x)\frac{f(t)-f(x)}{t-x}$$
$$= f(x)g'(x) + f'(x)g(x). \qquad \square$$

Corollary 5.7 (Power rule, version 1). *Suppose $f(x) = x^n$ for some $n \in \mathbb{N}$. Then, f is differentiable, and $f'(x) = nx^{n-1}$.*

Proof. The proof is by induction on n. For the base case $n = 1$, let $\mathrm{id}(x) = x$, and observe that $\lim_{t\to x} \frac{\mathrm{id}(t)-\mathrm{id}(x)}{t-x} = \lim_{t\to x} \frac{t-x}{t-x} = 1 = 1x^{1-1}$. (This is just a special case of Example 5.2.) For the inductive step, let $f(x) = x^n$. Let $g(x) = x^{n-1}$ and suppose (as the inductive hypothesis) that g is differentiable and $g'(x) = (n-1)x^{n-2}$ for every x. Because $f(x) = \mathrm{id}(x)g(x)$, we may conclude from Theorem 5.6 that f is differentiable, and

$$f'(x) = \mathrm{id}(x)g'(x) + \mathrm{id}'(x)g(x)$$
$$= x \cdot (n-1)x^{n-2} + 1 \cdot x^{n-1} = (n-1)x^{n-1} + x^{n-1} = nx^{n-1}. \qquad \square$$

Corollary 5.8. *Let $D \subseteq \mathbb{R}$, and suppose f is a function $D \to \mathbb{R}$ differentiable at some point x. For any $c \in \mathbb{R}$, the function cf is differentiable at x, and $(cf)'(x) = cf'(x)$.*

Proof. This follows from the product rule by taking g to be the constant function $x \mapsto c$. (It is also fairly straightforward to prove this corollary directly from the definition of a derivative, without needing to prove the product rule first.) $\square$

Corollary 5.9. *Polynomial functions are differentiable.*

Proof. Combine Theorem 5.5 and Corollaries 5.7 and 5.8. $\square$

Theorem 5.10 (Quotient rule). *Let $D \subseteq \mathbb{R}$, and suppose f, g are functions $D \to \mathbb{R}$ both differentiable at some point x, and $g(x) \neq 0$. Then the function f/g is differentiable at x, and $(f/g)'(x) = \frac{g(x)f'(x)-f(x)g'(x)}{(g(x))^2}$.*

Proof. If f and g are both differentiable at x and $g(x) \neq 0$, then

$$\lim_{t\to x} \frac{f(t)/g(t) - f(x)/g(x)}{t-x} = \lim_{t\to x} \frac{\frac{f(t)g(x)-f(x)g(t)}{g(t)g(x)}}{t-x}$$
$$= \lim_{t\to x} \frac{1}{g(t)g(x)} \frac{f(t)g(x) - f(x)g(x) + f(x)g(x) - f(x)g(t)}{t-x}$$
$$= \frac{1}{(g(x))^2} \lim_{t\to x} \frac{g(x)\big(f(t)-f(x)\big) - f(x)\big(g(t)-g(x)\big)}{t-x}$$
$$= \frac{1}{(g(x))^2} \lim_{t\to x} \left(g(x)\frac{f(t)-f(x)}{t-x} - f(x)\frac{g(t)-g(x)}{t-x} \right)$$
$$= \frac{g(x)f'(x) - f(x)g'(x)}{(g(x))^2}. \qquad \square$$

Corollary 5.11 (Power rule, version 2). *Suppose $f(x) = x^n$ for some $n \in \mathbb{Z}$. Then, f is differentiable, and $f'(x) = nx^{n-1}$.*

Proof. For $n \in \mathbb{N}$ this was already proved in Corollary 5.7. If $n = 0$, then $f(x) = 1$ and $\lim_{t\to x} \frac{f(t)-f(x)}{t-x} = \lim_{t\to x} \frac{1-1}{t-x} = 0 = 0x^{0-1}$ for all x. For the remaining case, suppose $n \in \mathbb{Z}$ and $n < 0$, and let $f(x) = x^n$. Then, $f(x) = 1/x^{|n|}$, and, applying Corollary 5.7 and Theorem 5.10, we conclude that f is differentiable and

$$f'(x) = \frac{x^{|n|} \cdot 0 - 1 \cdot |n|x^{|n|-1}}{(x^{|n|})^2} = -|n|x^{|n|-1-2|n|} = nx^{n-1}. \qquad \square$$

Corollary 5.12. *Every rational function is differentiable at every point of its domain.*

Proof. Combine the quotient rule with Corollary 5.9. □

Theorem 5.13 (Chain rule). *Suppose that $D \subseteq \mathbb{R}$ and that f is a function $D \to \mathbb{R}$; also, suppose $f[D] \subseteq E \subseteq \mathbb{R}$, and g is a function $E \to \mathbb{R}$. If f is differentiable at some point x and g differentiable at $f(x)$, then the composite function $g \circ f$ is differentiable at x, and $(g \circ f)'(x) = g'(f(x))f'(x)$.*

Proof. Let $y = f(x)$. Define a new expression as follows:

$$u(t) = \begin{cases} \frac{f(t)-f(x)}{t-x} - f'(x) & \text{if } t \neq x \\ 0 & \text{if } t = x. \end{cases}$$

Because f is differentiable at x, we have $\lim_{t\to x} u(t) = 0$, which means that u is continuous at x. Similarly, let

$$v(s) = \begin{cases} \frac{g(s)-g(y)}{s-y} - g'(y) & \text{if } s \neq y \\ 0 & \text{if } s = y. \end{cases}$$

As with u, the fact that g is differentiable at y implies v is continuous at y. Rearranging the previous equations gives

$$\begin{aligned} f(t) - f(x) &= (t-x)\big(u(t) + f'(x)\big), \quad \text{and} \\ g(s) - g(y) &= (s-y)\big(v(s) + g'(y)\big). \end{aligned}$$

Combining these expressions (with $s = f(t)$) gives

$$\begin{aligned} (g \circ f)(t) - (g \circ f)(x) &= g(f(t)) - g(y) \\ &= \big(f(t) - y\big)\big(v(f(t)) + g'(y)\big) \\ &= (t-x)\big(u(t) + f'(x)\big)\big(v(f(t)) + g'(y)\big). \end{aligned}$$

Because v is continuous at $y=f(x)$ and f is continuous at x (by Theorem 5.3), the composite function $v\circ f$ is continuous at x. It follows that $\lim_{t\to x} v(f(t)) = v(f(x)) = 0$. Therefore,

$$\begin{aligned}\lim_{t\to x} & \frac{(g\circ f)(t)-(g\circ f)(x)}{t-x} \\ &= \lim_{t\to x}\big(u(t)+f'(x)\big)\big(v(f(t))+g'(y)\big) \\ &= \big(u(x)+f'(x)\big)\big(0+g'(y)\big) = f'(x)g'(f(x)).\end{aligned}$$

□

Theorem 5.14 (Inverse function rule). *Suppose $a, b \in \mathbb{R}$ with $a<b$ and f is a continuous injective function $(a, b)\to\mathbb{R}$. Let g be the inverse of f on (a, b), i.e., g is the (unique) function such that $g\circ f(t) = f\circ g(t) = t$ for all $t\in(a, b)$. If f is differentiable at some point $x\in(a, b)$ and $f'(x)\neq 0$, then g is differentiable at $f(x)$, and $g'(f(x)) = 1/f'(x)$.*

Despite its somewhat complicated-looking statement, the idea behind this theorem is fairly simple. If f and g are inverses of one another on some open set U, then the graph of g is obtained from the graph of f simply by reflecting it across the line $y=x$.

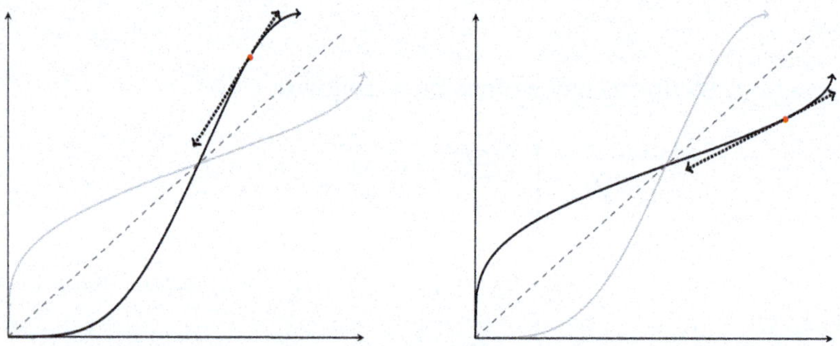

As the picture suggests, if the point $(x, f(x))$ is on the graph of f, then the point $(f(x), x)$ is on the graph of g, and the slopes of the corresponding tangent lines to these graphs are reciprocals of one another. This is more or less what Theorem 5.14 says. It also requires that f be invertible on some interval containing x, but of course this must be true if we wish to say anything meaningful about the derivative of its inverse.

Our proof of Theorem 5.14 is not as short or easy as this picture might suggest. Before embarking on the proof, let us point out that the real content of the theorem is not merely the formula $g'(f(x)) = 1/f'(x)$, but the fact that g is differentiable in the first place. In fact, if we assume g is differentiable, then the formula is easy to derive. Because $g\circ f = \mathrm{id}$, the chain rule says

$$1 = \mathrm{id}'(x) = (g\circ f)'(x) = g'(f(x))f'(x),$$

which implies that $g'(f(x)) = 1/f'(x)$. Before showing g is differentiable, let us first show it is continuous.

Definition 5.15. Let $D \subseteq \mathbb{R}$. A function $f : D \to \mathbb{R}$ is *strictly increasing* if $f(a) < f(b)$ whenever $a, b \in D$ and $a < b$. It is *strictly decreasing* if $f(a) > f(b)$ whenever $a, b \in D$ and $a < b$.

Lemma 5.16. *Suppose $a, b \in \mathbb{R}$ with $a < b$, and let f be a continuous injection $[a, b] \to \mathbb{R}$. Then, f is either strictly increasing or strictly decreasing.*

Proof. Let $x, y \in (a, b]$ with $a < x < y < z \leq b$. Aiming for a contradiction, assume $f(x) < f(y) > f(z)$. Fix λ with $\max\{f(x), f(z)\} < \lambda < f(y)$. Using the fact that f is continuous, the Intermediate Value Theorem implies that f maps a point in (x, y) to λ and that f maps a point in (y, z) to λ, contradicting the injectivity of f. By essentially the same argument, the assumption $f(x) > f(y) < f(z)$ also leads to a contradiction. Because $f(x) = f(y)$ and $f(y) = f(z)$ are ruled out by the injectivity of f, this leaves only two options: (1) $f(x) < f(y) < f(z)$ or (2) $f(x) > f(y) > f(z)$.

If $f(a) < f(b)$, then setting $x = a$ and $z = b$ in the previous paragraph, we have $f(a) < f(y) < f(b)$ for any $y \in (a, b)$. Given any $x, y \in (a, b)$ with $x < y$, we have $f(a) < f(y)$ (by the previous sentence), which implies $f(a) < f(x) < f(y)$ (by the previous paragraph). Thus, f is strictly increasing. A similar argument shows that if $f(a) > f(b)$, then f is strictly decreasing. □

Lemma 5.17. *Suppose $a, b \in \mathbb{R}$ with $a < b$. If $f : [a, b] \to \mathbb{R}$ is strictly increasing, then f^{-1} is a continuous, strictly increasing function $[f(a), f(b)] \to \mathbb{R}$. If $f : [a, b] \to \mathbb{R}$ is strictly decreasing, then f^{-1} is a continuous, strictly decreasing function $[f(b), f(a)] \to \mathbb{R}$.*

Proof. First, assume f is strictly increasing on $[a, b]$. Recall that continuous functions map connected sets to connected sets, and they map compact sets to compact sets. In particular, because $[a, b]$ is both compact and connected, so is $f\big[[a, b]\big]$. Furthermore, the fact that f is strictly increasing means that $f(a) = \inf f\big[[a, b]\big]$ and $f(b) = \sup f\big[[a, b]\big]$. It follows that $f\big[[a, b]\big] = [f(a), f(b)]$, i.e., f is a surjection onto $[f(a), f(b)]$. If $y, z \in [f(a), f(b)]$, then $f^{-1}(y), f^{-1}(z) \in [a, b]$, and the fact that f is strictly increasing means that if $f^{-1}(y) < f^{-1}(z)$, then $y < z$ and if $f^{-1}(y) > f^{-1}(z)$, then $y > z$. Thus, $y < z$ if and only if $f^{-1}(y) < f^{-1}(z)$, which means f^{-1} is strictly increasing.

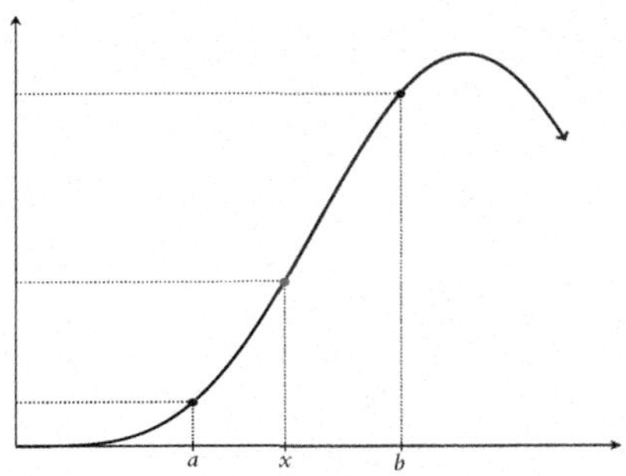

Thus, f^{-1} is a strictly increasing surjection $[f(a), f(b)] \to [a, b]$. Because it is surjective and strictly increasing, f^{-1} has no jump discontinuities. By Froda's theorem, this means f^{-1} is continuous. This finishes the proof in the case that f is strictly increasing on $[a, b]$. The argument for strictly decreasing functions is essentially identical. □

The previous two lemmas show that a continuous injection $[a, b] \to \mathbb{R}$ must be continuous. This enables us to subsequently prove Theorem 5.14 But first, let us observe that with a different domain or codomain, continuous injections do not generally have continuous inverses. This example shows that using some of the order-theoretic structure of $\mathbb{R}$ was really necessary for proving the previous two lemmas.

Example 5.18. Consider the function $(0, 1) \cup [2, 3) \to (-1, 1)$ with $f(t) = t$ if $t \in (0, 1)$ and $f(t) = 2 - t$ if $t \in [2, 3)$. This function is a continuous bijection $(0, 1) \cup [2, 3) \to (-1, 1)$, but it is not monotone and its inverse is not continuous.

Consider the function $f(t) = (\sin t, \cos t)$. This is a continuous bijection from $[0, 2\pi)$ to the unit circle $\{(x, y) \in \mathbb{R}^2 : x^2 + y^2 = 1\}$. But, f^{-1} is not continuous at $(1, 0)$.

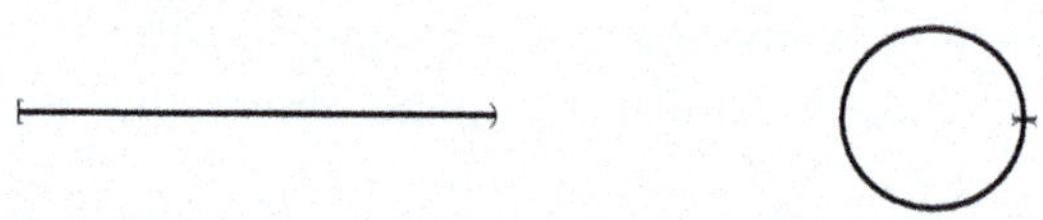

Proof of Theorem 5.14. Let $y = f(x)$ and let $\varepsilon > 0$. Our goal is to show that there is some $\beta > 0$ such that if $|s - y| < \beta$, then $\left|\frac{g(s)-g(y)}{s-y} - \frac{1}{f'(x)}\right| < \varepsilon$.

The function $t \mapsto 1/t$ is continuous at every point but 0. Because $f'(x) \neq 0$, there is some $\delta > 0$ such that if $|t - f'(x)| < \delta$, then $\left|\frac{1}{t} - \frac{1}{f'(x)}\right| < \varepsilon$.

Because $f'(x) = \lim_{t\to x} \frac{f(t)-f(x)}{t-x}$, there is also some $\gamma > 0$ such that if $|t - x| < \gamma$, then $\left|\frac{f(t)-f(x)}{t-x} - f'(x)\right| < \delta$.

Finally, because g is continuous at y by the previous two lemmas, there is some $\beta > 0$ such that if $|s - y| < \beta$, then $|g(s) - g(y)| < \gamma$.

Now, suppose $|s - y| < \beta$. This implies $|g(s) - g(y)| < \gamma$. By our choice of γ, this implies (using the fact that $f \circ g = \text{id}$) that

$$\left|\frac{f(g(s)) - f(g(y))}{g(s) - g(y)} - f'(x)\right| = \left|\frac{s - y}{g(s) - g(y)} - f'(x)\right| < \delta.$$

By our choice of δ, this implies $\left|\frac{g(s)-g(y)}{s-y} - \frac{1}{f'(x)}\right| < \varepsilon$.

Thus, there is a $\beta > 0$ such that $|s - y| < \beta$ implies $\left|\frac{g(s)-g(y)}{s-y} - \frac{1}{f'(x)}\right| < \varepsilon$. As ε was arbitrary, this shows that $\lim_{s\to y} \frac{g(s)-g(y)}{s-y} = 1/f'(x)$. □

Corollary 5.19 (Power rule, version 3). *Suppose $f(x) = x^p$ for some $p \in \mathbb{Q}$. Then, f is differentiable at all $x > 0$, and $f'(x) = px^{p-1}$.*

Proof. The case $p = 0$ is covered already by Corollary 5.11, so let us assume $p \in \mathbb{Q}$ and $p \neq 0$. Thus, $p = {}^m/_n$ for some $m \in \mathbb{Z}$ and $n \in \mathbb{N}$ with $m \neq 0$. Recall that $x^{m/n} = (x^m)^{1/n}$, where the function $t \mapsto t^{1/n}$ is defined as the inverse of the function $t \mapsto t^n$ on the open set $(0, \infty)$. In other words, the function $x \mapsto x^p$ is equal to $g^{-1} \circ f$, where $f(x) = x^m$ and $g(x) = x^n$. Now fix some $x > 0$. Then, $f(x) = x^m > 0$, and g is continuous and injective on the open set $(0, b)$, for any $b > f(x)$. By the inverse function rule, g^{-1} is differentiable at $f(x)$. Combining this with the chain rule and Corollary 5.11, the function $g^{-1} \circ f$ is differentiable at x, and its derivative is

$$\begin{aligned}(g^{-1} \circ f)'(x) &= (g^{-1})'(f(x)) \cdot f'(x) = \frac{1}{g'(g^{-1}(f(x)))} f'(x) \\ &= \frac{1}{g'(x^p)} f'(x) = \frac{1}{n(x^p)^{n-1}} m x^{m-1} = \frac{m}{n} \frac{x^{m-1}}{x^{\frac{m(n-1)}{n}}} \\ &= \frac{m}{n} x^{\frac{n(m-1)}{n} - \frac{m(n-1)}{n}} = \frac{m}{n} x^{\frac{m-n}{n}} = p x^{p-1}.\end{aligned}$$

□

The Mean Value Theorem

The main theorem of this section is the *Mean Value Theorem* (or the MVT for short). A version of this theorem can be traced all the way back to the 15th-century Indian mathematician Parameshvara, though in the general form given here the theorem is usually attributed to Cauchy (1823). Before proving it, we begin with a related and even more well-known fact: A differentiable function achieves its maximum or minimum when its derivative is 0.

Definition 5.20. Suppose $f : D \to \mathbb{R}$ is a differentiable function, and $A \subseteq D$.

We say that f *achieves a local maximum value on* A at a point $x \in A$ if there is some $\varepsilon > 0$ such that $f(t) \leq f(x)$ for all $t \in A$ with $|x - t| < \varepsilon$. In this case, $f(x)$ is called a *local maximum value for* f *on* A. In context, the "on A" is omitted if it is clear what A is intended, e.g., if $A = D = \mathbb{R}$.

Similarly, f *achieves a local minimum value on* A at a point $x \in A$ if there is some $\varepsilon > 0$ such that $f(t) \geq f(x)$ for all $t \in A$ with $|x - t| < \varepsilon$. In this case, $f(x)$ is called a *local minimum value for* f *on* A.

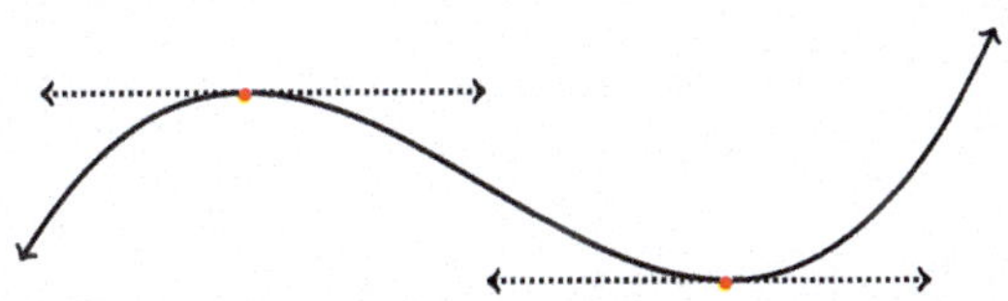

Theorem 5.21. *Suppose $a, b \in \mathbb{R}$ with $a < b$, and f is a differentiable function $[a, b] \to \mathbb{R}$. If f achieves either a local maximum or local minimum value on (a, b) at x, then $f'(x) = 0$.*

Proof. Suppose $x \in (a, b)$ and f achieves a local maximum value on (a, b) at x. Then, there is some $\varepsilon > 0$ such that $f(t) - f(x) \leq 0$ whenever $t \in (x - \varepsilon, x + \varepsilon) \cap (a, b)$. It follows that the limit $\lim_{t \to x^-} \frac{f(t)-f(x)}{t-x}$, if it exists, is nonnegative (because the numerator and denominator are both nonpositive), and the limit $\lim_{t \to x^+} \frac{f(t)-f(x)}{t-x}$, if it exists, is nonpositive (because the numerator is nonpositive and the denominator is nonnegative). But f is differentiable, so both these limits do exist, and are both equal to $f'(x) = \lim_{t \to x} \frac{f(t)-f(x)}{t-x}$. It follows that $f'(x)$ is both nonnegative and nonpositive, i.e., $f'(x) = 0$. A similar argument shows $f'(x) = 0$ if f achieves a local maximum at x. □

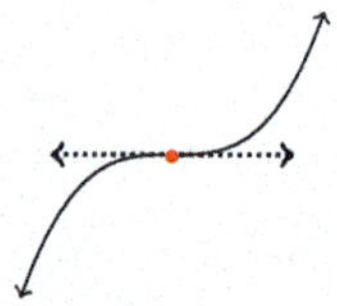

Note that the converse of this theorem is not true. For example, the function $f(x) = x^3$ has derivative $f'(x) = 3x^2$, and in particular $f'(0) = 0$. But this function has neither a local maximum nor a local minimum at $x = 0$.

Theorem 5.22 (Mean Value Theorem). *Suppose $a, b \in \mathbb{R}$ with $a < b$, and suppose f is a real-valued function that is continuous on $[a, b]$ and differentiable on (a, b). Then, there is some $x \in (a, b)$ such that $f'(x) = \frac{f(b)-f(a)}{b-a}$.*

In other words, the MVT states that the average rate of change of f between a and b, which is $\frac{f(b)-f(a)}{b-a}$, must be equal to f' somewhere between a and b (possibly at more than one point, as in the next graph).

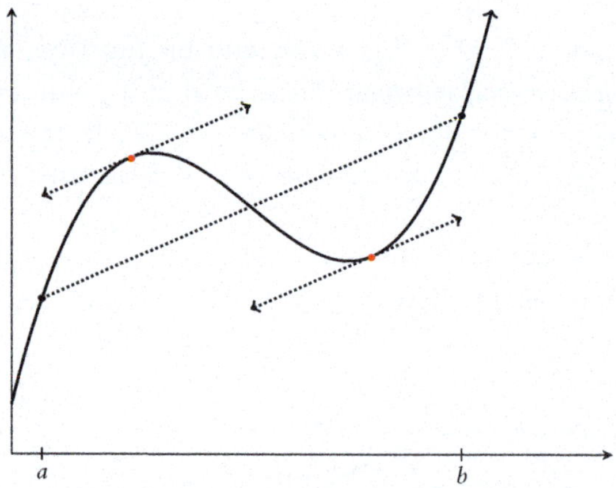

Proof of Theorem 5.22. Roughly, the idea of this proof is to subtract a linear function from f such that the new function (which we call g) has average rate of change 0 between a and b. The points x satisfying the MVT for f will be precisely those points with $g'(x) = 0$, and we can find such points between a and b using Theorem 5.21.

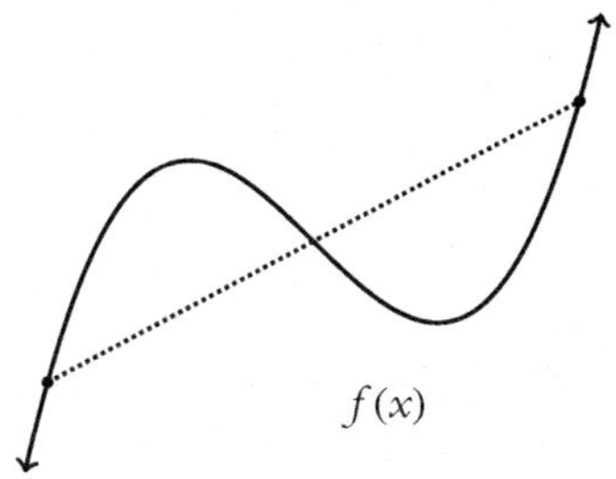

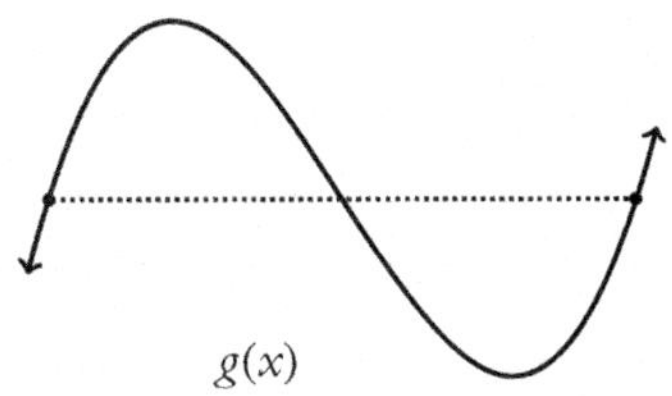

More precisely, define

$$g(t) = f(t) - \frac{f(b)-f(a)}{b-a}t.$$

Because f is continuous on $[a, b]$, g is, too. Because f is differentiable on (a, b), g is, too, and $g'(x) = f'(x) - \frac{f(b)-f(a)}{b-a}$. Furthermore,

$$\begin{aligned} g(a) &= f(a) - \frac{f(b)-f(a)}{b-a}a = \frac{(b-a)f(a)-af(b)+af(a)}{b-a} = \frac{bf(a)-af(b)}{b-a}, \\ g(b) &= f(b) - \frac{f(b)-f(a)}{b-a}b = \frac{(b-a)f(b)-bf(b)+bf(a)}{b-a} = \frac{bf(a)-af(b)}{b-a}. \end{aligned}$$

In particular, $g(a) = g(b)$.

We wish to find a point $x \in (a, b)$ where $g'(x) = 0$. If g is constant on $[a, b]$, then $g'(x) = 0$ for all $x \in (a, b)$. If g is not constant on $[a, b]$, then because $g(a) = g(b)$, there is some $s \in (a, b)$ with either $g(s) > g(a), g(b)$ or $g(s) < g(a), g(b)$.

Suppose that $g(s) > g(a), g(b)$. By the Extreme Value Theorem, because g is continuous on $[a, b]$, there is some $x \in [a, b]$ such that $g(x) = \sup\{g(t) : t \in [a, b]\}$. Because $g(x) \geq g(s) > g(a), g(b)$, clearly, $x \neq a, b$. Hence, $x \in (a, b)$, and f achieves a local maximum at x. Using the fact that g is differentiable on (a, b), Theorem 5.21 implies $g'(x) = 0$.

Supposing instead that $g(s) < g(a), g(b)$, a similar argument (but using a local minimum for g instead of a local maximum) shows that there is some $x \in (a, b)$ with $g'(x) = 0$.

Either way, there is some $x \in (a, b)$ with $g'(x) = 0$. Because $g'(x) = f'(x) - \frac{f(b)-f(a)}{b-a}$, this means $f'(x) = \frac{f(b)-f(a)}{b-a}$. □

Corollary 5.23. *Suppose $a, b \in \mathbb{R}$ with $a < b$, and suppose f is a real-valued function that is continuous on $[a, b]$ and differentiable on (a, b). Then,*

(1) *$f'(x) \geq 0$ for all $x \in (a, b)$ if and only if f is nondecreasing.*

(2) *$f'(x) \leq 0$ for all $x \in (a, b)$ if and only if f is nonincreasing.*

(3) *$f'(x) = 0$ for all $x \in (a, b)$ if and only if f is constant.*

Proof. For the "if" direction of (1), suppose f is nondecreasing. Then, $\frac{f(t)-f(x)}{t-x}$ has a nonnegative numerator and denominator when $t > x$, and it has a nonpositive numerator and denominator when $t < x$. In either case, $\frac{f(t)-f(x)}{t-x} \geq 0$ for all $t \neq x$, and therefore $f'(x) = \lim_{t \to x} \frac{f(t)-f(x)}{t-x} \geq 0$.

For the "only if" direction of (1), we prove the contrapositive. If f is not nondecreasing on $[a, b]$, then there are some $x, y \in [a, b]$ such that $x < y$ and $f(x) > f(y)$. Applying the Mean Value Theorem to f on the interval $[x, y]$, there is some $z \in (x, y)$ such that $f'(z) = \frac{f(x)-f(y)}{x-y} < 0$.

A similar argument is used to prove (2). Finally, (3) follows by combining (1) and (2): $f'(x) = 0$ for all $x \in (a, b)$ if and only if $f'(x) \leq 0$ and $f'(x) \geq 0$ for all $x \in (a, b)$, if and only if f is both nondecreasing and nonincreasing, i.e., constant. □

Must a derivative be continuous?

If f is differentiable on an interval (a, b), then we know f is also continuous on (a, b). But what about $f'(x)$? Must this function be continuous, too? Part of the answer to this question is: Derivatives are not necessarily continuous, but they do share at least one important property with continuous functions, namely, they satisfy the conclusion of the Intermediate Value Theorem. Proving this is the topic of this section. (Another part of the answer, which we delay until after studying function spaces in Chap. 8, is that a derivative cannot be too discontinuous: The set of its points of discontinuity must be meager.)

One attempt at getting a function f with a noncontinuous derivative is to take a noncontinuous function g and integrate it. After all, everyone knows that if g is the integral of f then f is the derivative of g. Right?

Not exactly. The first fundamental theorem of calculus, covered in the next chapter, states that a continuous function is the derivative of its integral. However, integrating a noncontinuous function can result in something not even differentiable.

This kind of example does not address the question stated above: if f is differentiable, then how close is f' to being continuous? We are looking for a differentiable function $f(x)$ such that the everywhere-defined function $f'(x)$ is not continuous. One example of such a function is as follows.

Example 5.24. Consider the function

$$f(x) = \begin{cases} x^2 \sin(1/x) & \text{if } x \neq 0 \\ 0 & \text{if } x = 0. \end{cases}$$

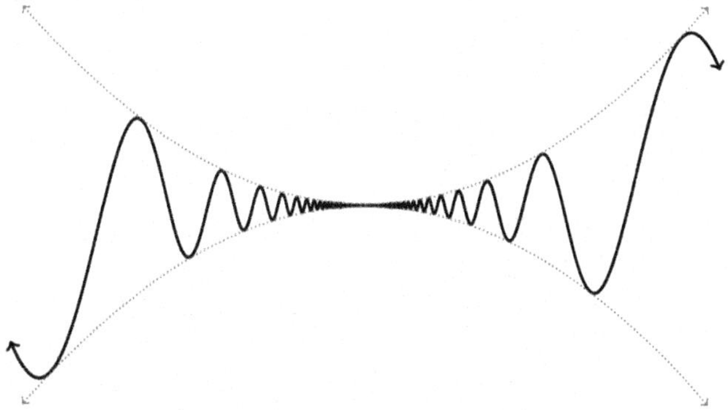

This function is differentiable. Using the chain rule, quotient rule, and product rule, it is clear that f is differentiable at x if $x \neq 0$. (At least, this is clear assuming the function $x \mapsto \sin(x)$ is differentiable. Proving this is one of the exercises.) It remains to show f is differentiable at $x = 0$.

To that f is differentiable at 0, let us verify the definition directly. Let $\varepsilon > 0$. Because $-1 \leq \sin(s) \leq 1$ for all s, if $|t| < \varepsilon$ then

$$\left|\frac{f(t)-f(0)}{t-0}\right| = \left|\frac{t^2 \sin\left(1/t\right)}{t}\right| = \left|t\sin\left(1/t\right)\right| \leq |t|.$$

It follows that $\lim_{t\to 0}\left|\frac{f(t)-f(0)}{t-0}\right| = 0$, so $f'(0)$ exists, and $f'(0) = 0$. Combining this with the chain rule, quotient rule, and product rule, and the fact that the derivative of the sine function is the cosine function (proved as a homework exercise), we get

$$f'(x) = \begin{cases} 2x\sin\left(1/x\right) - \cos\left(1/x\right) & \text{if } x \neq 0 \\ 0 & \text{if } x = 0. \end{cases}$$

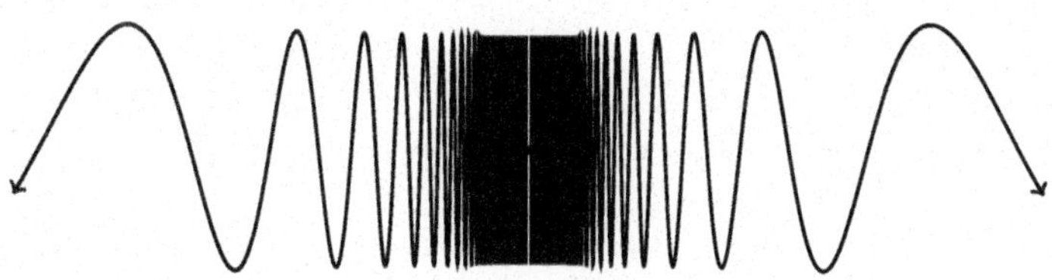

This function has a discontinuity of the second kind at $x = 0$.

According to Corollary 5.27, a derivative can never have a discontinuity of the first kind. In other words, if a derivative is discontinuous, then it must have a discontinuity of the second kind. This means that the discontinuous derivative from Example 5.24, though it may look somewhat complicated, is really the simplest example of a discontinuous derivative that one could hope for.

Recall the statement of the Intermediate Value Theorem:

(IVT) Suppose $a, b \in \mathbb{R}$ with $a < b$, and suppose f is a continuous function $[a, b] \to \mathbb{R}$. For every $y \in \mathbb{R}$ that is between $f(a)$ and $f(b)$, there is some x between a and b such that $f(x) = y$.

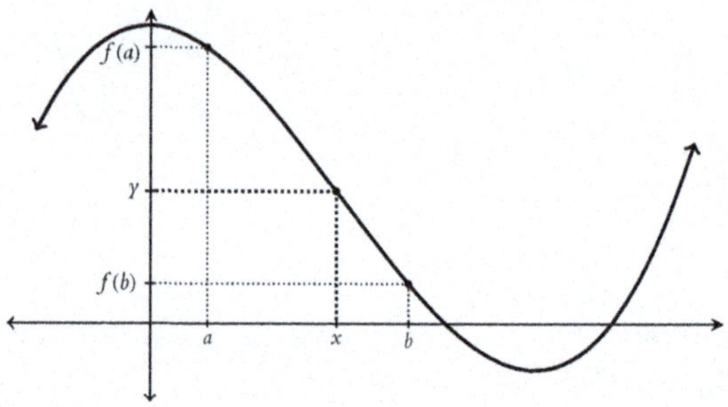

Definition 5.25. A function $f : \mathbb{R} \to \mathbb{R}$ is said to *satisfy the conclusion of the IVT* if for every $a, b \in \mathbb{R}$ with $a < b$, for every $y \in \mathbb{R}$ between $f(a)$ and $f(b)$ there is some x between a and b such that $f(x) = y$.

Sometimes such functions are called *Darboux* functions. The IVT states that all continuous functions $\mathbb{R} \to \mathbb{R}$ satisfy the conclusion of the IVT. However, the converse is not true: a function may satisfy the conclusion of the IVT and yet not be continuous. A classic example of this phenomenon is the function $f'(x)$ from Example 5.24. A careful look at the graph of this function may (or may not) be enough to conclude that it satisfies the conclusion of the IVT. The following theorem provides a proof that it does, and so does the derivative of every other differentiable function.

Theorem 5.26 (Darboux's Theorem). *Suppose $f : \mathbb{R} \to \mathbb{R}$ is a differentiable function. Then its derivative $f'(x)$ satisfies the conclusion of the IVT.*

Proof. Let $f : \mathbb{R} \to \mathbb{R}$ be differentiable, and let $a, b \in \mathbb{R}$ with $a < b$. Supposing that $f'(a) < f'(b)$, we will show that for every $\lambda \in (f'(a), f'(b))$ there is some $x \in (a, b)$ such that $f'(x) = \lambda$. (The proof of the other case, where $f'(a) > f'(b)$, is essentially the same.)

Fix some $\lambda \in \mathbb{R}$ such that $f'(a) < \lambda < f'(b)$. Define $g : \mathbb{R} \to \mathbb{R}$ by setting

$$g(t) = f(t) - \lambda t$$

for all t. Because f is differentiable, g is, too, and $g'(t) = f'(t) - \lambda$ for all t. In particular, $g'(a) = f'(a) - \lambda < 0$, and $g'(b) = f'(b) - \lambda > 0$.

Because $g'(a) = \lim_{t \to a} \frac{g(t) - g(a)}{t - a} < 0$, there is some $\varepsilon_a > 0$ such that $\frac{g(t) - g(a)}{t - a} < 0$ for all t with $|a - t| < \varepsilon_a$. In particular, if $a < t < a + \varepsilon_a$ then $g(t) - g(a) < 0$ (because $\frac{g(t) - g(a)}{t - a} < 0$ and $t - a > 0$), which means $g(t) < g(a)$. Thus $g(a) \neq \inf\{g(t) : t \in [a, b]\}$, or in other words, the minimum value of g on $[a, b]$ is not attained at $x = a$. (This is clear if one thinks of $g'(a)$ as the slope of the graph of g in a microscopic region around $(a, g(a))$.)

Similarly, because $g'(b) > 0$ there is some $\varepsilon_b > 0$ such that if $b - \varepsilon_b < t < b$ then $g(t) < g(b)$. This means that the minimum value of g on $[a, b]$ is not attained at $x = b$.

Nonetheless, because g is continuous (by Theorem 5.3) the Extreme Value Theorem implies that the minimum value of g on $[a, b]$ is attained at some $x \in [a, b]$. Therefore the minimum value of g on $[a, b]$ is attained at some $x \in (a, b)$. This means g achieves a local minimum value at x. By Theorem 5.21, this implies $g'(x) = 0$. Because $g'(x) = f'(x) - \lambda$, this implies $f'(x) = \lambda$. □

Corollary 5.27. *Suppose $f : \mathbb{R} \to \mathbb{R}$ is a differentiable function. Then its derivative $f'(x)$ does not have any discontinuities of the first kind.*

Proof. By Exercise 11 from Chap. 4, if a function has a discontinuity of the first kind at some point, then it does not satisfy the conclusion of the IVT. The corollary follows from this fact plus Darboux's Theorem. □

A continuous, nowhere differentiable function

We end this chapter with an example of a function that is continuous at every point, but not differentiable at any point. The first example of such a function was concocted by Weierstrass in 1872.

Theorem 5.28. *There is a continuous function $f : \mathbb{R} \to \mathbb{R}$ that is not differentiable at any point.*

Proof. To begin, define a function $\phi : \mathbb{R} \to \mathbb{R}$ by setting

$$\phi(x) = \inf\{|2n - x| : n \in \mathbb{Z}\}$$

for all $x \in \mathbb{R}$. In other words, $\phi(x)$ is the distance from x to the nearest even integer.

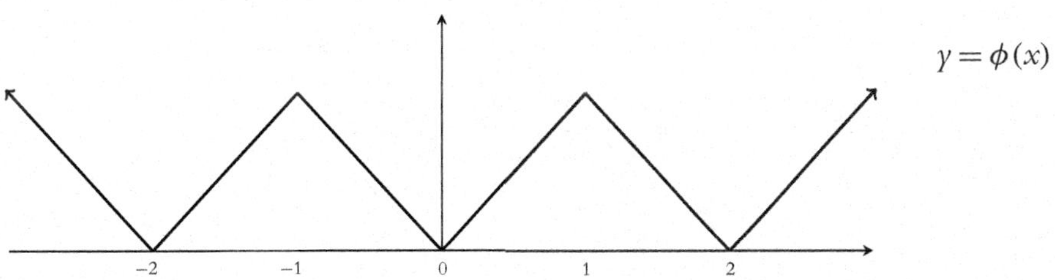

Next, define a function $f : \mathbb{R} \to \mathbb{R}$ by setting

$$f(x) = \sum_{k=0}^{\infty} \left(\frac{3}{4}\right)^k \phi(4^k x)$$

for each $x \in \mathbb{R}$. This sum converges for every $x \in \mathbb{R}$ by the Comparison Test, because all the terms of the sum are nonnegative, and the sum is termwise bounded above by a geometric series:

$$f(x) = \sum_{k=0}^{\infty} \left(\frac{3}{4}\right)^k \phi(4^k x) \le \sum_{k=0}^{\infty} \left(\frac{3}{4}\right)^k = 4.$$

Thus f is a well-defined function. We claim that this function is continuous at every $x \in \mathbb{R}$, but not differentiable at any $x \in \mathbb{R}$.

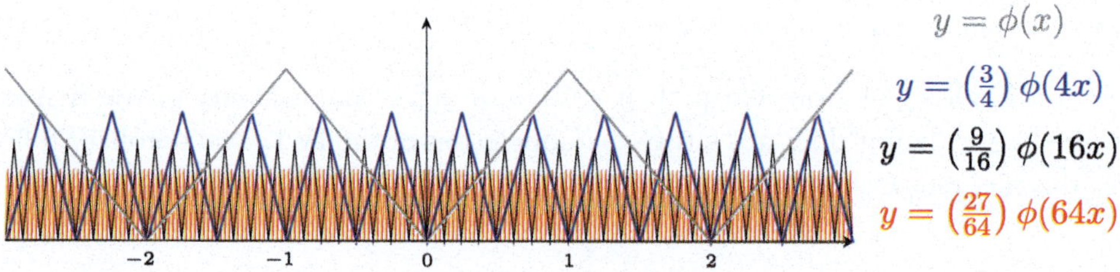

For each $n \in \mathbb{N}$, let $f_n(x)$ denote the finite partial sum

$$f_n(x) = \sum_{k=0}^{n} \left(\frac{3}{4}\right)^k \phi(4^k x)$$

of the first $n + 1$ terms in the series defining $f(x)$.

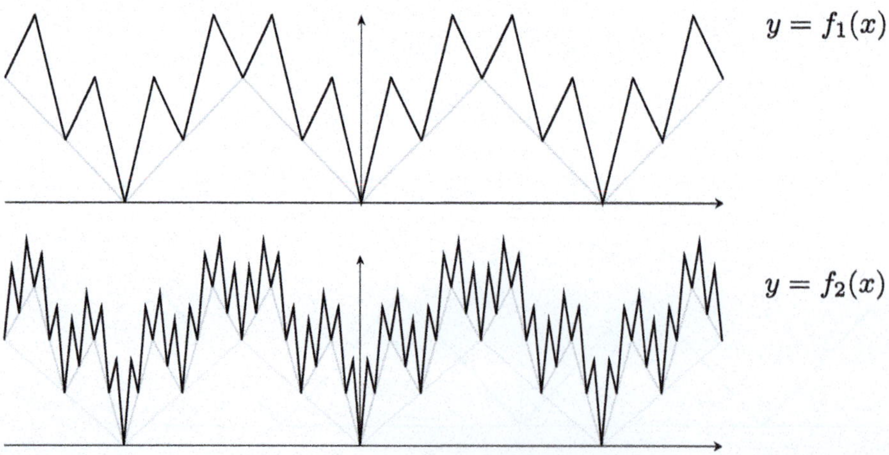

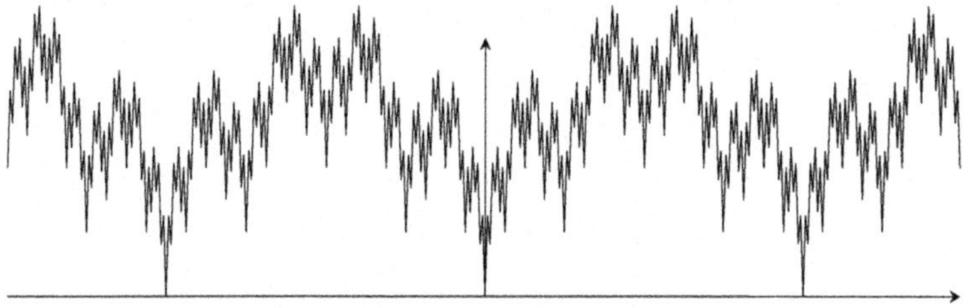

To prove that f is continuous, let $\varepsilon > 0$. Fix some $N \in \mathbb{N}$ large enough that $\sum_{k=N+1}^{\infty} \left(\frac{3}{4}\right)^k < \varepsilon/2$. Note that the finite partial sum $f_N(x)$ is a uniformly continuous function of x, because it is a sum of the finitely many uniformly continuous functions $\phi(x), \left(\frac{4}{3}\right)\phi(4x), \ldots, \left(\frac{4}{3}\right)^N \phi(4^N x)$. Thus there is some $\delta > 0$ such that if $|x - y| < \delta$ then $|f_N(x) - f_N(y)| < \varepsilon/2$. But then

$$\begin{aligned}
&|f(x) - f(y)| \\
&= \left| \left(f_N(x) + \sum_{k=N+1}^{\infty} \left(\frac{3}{4}\right)^k \phi(4^k x) \right) - \left(f_N(y) + \sum_{k=N+1}^{\infty} \left(\frac{3}{4}\right)^k \phi(4^k y) \right) \right| \\
&= \left| \left(f_N(x) - f_N(y) \right) + \sum_{k=N+1}^{\infty} \left(\frac{3}{4}\right)^k \left(\phi(4^k x) - \phi(4^k y) \right) \right| \\
&\leq \left| \left(f_N(x) - f_N(y) \right) + \sum_{k=N+1}^{\infty} \left(\frac{3}{4}\right)^k \right| \\
&\leq \left| f_N(x) - f_N(y) \right| + \sum_{k=N+1}^{\infty} \left(\frac{3}{4}\right)^k \\
&\leq \varepsilon/2 + \varepsilon/2 = \varepsilon
\end{aligned}$$

whenever $|x - y| < \delta$. As ε was arbitrary, this shows that f is (uniformly) continuous.

To finish the proof, it remains to show that f is not differentiable at any point. Fix $x \in \mathbb{R}$.

Claim. *For every $m \in \mathbb{N}$, there is some $t_m \in \mathbb{R}$ with $|x - t_m| = \frac{1}{2 \cdot 4^m}$ and*

$$\left| \frac{f(x) - f(t_m)}{x - t_m} \right| \geq \frac{3^m + 1}{2}.$$

Proof of Claim. Fix $m \in \mathbb{N}$. Observe that for every $n \in \mathbb{Z}$, the restriction of the function $\phi(x)$ to the interval $[n, n+1]$ is a line with slope ± 1. It follows that for every $n \in \mathbb{Z}$, the restriction of $\phi(4^m x)$ to $\left[\frac{n}{4m}, \frac{n+1}{4m}\right]$ is a line of slope $\pm 4^m$. Fix some $n \in \mathbb{Z}$ such that $x \in \left[\frac{n}{4m}, \frac{n+1}{4m}\right]$.

If x is in the left half of this interval, i.e. $x \in \left[\frac{n}{4m}, \frac{n}{4m} + \frac{1}{2 \cdot 4^m}\right]$, then because the interval has length $\frac{1}{4^m}$, we have $x + \frac{1}{2 \cdot 4^m} \in \left[\frac{n}{4m}, \frac{n+1}{4m}\right]$. In this case, let $t_m = x + \frac{1}{2 \cdot 4^m}$. Otherwise, $x \in$

$\left(\frac{n}{4m} + \frac{1}{2 \cdot 4^m}, \frac{n}{4m}\right]$ (the right half of the interval); in this case we have $x - \frac{1}{2 \cdot 4^m} \in \left[\frac{n}{4m}, \frac{n+1}{4m}\right]$, and we set $t_m = x - \frac{1}{2 \cdot 4^m}$.

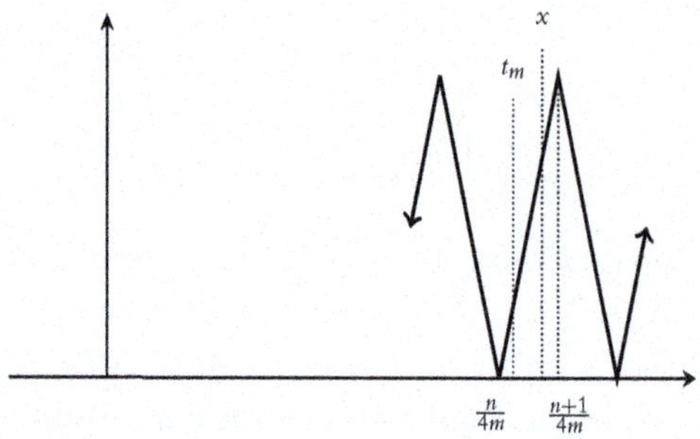

Either way, there is some t_m with $|x - t_m| = \frac{1}{2 \cdot 4^m}$, such that x and t_m both lie in the interval $\left[\frac{n}{4m}, \frac{n+1}{4m}\right]$. Because $\phi(4^m x)$ restricts to a line of slope $\pm 4^m$ on this interval, this means

$$|\phi(4^m x) - \phi(4^m t_m)| = 4^m |x - t_m| = 4^m \cdot \frac{1}{2 \cdot 4^m} = \frac{1}{2}.$$

More generally, suppose $k \leq m$. Then the interval $\left[\frac{n}{4m}, \frac{n+1}{4m}\right]$ containing x and t_m is strictly contained in an interval of the form $\left[\frac{\ell}{4k}, \frac{\ell+1}{4k}\right]$, for some $\ell \in \mathbb{N}$. In particular, x and t_m are both contained in an interval on which the function $\phi(4^k x)$ restricts to a line of slope $\pm 4^k$. Therefore

$$\left|\phi(4^k x) - \phi(4^k t_m)\right| = 4^k |x - t_m| = 4^k \cdot \frac{1}{2 \cdot 4^m} = \frac{1}{2} 4^{k-m}.$$

On the other hand, suppose $k > m$. Then the function $\phi(4^k x)$ is periodic with period $\frac{2}{4^k} = \frac{1}{2} \frac{1}{4^{k-1}}$. And $m < k$ implies $|x - t_m| = \frac{1}{2 \cdot 4^m}$ is an integer multiple of this period. Hence $\phi(4^k x) = \phi(4^k t_m)$. In summary,

$$\left|\phi(4^k x) - \phi(4^k t_m)\right| = \begin{cases} \frac{1}{2 \cdot 4^{m-k}} & \text{if } k \leq m, \\ 0 & \text{if } k > m. \end{cases}$$

Using this information, we can compute

$$\begin{aligned} &\left|\frac{f(x) - f(t_m)}{x - t_m}\right| \\ &\quad = \frac{1}{|x - t_m|} \left| \sum_{k=0}^{\infty} \left(\frac{3}{4}\right)^k \phi(4^k x) - \sum_{k=0}^{\infty} \left(\frac{3}{4}\right)^k \phi(4^k t_m) \right| \\ &\quad = 2 \cdot 4^m \left| \sum_{k=0}^{\infty} \left(\frac{3}{4}\right)^k \left(\phi(4^k x) - \phi(4^k t_m)\right) \right| \end{aligned}$$

$$= 2 \cdot 4^m \left| \sum_{k=0}^{m} \left(\frac{3}{4}\right)^k \left(\phi(4^k x) - \phi(4^k t_m)\right) \right|$$

$$\geq 2 \cdot 4^m \left(\frac{3}{4}\right)^m |\phi(4^m x) - \phi(4^m t_m)|$$

$$- 2 \cdot 4^m \sum_{k=0}^{m-1} \left(\frac{3}{4}\right)^k \left|\phi(4^k x) - \phi(4^k t_m)\right|$$

$$= 2 \cdot 4^m \left(\frac{3}{4}\right)^m \cdot \frac{1}{2} - 2 \cdot 4^m \sum_{k=0}^{m-1} \left(\frac{3}{4}\right)^k \frac{1}{2 \cdot 4^{m-k}}$$

$$= 3^m - \sum_{k=0}^{m-1} 3^k = \frac{3^m + 1}{2}. \qquad \square$$

For each $m \in \mathbb{N}$, let t_m be the number described in the preceding claim. Then $\langle t_m : m \in \mathbb{N} \rangle$ is a sequence of numbers with $\lim_{m \to \infty} t_m = x$, but (by the claim) $\lim_{m \to \infty} \left| \frac{f(x) - f(t_m)}{x - t_m} \right| = \infty$. Thus $\lim_{m \to \infty} \frac{f(x) - f(t_m)}{x - t_m}$ does not converge to a real number, and this means that f is not differentiable at x. (We have not proved the limit does not converge in the extended reals, and in fact it does converge to $\pm\infty$ for some, but not all, values of x. But the definition of differentiability requires this limit to converge to a real number.) As x was arbitrary, this concludes the proof of the theorem. $\square$

Exercises

1. Let $D \subseteq \mathbb{R}$, let $f : D \to \mathbb{R}$ be a function, and suppose $x \in \text{Int}(D)$. Prove that f is differentiable at x if and only if $\lim_{h \to 0} \frac{f(x+h) - f(x)}{h}$ exists, and if this limit does exist then it is equal to $f'(x)$.
2. Prove that the function

$$f(x) = \begin{cases} x^p \sin(1/x) & \text{if } x \neq 0 \\ 0 & \text{if } x = 0 \end{cases}$$

is differentiable at 0 if and only if $p > 1$.

3. Suppose that $f : \mathbb{R} \to \mathbb{R}$ is a function and that $|f(x) - f(y)| \leq (x - y)^2$ for all $x, y \in \mathbb{R}$. Prove that f is a constant function.
4. Suppose $f : \mathbb{R} \to \mathbb{R}$ is differentiable. Prove that if f has a bounded derivative (meaning there is some fixed $B > 0$ such that $|f'(x)| < B$ for all x) then f is uniformly continuous.
5. In contrast to the previous exercise, find an example of a differentiable, uniformly continuous function $f : \mathbb{R} \to \mathbb{R}$ whose derivative is not bounded.

6. Given $t \in \mathbb{R}$, consider the function $t \mapsto \vec{r}(t)$ that sends t to the point on the graph of the unit circle $x^2 + y^2 = 1$ that is reached by traveling counterclockwise along the circumference of the circle exactly t units from the point $(1, 0)$.

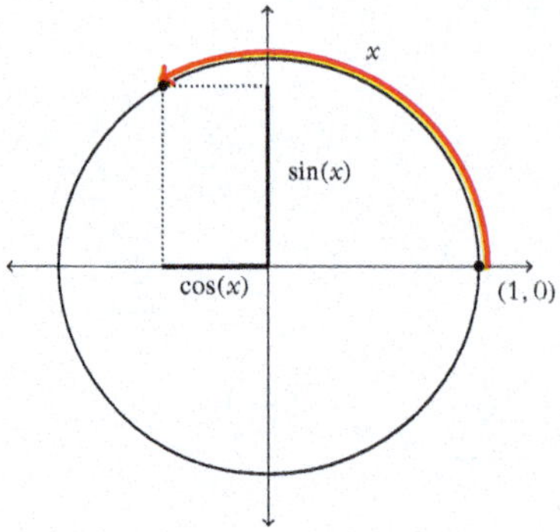

Define $\sin(t)$ to be the Y-coordinate of $\vec{r}(t)$, and define $\cos(t)$ to be the X-coordinate of $\vec{r}(t)$. Prove that the derivative of the function $f(x) = \sin(x)$ is the function $f'(x) = \cos(x)$, and prove the function $x \mapsto \cos(x)$ is continuous.

7. Consider the following statement:

(∗) Suppose f is a function $\mathbb{R} \to \mathbb{R}$, f is differentiable at some $a \in \mathbb{R}$, and $f'(a) > 0$. Then there is some $\varepsilon > 0$ such that f is strictly increasing on the interval $(a - \varepsilon, a + \varepsilon)$ (which means that if $a - \varepsilon < s < t < a + \varepsilon$ then $f(s) < f(t)$).

Decide whether (∗) is true or false. If it is true then provide a proof. If it is false then provide a counterexample.

8. Suppose f and g are differentiable functions $[a, b] \to \mathbb{R}$, and that $f'(x) = g'(x)$ for all $x \in (a, b)$. Show that f and g differ by a constant.

9. Let us say that a function $f : \mathbb{R} \to \mathbb{R}$ is *convex* if

$$f(t) \leq f(a) + (t - a)\frac{f(b) - f(a)}{b - a}$$

for $a, t, b \in \mathbb{R}$ with $a < t < b$. (In other words, the linear function connecting $(a, f(a))$ with $(b, f(b))$ is $\geq f(t)$ on the interval (a, b)).

Prove that if $f : \mathbb{R} \to \mathbb{R}$ is differentiable, then f is convex if and only if its derivative $f'(x)$ is nondecreasing.

10. Find an example of a continuous function $f : \mathbb{R} \to \mathbb{R}$ that is differentiable at exactly one point.
(*Hint*: Modify the function from Theorem 5.28.)

11. More generally, given a closed set $F \subseteq \mathbb{R}$, find an example of a continuous function $f : \mathbb{R} \to \mathbb{R}$ that is differentiable at x if and only if $x \in F$.

12. Find an example of a function $f : \mathbb{R} \to \mathbb{R}$ that is continuous at only one point, and is also differentiable at that point.

13. Define a function $[0, 1] \to \mathbb{R}$ that is continuous at every $x \in [0, 1]$, and is differentiable at x if and only if x is irrational.

14. Suppose that f and g are differentiable functions $\mathbb{R} \to \mathbb{R}$, and that $\lim_{x\to\infty} f(x) = \infty$ and $\lim_{x\to\infty} g(x) = \infty$. Suppose further that $\lim_{x\to\infty} f'(x)$ and $\lim_{x\to\infty} g'(x)$ both exist (as real numbers), and that $\lim_{x\to\infty} g'(x) \neq 0$. Prove that

$$\lim_{x\to\infty} \frac{f(x)}{g(x)} = \frac{\lim_{x\to\infty} f'(x)}{\lim_{x\to\infty} g'(x)}.$$

This is one instance of the principle known as *L'Hôpital's rule.*

15. Suppose again that f and g are differentiable functions $\mathbb{R} \to \mathbb{R}$, and that $\lim_{x\to\infty} f(x) = \infty$ and $\lim_{x\to\infty} g(x) = \infty$. Suppose further that $\lim_{x\to\infty} f'(x)$ exists (as a real) and $\lim_{x\to\infty} g'(x) = \infty$. Prove that

$$\lim_{x\to\infty} \frac{f(x)}{g(x)} = 0.$$

This is another instance of *L'Hôpital's rule.*

16. Let $c \in \mathbb{R}$. Suppose $f : \mathbb{R} \to \mathbb{R}$ is differentiable on (c, ∞), and its derivative f' has a vertical asymptote at c. Does this imply that f also has a vertical asymptote at c? If so then prove it, and if not then find a counterexample.

17. Suppose f is a differentiable function $\mathbb{R} \to \mathbb{R}$ such that $\lim_{x\to\infty} f(x)$ and $\lim_{x\to\infty} f'(x)$ both exist (as reals). Prove that $\lim_{x\to\infty} f'(x) = 0$.

The last two exercises in this set of notes show, in contrast with Theorem 5.26, two ways in which derivatives do not necessarily resemble continuous functions.

18. Give an example of a differentiable function $f : \mathbb{R} \to \mathbb{R}$ such that $f'(x)$ is not bounded on the interval $[0, 1]$.

19. A function $f : \mathbb{R} \to \mathbb{R}$ is said to *satisfy the conclusion of the EVT* if for every $a, b \in \mathbb{R}$ with $a < b$, there are some $x, y \in [a, b]$ such that $f(x) = \sup\{f(t) : t \in [a, b]\}$ and $f(y) = \inf\{f(t) : t \in [a, b]\}$. Provide an example of a differentiable function $\mathbb{R} \to \mathbb{R}$ whose derivative f' is bounded, but f' does not satisfy the conclusion of the EVT.

CHAPTER 6

Integration

Aspects of the theory of integration can be seen in the work of Archimedes and other ancient Greek mathematicians. A unified calculus, relating integration and differentiation, emerged much later, in the 1600s, with the work of Gregory, Barrow, Newton, and Leibniz. A rigorous definition of integration, and much of the mathematical theory surrounding it, came later still, beginning with the work of Bernhard Riemann in the mid-1800s. Riemann's definition of the integral was later simplified by Gaston Darboux, who contributed the following definition.

Definition and preliminaries

Definition 6.1. Let $a, b \in \mathbb{R}$ with $a < b$. A *partition* of $[a, b]$ is a finite sequence $\langle x_0, x_1, \ldots, x_n \rangle$ such that $a = x_0 < x_1 < \cdots < x_n = b$. For each $i = 1, 2, \ldots, n$, we define $\Delta x_i = x_i - x_{i-1}$.

Now, suppose f is a bounded function $[a, b] \to \mathbb{R}$ (i.e., the image of f is a bounded set). If $P = \langle x_0, x_1, \ldots, x_n \rangle$ is a partition of $[a, b]$, then we define

$$U(P,f) = \sum_{i=1}^{n} M_i^f \Delta x_i, \quad \text{where } M_i^f = \sup\{f(t) \colon x_{i-1} \le t \le x_i\}, \text{ and}$$

$$L(P,f) = \sum_{i=1}^{n} m_i^f \Delta x_i, \quad \text{where } m_i^f = \inf\{f(t) \colon x_{i-1} \le t \le x_i\}.$$

Next, define

$$\overline{\int_a^b} f(x)\,dx = \inf\{U(P,f) \colon P \text{ is a partition of } [a,b]\}, \text{ and}$$

$$\underline{\int_a^b} f(x)\,dx = \sup\{L(P,f) \colon P \text{ is a partition of } [a,b]\}.$$

If $\overline{\int_a^b} f(x)\,dx = \underline{\int_a^b} f(x)\,dx$, then we say f is *Riemann–Darboux integrable*, or simply *Riemann integrable*, on $[a, b]$. When this is the case, we write

$$\int_a^b f(x)\,dx = \overline{\int_a^b} f(x)\,dx = \underline{\int_a^b} f(x)\,dx,$$

and we say that $\int_a^b f(x)\,dx$ is the *integral of f from a to b*.

Foundations of Real Analysis
https://doi.org/10.1016/B978-0-44-348527-5.00010-X

A function with domain $D \supseteq [a, b]$ is said to be Riemann integrable on $[a, b]$ if its restriction to $[a, b]$ satisfies this definition. If the interval $[a, b]$ is clear from context, or if a function f is integrable on any interval, we simply say f is *Riemann integrable.* ☕

Before moving on, let us check that the various quantities used in this definition, the M_i^f and m_i^f, and $\overline{\int_a^b} f(x)\,dx$ and $\underline{\int_a^b} f(x)\,dx$, are well-defined. These quantities are all defined as the infimum/supremum of certain sets, so in order to show that they exist, we need to show that these sets are non-empty and have a lower/upper bound.

Lemma 6.2. *Suppose $[a, b]$ is an interval in $\mathbb{R}$, $P = \langle x_0, x_1, \ldots, x_n \rangle$ is a partition of $[a, b]$, and f is a bounded function $[a, b] \to \mathbb{R}$. For each $i = 1, 2, \ldots, n$, M_i^f and m_i^f are well-defined. The quantities $\overline{\int_a^b} f(x)\,dx$ and $\underline{\int_a^b} f(x)\,dx$ are also well-defined.*

Proof. Because f is a bounded function, there is some $B \in \mathbb{R}$ such that $|f(x)| \leq B$ for all $x \in [a, b]$. Let $P = \langle x_0, \ldots, x_n \rangle$ be a partition of $[a, b]$. For each $i \leq n$, the set $\{f(t) : x_{i-1} \leq t \leq x_i\}$ is clearly non-empty, and it is also bounded above by B; therefore, $M_i^f = \sup\{f(t) : x_{i-1} \leq t \leq x_i\}$ is well-defined, and $M_i^f \leq B$. Similarly, $m_i^f = \inf\{f(t) : x_{i-1} \leq t \leq x_i\}$ is well-defined and $m_i^f \geq -B$. It follows that

$$L(P, f) = \sum_{i=1}^{n} m_i^f \Delta x_i \leq \sum_{i=1}^{n} M_i^f \Delta x_i \leq \sum_{i=1}^{n} B \Delta x_i = B \sum_{i=1}^{n} \Delta x_i = B(b - a),$$

and, similarly, $U(P, f) \geq -B(b - a)$.

As P was an arbitrary partition of $[a, b]$, this shows that $B(b - a)$ is an upper bound for $\{L(P, f) : P \text{ is a partition of } [a, b]\}$. As this set is clearly non-empty, this means $\underline{\int_a^b} f(x)\,dx = \sup\{L(P, f) : P \text{ is a partition of } [a, b]\}$ is well-defined (and is at most $B(b - a)$). Similarly, $-B(b - a)$ is a lower bound for $\{U(P, f) : P \text{ is a partition of } [a, b]\}$, and this means $\overline{\int_a^b} f(x)\,dx = \inf\{U(P, f) : P \text{ is a partition of } [a, b]\}$ is well-defined. □

The idea behind Definition 6.1 is of course that $\int_a^b f(x)\,dx$ should be defined to be equal to the signed area between the curve $y = f(x)$ and the x-axis. We cannot simply define $\int_a^b f(x)\,dx$ to be this area, however, because at this point, such a definition would be circular: We have not defined the "area" of an arbitrary planar region, and have no way of computing planar areas (other than the polygons and circles of basic geometry), except via integration.

It is worth noting that there are other ways of defining integration, not just Riemann–Darboux integration. The most well-known of these (but not the only one) is Lebesgue integration. All the different ways of defining integration provide a continuous analogue of summation, computing the signed area under a curve via some limiting process. They all agree on the value of the integrals of basic functions, though they may disagree on the integrability of more complicated functions. In this chapter, we study only Riemann–Darboux integration, which is in many ways the most basic.

Definition 6.1 encapsulates a logical approach to computing the area in question. For any partition P of $[a, b]$, $U(P,f)$ is an overestimation of the area, and $L(P,f)$ is an underestimation.

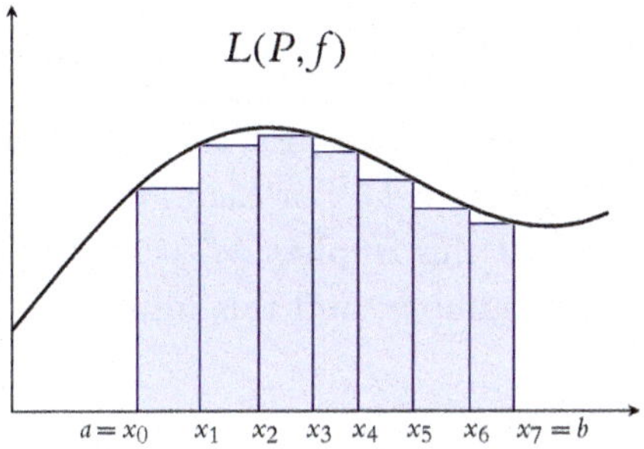

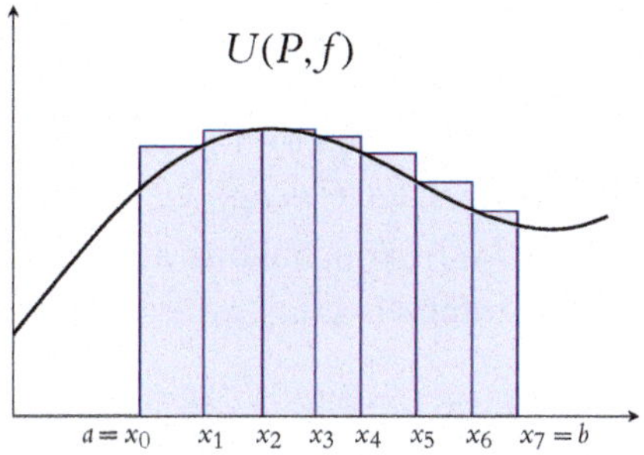

As we put more and more points into P, the overestimation $U(P,f)$ should grow closer and closer to the true area (assuming such a thing exists). The infimum of all such overestimations, $\overline{\int_a^b} f(x)\,dx$, is therefore an upper bound for the true area: It is our best guess at the true area, as approximated from above. Similarly, $\underline{\int_a^b} f(x)\,dx$ is a lower bound for the true area, and is our best guess at the true area as approximated from below. If these upper and lower bounds coincide, their common value must be the true area.

Example 6.3. Define $f : [0, 1] \to \mathbb{R}$ as follows:

$$f(x) = \begin{cases} 1/\sqrt{x} & \text{if } x > 0 \\ 0 & \text{if } x = 0. \end{cases}$$

This function is not Riemann integrable on $[0, 1]$ because it is not bounded: f fails to meet one of the basic requirements in Definition 6.1.

However, one can argue that it still makes sense to assign an area to the region between the x-axis and the graph of this function. In fact, integration can still give us the area indirectly, getting around the unboundedness issue with something known as *improper integration*. Observing that $\int_t^1 f(x)\,dx$ is well-defined for all $t \in (0, 1)$, we may define

$$\int_0^1 f(x)\,dx = \lim_{t \to 0^+} \int_t^1 f(x)\,dx = 2.$$

This is a kind of trick for finding the integral of f from 0 to 1, even though it is not quite "integrable" in the sense of Definition 6.1. Improper integration is explored further in the exercises at the end of the chapter.

The next example shows a function that fails to be Riemann integrable in a more fundamental, unfixable way.

Example 6.4. Recall the Dirichlet function

$$f(x) = \begin{cases} 1 & \text{if } x \in \mathbb{Q} \\ 0 & \text{if } x \notin \mathbb{Q}. \end{cases}$$

We claim that this function is not Riemann integrable on $[a, b]$, for any $a, b \in \mathbb{R}$ with $a < b$. To see this, let $P = \langle x_0, x_1, \ldots, x_n \rangle$ be a partition of $[a, b]$. For each $i = 1, 2, \ldots, n$, the interval $[x_{i-1}, x_i]$ contains a rational number, and this implies $M_i^f = 1$. Similarly, each of the intervals $[x_{i-1}, x_i]$ contains an irrational number, and this implies $m_i^f = 0$. Therefore

$$L(P,f) = \sum_{i=1}^{n} m_i^f \Delta x_i = \sum_{i=1}^{n} 0 \cdot \Delta x_i = 0, \quad \text{while}$$

$$U(P,f) = \sum_{i=1}^{n} M_i^f \Delta x_i = \sum_{i=1}^{n} 1 \cdot \Delta x_i = \sum_{i=1}^{n} x_i - x_{i-1} = x_n - x_0 = b - a.$$

It follows that $\overline{\int_a^b} f(x)\,dx = b - a$, while $\underline{\int_a^b} f(x)\,dx = 0$. ☕

The Cauchy criterion

Definition 6.5. If P and Q are partitions of an interval $[a, b]$, we say Q *refines* P if every point in P is also in Q (i.e., P is a subsequence of Q). ☕

Lemma 6.6. *Suppose f is a bounded function $[a, b] \to \mathbb{R}$ and that P and Q are partitions of $[a, b]$. If Q refines P, then*

$$L(P,f) \le L(Q,f) \le U(Q,f) \le U(P,f).$$

Proof. This lemma is best explained with the following illustrations:

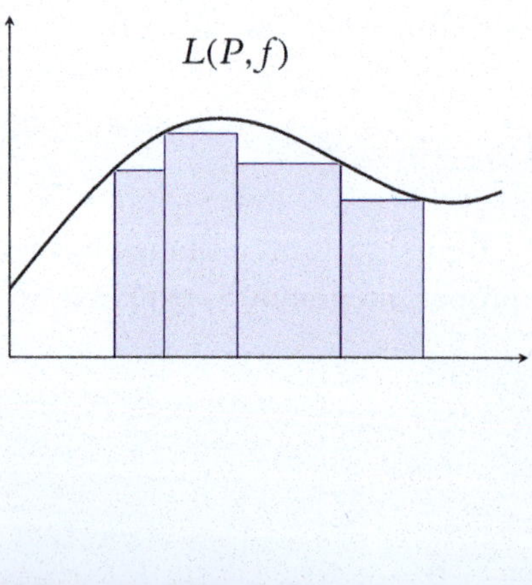

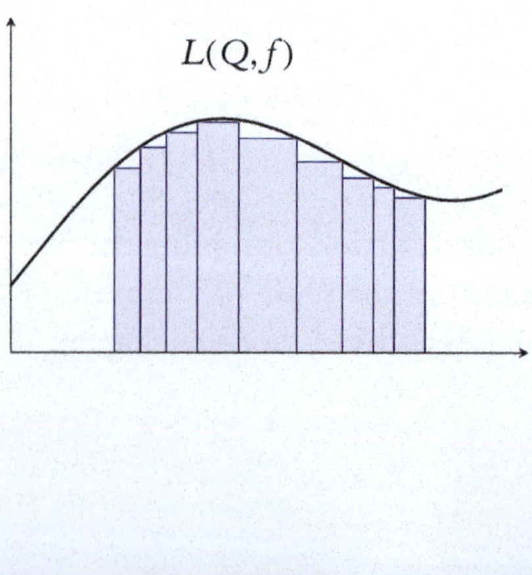

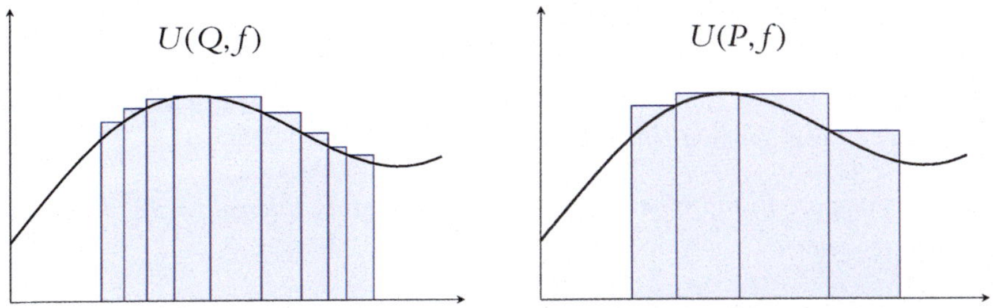

Let us now make the idea in these pictures a bit more precise. Let $P = \langle x_0, x_1, \ldots, x_n \rangle$ be a partition of $[a, b]$, and let $Q = \langle y_0, y_1, \ldots, y_m \rangle$ be a refinement of P. That is, there is an increasing sequence $k_0, k_1, \ldots, k_n$ of integers such that $k_0 = 0$, $k_n = m$, and $y_{k_i} = x_i$ for all $i \leq n$.

Within this proof, let us denote

$$M_i^{f,P} = \sup\{f(t) \colon x_{i-1} \leq t \leq x_i\} \quad \text{and} \quad m_i^{f,P} = \inf\{f(t) \colon x_{i-1} \leq t \leq x_i\}$$

whenever $1 \leq i \leq n$, and

$$M_j^{f,Q} = \sup\{f(t) \colon y_{j-1} \leq t \leq y_j\} \quad \text{and} \quad m_j^{f,Q} = \inf\{f(t) \colon y_{j-1} \leq t \leq y_j\}$$

whenever $1 \leq j \leq m$. Also, let $\Delta y_j = y_j - y_{j-1}$ whenever $1 \leq j \leq m$.

If $i \leq n$ and $k_{i-1} < j \leq k_i$, then $M_j^{f,Q} \leq M_i^{f,P}$. This is because $[y_{j-1}, y_j] \subseteq [x_{i-1}, x_i]$, which implies

$$M_i^{f,P} = \sup\{f(t) \colon x_{i-1} \leq t \leq x_i\} \leq \sup\{f(t) \colon y_{j-1} \leq t \leq y_j\} = M_j^{f,Q}.$$

It follows that

$$\begin{aligned} U(Q,f) &= \sum_{j=1}^{m} M_j^{f,Q} \Delta y_j = \sum_{i=1}^{n} \sum_{j=k_{i-1}+1}^{k_i} M_j^{f,Q} \Delta y_j \\ &\leq \sum_{i=1}^{n} \sum_{j=k_{i-1}+1}^{k_i} M_i^{f,P} \Delta y_j = \sum_{i=1}^{n} M_i^{f,P} \sum_{j=k_{i-1}+1}^{k_i} y_j - y_{j-1} \\ &= \sum_{i=1}^{n} M_i^{f,P} (y_{k_i} - y_{k_{i-1}}) = \sum_{i=1}^{n} M_i^{f,P} \Delta x_i = U(P,f). \end{aligned}$$

The proof that $L(P,f) \leq L(Q,f)$ is essentially the same. It remains to show that $L(Q,f) \leq U(Q,f)$, which is relatively easy to see in the following,

$$L(Q,f) = \sum_{j=1}^{m} m_j^{f,Q} \Delta y_j \leq \sum_{j=1}^{m} M_j^{f,Q} \Delta y_j = U(Q,f),$$

because $m_j^{f,Q} \le M_j^{f,Q}$ for every $j \le m$. □

Definition 6.7. If P and P' are two partitions of $[a, b]$ and Q is a refinement of both of them, we say Q is a *common refinement* of P and P'.

Lemma 6.8. *Suppose P and P' are any two partitions of an interval $[a, b]$. Then, P and P' have a common refinement.*

Proof. Let S_P denote the (unordered) set of numbers in the sequence P, and let $S_{P'}$ denote, similarly, the set of numbers in P'. Then define Q to be the sequence listing the elements of $S_P \cup S_{P'}$ in increasing order. □

Lemma 6.9. *Let f be a bounded function $[a, b] \to \mathbb{R}$ for some interval $[a, b]$. If P and P' are any two partitions of $[a, b]$, then $L(P,f) \le U(P',f)$.*

Proof. By Lemma 6.8, there is a common refinement Q of P and P'. By Lemma 6.6, $L(P,f) \le L(Q,f) \le U(Q,f) \le U(P',f)$. □

Theorem 6.10. *Let f be a bounded function $[a, b] \to \mathbb{R}$ for some interval $[a, b]$. Then, $\underline{\int_a^b} f(x)\,dx \le \overline{\int_a^b} f(x)\,dx$.*

Proof. Let Q be a partition of $[a, b]$. By Lemma 6.9, $U(Q,f)$ is an upper bound for $\{L(P,f) : P \text{ is a partition of } [a, b]\}$. Therefore,

$$U(Q,f) \ge \sup\{L(P,f) : P \text{ is a partition of } [a, b]\} = \underline{\int_a^b} f(x)\,dx.$$

Because Q was an arbitrary partition of $[a, b]$, this means $\underline{\int_a^b} f(x)\,dx$ is a lower bound for $\{U(Q,f) : Q \text{ is a partition of } [a, b]\}$. Therefore,

$$\underline{\int_a^b} f(x)\,dx \le \inf\{U(Q,f) : Q \text{ is a partition of } [a, b]\} = \overline{\int_a^b} f(x)\,dx$$

as claimed. □

This section ends with a criterion for Riemann integrability, known as the *Cauchy criterion*. In practice, one never actually checks that the conditions of Definition 6.1 hold for a function: To prove Riemann integrability, one always checks the Cauchy criterion instead.

Definition 6.11. Let $a, b \in \mathbb{R}$ with $a < b$. A bounded function $f : [a, b] \to \mathbb{R}$ satisfies the *Cauchy criterion* if for every $\varepsilon > 0$, there is a partition P of $[a, b]$ such that $U(P,f) - L(P,f) < \varepsilon$.

Theorem 6.12. *Suppose $[a, b]$ is an interval in $\mathbb{R}$ and f is a bounded function $[a, b] \to \mathbb{R}$. Then, f is Riemann integrable on $[a, b]$ if and only if f satisfies the Cauchy criterion.*

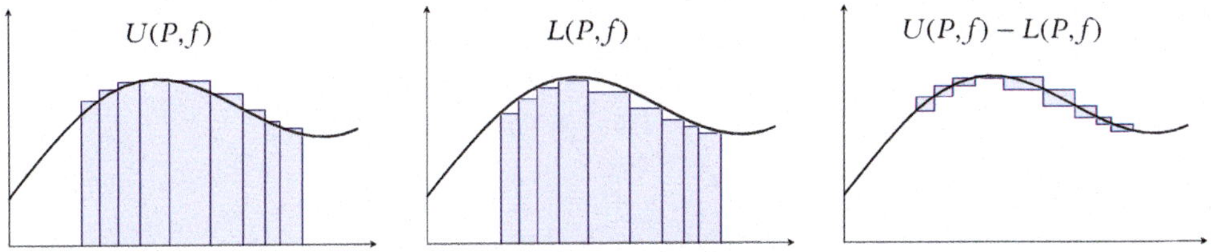

Proof. For the "only if" direction, suppose that f is Riemann integrable. Let $\varepsilon > 0$. That f is Riemann integrable means $\underline{\int_a^b} f(x)\,dx = \overline{\int_a^b} f(x)\,dx$. By the definition of $\underline{\int_a^b} f(x)\,dx$, there is some partition P_L of $[a, b]$ such that $0 \le \underline{\int_a^b} f(x)\,dx - L(P_L, f) < \varepsilon/2$. Similarly, there is some partition P_U of $[a, b]$ such that $0 \le U(P_U, f) - \overline{\int_a^b} f(x)\,dx < \varepsilon/2$. Because $\underline{\int_a^b} f(x)\,dx = \overline{\int_a^b} f(x)\,dx$, it follows that $U(P_U, f) - L(P_L, f) < \varepsilon$.

Applying Lemma 6.8, let P be a common refinement of P_L and P_U. By Lemma 6.6, $L(P_L, f) \le L(P, f) \le U(P, f) \le U(P_U, f)$. Because $U(P_U, f) - L(P_L, f) < \varepsilon$, it follows that $U(P, f) - L(P, f) < \varepsilon$.

For the "if" direction, suppose f satisfies the Cauchy criterion. Let $\varepsilon > 0$, and let P be a partition of $[a, b]$ such that $U(P, f) - L(P, f) < \varepsilon$. By the definition of $\underline{\int_a^b} f(x)\,dx$ and $\overline{\int_a^b} f(x)\,dx$ and by Theorem 6.10,

$$L(P,f) \le \underline{\int_a^b} f(x)\,dx \le \overline{\int_a^b} f(x)\,dx \le U(P,f).$$

It follows that $\left|\overline{\int_a^b} f(x)\,dx - \underline{\int_a^b} f(x)\,dx\right| < U(P, f) - L(P, f) < \varepsilon$. As ε was arbitrary, this means $\underline{\int_a^b} f(x)\,dx = \overline{\int_a^b} f(x)\,dx$, which means that f is Riemann integrable. □

The following fact is often useful for verifying the Cauchy criterion.

Proposition 6.13. *Let f be a bounded function $[a, b] \to \mathbb{R}$, and let $P = \langle x_0, x_1, \ldots, x_n\rangle$ be a partition of $[a, b]$. For each $i \le n$, $M_i^f - m_i^f = \omega_f([x_{i-1}, x_i]) = \sup\{f(t) - f(s) \colon s, t \in [x_{i-1}, x_i]\}$. Consequently, $U(P, f) - L(P, f) = \sum_{i=1}^n \omega_f([x_{i-1}, x_i])\Delta x_i$.*

Proof. Clearly, $U(P, f) - L(P, f) = \sum_{i=1}^n (M_i^f - m_i^f)\Delta x_i$, so we must show

$$M_i^f - m_i^f = \omega_f([x_{i-1}, x_i])$$

for every $i \le n$. Fix $i \le n$, and let $\varepsilon > 0$. There is some $s_0 \in [x_{i-1}, x_i]$ with $f(s_0) < m_i^f + \varepsilon/2$ and some $t_0 \in [x_{i-1}, x_i]$ with $f(t_0) > M_i^f - \varepsilon/2$. Thus,

$$\omega_f([x_{i-1}, x_i]) = \sup\{f(t) - f(s) \colon s, t \in [x_{i-1}, x_i]\}$$

$$\geq f(t_0) - f(s_0) > (M_i^f - m_i^f) - \varepsilon.$$

On the other hand, there exists some $s_1, t_1 \in [x_{i-1}, x_i]$ with $f(t_1) - f(s_1) > \omega_f([x_{i-1}, x_i]) - \varepsilon$. But $f(s_0) \geq m_i^f$ and $f(t_0) \leq M_i^f$, so

$$M_i^f - m_i^f \geq f(t_1) - f(s_1) > \omega_f([x_{i-1}, x_i]) - \varepsilon.$$

Because $\varepsilon > 0$ was arbitrary, $M_i^f - m_i^f = \omega_f([x_{i-1}, x_i])$ as claimed. □

Some integrable functions

Theorem 6.14. *Continuous functions are Riemann integrable.*

Proof. Fix $a, b \in \mathbb{R}$ with $a < b$. Note that f is bounded on $[a, b]$ by the Extreme Value Theorem (EVT) because f is continuous and $[a, b]$ is compact.

Let $\varepsilon > 0$. Because $[a, b]$ is compact, f is uniformly continuous. Thus, there is some $\delta > 0$ such that whenever $x, y \in [a, b]$ and $|x - y| < \delta$, then $|f(x) - f(y)| < \frac{\varepsilon}{b-a}$. Let $P = \langle x_0, x_1, \ldots, x_n \rangle$ be a partition of $[a, b]$ such that $\Delta x_i < \delta$ for all $i \leq n$.

We claim that $M_i - m_i < \frac{\varepsilon}{b-a}$ for all $i \leq n$. Indeed, applying the EVT again, there are some $c_i, d_i \in [x_{i-1}, x_i]$ such that

$$f(c_i) = \inf \{f(t) : t \in [x_{i-1}, x_i]\} = m_i^f, \text{ and}$$
$$f(d_i) = \sup \{f(t) : t \in [x_{i-1}, x_i]\} = M_i^f.$$

But $|c_i - d_i| < \Delta x_i < \delta$, by our choice of P, so this implies $M_i^f - m_i^f = |f(c_i) - f(d_i)| < \frac{\varepsilon}{b-a}$ by our choice of δ.

Using the previous paragraph,

$$\begin{aligned} U(P,f) - L(P,f) &= \sum_{i=1}^{n} M_i^f \Delta x_i - \sum_{i=1}^{n} m_i^f \Delta x_i = \sum_{i=1}^{n} (M_i^f - m_i^f) \Delta x_i \\ &< \sum_{i=1}^{n} \frac{\varepsilon}{b-a} \Delta x_i = \frac{\varepsilon}{b-a} \sum_{i=1}^{n} \Delta x_i = \frac{\varepsilon}{b-a} (b-a) = \varepsilon. \end{aligned}$$

Thus, f satisfies the Cauchy criterion. By Theorem 6.12, this means f is Riemann integrable. □

The Cauchy criterion requires that a certain collection of rectangles have total area $< \varepsilon$. Essentially, the preceding proof works by showing that the rectangles in question can be made to have maximum height $\varepsilon/b-a$, which works because their total width is $b - a$ and

$$\begin{aligned} \text{total area} &= (\text{average height}) \times (\text{total width}) \\ &\leq (\text{maximum height}) \times (\text{total width}). \end{aligned}$$

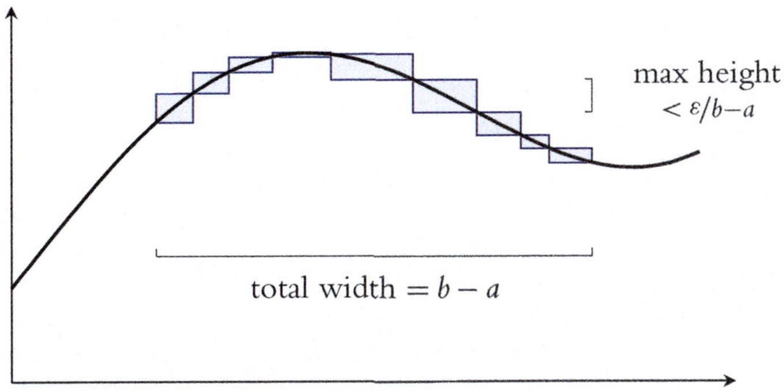

The next proof works on the opposite principle, namely, that

$$\begin{aligned}\text{total area} &= (\text{total height}) \times (\text{average width})\\ &\leq (\text{total height}) \times (\text{maximum width}).\end{aligned}$$

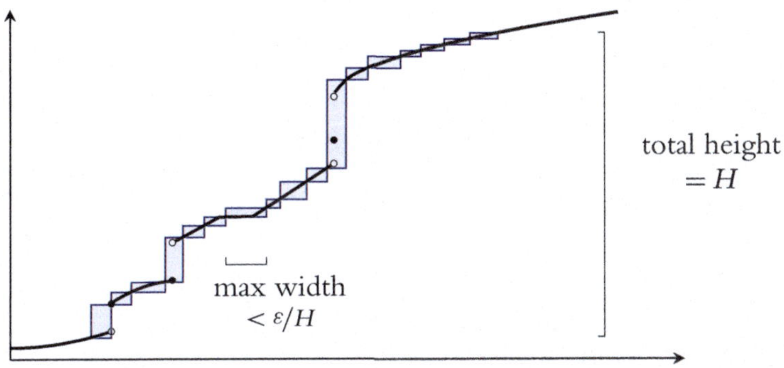

Theorem 6.15. *Monotone functions are Riemann integrable.*

Proof. Fix $a, b \in \mathbb{R}$ with $a < b$, and let $f : [a, b] \to \mathbb{R}$ be nondecreasing. Note that f is bounded on $[a, b]$ by its values at the endpoints, that is, f is bounded below by $f(a)$ and bounded above by $f(b)$.

Let $\varepsilon > 0$. Let $H = f(b) - f(a) + 1$. (We will divide something by H in a moment; the point of the "+1" is to ensure $H > 0$, even in the case where f is a constant function.) Let $P = \langle x_0, x_1, \ldots, x_n \rangle$ be a partition of $[a, b]$ such that $\Delta x_i < \varepsilon/H$ for all $i \leq n$. Because f is nondecreasing,

$$\begin{aligned}f(x_{i-1}) &= \inf\{f(t) : t \in [x_{i-1}, x_i]\} = m_i^f, \text{ and}\\ f(x_i) &= \sup\{f(t) : t \in [x_{i-1}, x_i]\} = M_i^f\end{aligned}$$

for every $i \leq n$. It follows that

$$U(P, f) - L(P, f) = \sum_{i=1}^{n} M_i^f \Delta x_i - \sum_{i=1}^{n} m_i^f \Delta x_i = \sum_{i=1}^{n} (M_i^f - m_i^f) \Delta x_i$$

$$< \sum_{i=1}^{n}(f(x_i) - f(x_{i-1}))\frac{\varepsilon}{H} = \frac{\varepsilon}{H}\sum_{i=1}^{n} f(x_i) - f(x_{i-1})$$
$$= \frac{\varepsilon}{H}(f(b) - f(a)) = \frac{\varepsilon}{H}H = \varepsilon.$$

Thus, f satisfies the Cauchy criterion and therefore is Riemann integrable.

The proof for nonincreasing functions is essentially the same. □

In Theorem 6.14, we were able to control the height of our rectangles, while their width was fixed at $b - a$, and in Theorem 6.15 we were able to control the width of our rectangles, while their total height was fixed at $|f(b) - f(a)|$. The next proof combines these two ideas. In it, the rectangles are sorted into two groups, a "height control" group and a "width control" group. This same idea is used later on, in a more sophisticated way, to prove the Lebesgue–Vitali Theorem, one of the main results in this chapter.

Theorem 6.16. *Bounded functions with finitely many points of discontinuity are Riemann integrable.*

Proof. Fix $a, b \in \mathbb{R}$ with $a < b$, and let $f : [a, b] \to \mathbb{R}$ be a bounded function with finitely many points of discontinuity. Let $d_1, d_2, \ldots, d_k$ denote the points where f fails to be continuous. Because f is bounded on the interval $[a, b]$, $\sup\{f(t) : t \in [a, b]\}$ and $\inf\{f(t) : t \in [a, b]\}$ both exist. Let $H = \sup\{f(t) : t \in [a, b]\} - \inf\{f(t) : t \in [a, b]\} + 1$.

Let $\varepsilon > 0$. By making ε even smaller if necessary, we may (and do) assume that $|d_i - d_j| > \varepsilon/2kH$ whenever $i, j \leq k$ and $i \neq j$.

For each $j \leq k$, let $U_j = \left(d_j - \frac{\varepsilon}{4kH}, d_j + \frac{\varepsilon}{4kH}\right)$, the small open interval of length $\varepsilon/2kH$ centered at d_j. By our assumption about ε in the previous paragraph, these intervals, and even their closures, are disjoint.

Observe that f is continuous on the set $[a, b] \setminus \bigcup_{j \leq k} U_j$. This set is closed and bounded, hence compact. So in fact, f is uniformly continuous on this set. Fix $\delta > 0$ such that if $x, y \in [a, b] \setminus \bigcup_{j \leq k} U_j$ and $|x - y| < \delta$, then $|f(x) - f(y)| < \varepsilon/2(b-a)$.

Let $P = \langle x_0, x_1, \ldots, x_n \rangle$ be a partition of $[a, b]$ such that

(1) $\Delta x_i < \delta$ for all $i \leq n$, and

(2) If the endpoints of U_i (namely, $d_i - \frac{\varepsilon}{4kH}$ and $d_i + \frac{\varepsilon}{4kH}$) are in $[a, b]$, then they are in P.

It is clear that such partitions exist. We claim that $U(P, f) - L(P, f) < \varepsilon$.

To see this, divide the members of P into two groups: let A denote the set of all those $i \leq n$ such that there is some $j \leq k$ with x_{i-1} and x_i both in $\overline{U}_j = \left[d_j - \frac{\varepsilon}{4kH}, d_j + \frac{\varepsilon}{4kH}\right]$, and let $B = \{1, 2, \ldots, n\} \setminus A$.

If $i \in B$, then we claim $M_i^f - m_i^f < \varepsilon/2(b-a)$. To see this, first note that $[x_{i-1}, x_i]$ is disjoint from all the U_i. (This follows from the definition of A, plus the fact that the closures of the U_i are nonoverlapping.) Thus, f is continuous on $[x_{i-1}, x_i]$, and by the EVT, there are some $c_i, d_i \in [x_{i-1}, x_i]$ such that

$$f(c_i) = \inf\{f(t) : t \in [x_{i-1}, x_i]\} = m_i^f \text{ , and}$$

$$f(d_i) = \sup\{f(t) : t \in [x_{i-1}, x_i]\} = M_i^f.$$

Furthermore, $c_i, d_i \in [x_{i-1}, x_i] \subseteq [a, b] \setminus \bigcup_{j\le k} U_j$, and $|c_i - d_i| \le \Delta x_i < \delta$ by part (1) of our choice of P. This implies $M_i^f - m_i^f = |f(c_i) - f(d_i)| < \varepsilon/2(b-a)$.

If $i \in A$, then we cannot put any bounds on $M_i^f - m_i^f$ except the trivial bound $M_i^f - m_i^f < H$. However, observe that if $i \in A$, then there is some $j \le k$ such that $[x_{i-1}, x_i] \subseteq \left[d_j - \frac{\varepsilon}{4kH}, d_j + \frac{\varepsilon}{4kH}\right]$. It follows that

$$\sum_{i\in A} \Delta x_i \le \sum_{j\le k} \left(d_j + \frac{\varepsilon}{4kH}\right) - \left(d_j - \frac{\varepsilon}{4kH}\right) = k\frac{\varepsilon}{2kH} = \frac{\varepsilon}{2H}.$$

Putting these pieces together, we get

$$\begin{aligned} U(P,f) - L(P,f) &= \sum_{i=1}^{n} (M_i^f - m_i^f)\Delta x_i \\ &= \sum_{i\in A} (M_i^f - m_i^f)\Delta x_i + \sum_{i\in B} (M_i^f - m_i^f)\Delta x_i \\ &< \sum_{i\in A} H\Delta x_i + \sum_{i\in B} \frac{\varepsilon}{2(b-a)}\Delta x_i \\ &= H\sum_{i\in A} \Delta x_i + \frac{\varepsilon}{2(b-a)} \sum_{i\in B} \Delta x_i \\ &\le H\frac{\varepsilon}{2H} + \frac{\varepsilon}{2(b-a)}(b-a) = \varepsilon. \end{aligned}$$

Thus f satisfies the Cauchy criterion, and is Riemann integrable. □

Combining integrable functions

Next, we establish some basic rules about Riemann–Darboux integration, which are already familiar to anyone who has taken a basic calculus course. Note that the next theorem is somewhat obvious if one thinks of $\int_a^b f(x)\,dx$ as the area under the curve of $f(x)$ between the lines $x = a$ and $x = b$: We are simply saying that splitting an area into two pieces along the line $x = c$ results in two areas that together add up to the original.

Theorem 6.17. *Suppose f is a Riemann integrable function $[a, b] \to \mathbb{R}$. If $c \in (a, b)$, then f is Riemann integrable on both $[a, c]$ and $[c, b]$, and furthermore*

$$\int_a^b f(x)\,dx = \int_a^c f(x)\,dx + \int_c^b f(x)\,dx.$$

Proof. Let $\varepsilon > 0$. Let $P = \langle x_0, x_1, \ldots, x_n\rangle$ be a partition of $[a, b]$ such that $U(P,f) - L(P,f) < \varepsilon$. By refining P if necessary, we may (and do) assume that c is in P. Let $k \in \{1, 2, \ldots, n-1\}$ denote the index with $x_k = c$.

Let $P|[a, c]$ denote the partition $\langle x_0, x_1, \ldots, x_k \rangle$ of $[a, c]$. Then,

$$U(P|[a,c],f) - L(P|[a,c],f) = \sum_{i=1}^{k}(M_i^f - m_i^f)\Delta x_i$$
$$\le \sum_{i=1}^{n}(M_i^f - m_i^f)\Delta x_i = U(P,f) - L(P,f) < \varepsilon.$$

Thus, $f|[a, c]$ satisfies the Cauchy criterion, and is Riemann integrable on $[a, c]$. The proof that f is Riemann integrable on $[c, b]$ is similar, using $P|[c, b]$ instead of $P|[a, c]$.

Furthermore, observe that

$$L(P|[a,c],f) + L(P|[c,b],f) = \sum_{i=1}^{k} m_i^f \Delta x_i + \sum_{i=k+1}^{n} m_i^f \Delta x_i = L(P,f),$$

and similarly $U(P|[a, c], f) + U(P|[c, b], f) = U(P, f)$. Thus,

$$L(P,f) = L(P|[a,c],f) + L(P|[c,b],f) \le \int_a^c f(x)\,dx + \int_c^b f(x)\,dx$$
$$\le U(P|[a,c],f) + U(P|[c,b],f) = U(P,f),$$

while at the same time

$$L(P,f) \le \int_a^b f(x)\,dx \le U(P,f) \quad \text{and} \quad U(P,f) - L(P,f) < \varepsilon.$$

As ε was arbitrary, it follows that $\int_a^b f(x)dx = \int_a^c f(x)dx + \int_c^b f(x)dx$. □

Corollary 6.18. *Suppose f is a Riemann integrable function $[a, b] \to \mathbb{R}$. If $[c, d]$ is an interval with $[c, d] \subseteq [a, b]$, then f is Riemann integrable on $[c, d]$.*

Proof. By one application of the previous theorem, f is integrable on $[a, d]$, and then by a second application, f is integrable on $[c, d]$. □

Theorem 6.19. *Suppose f and g are integrable functions $[a, b] \to \mathbb{R}$.*

1. *The function $x \mapsto cf(x)$ is Riemann integrable for any $c \in \mathbb{R}$, and furthermore $\int_a^b cf(x)\,dx = c\int_a^b f(x)\,dx$.*
2. *The function $x \mapsto f(x) + g(x)$ is Riemann integrable, and furthermore $\int_a^b f(x) + g(x)\,dx = \int_a^b f(x)\,dx + \int_a^b g(x)\,dx$.*
3. *The function $x \mapsto f(x)g(x)$ is Riemann integrable.*
4. *Suppose there is some $c > 0$ such that $|f(x)| > c$ for all $x \in [a, b]$. Then, the function $x \mapsto 1/f(x)$ is Riemann integrable.*

Proof. For (1), let us first consider the case $c > 0$. For any partition $P = \langle x_0, x_1, \ldots, x_n \rangle$ of $[a, b]$, observe that

$$\begin{aligned} M_i^{cf} &= \sup\{cf(t)\colon t \in [x_{i-1}, x_i]\} = c \cdot \sup\{f(t)\colon t \in [x_{i-1}, x_i]\} = cM_i^f, \\ m_i^{cf} &= \inf\{cf(t)\colon t \in [x_{i-1}, x_i]\} = c \cdot \inf\{f(t)\colon t \in [x_{i-1}, x_i]\} = cm_i^f. \end{aligned}$$

It follows that

$$\begin{aligned} U(P, cf) &= \sum_{i=1}^{n} M_i^{cf} \Delta x_i = c\sum_{i=1}^{n} M_i^f \Delta x_i = cU(P, f), \text{ and} \\ L(P, cf) &= \sum_{i=1}^{n} m_i^{cf} \Delta x_i = c\sum_{i=1}^{n} m_i^f \Delta x_i = cL(P, f). \end{aligned}$$

Now, let $\varepsilon > 0$. Because $f(x)$ is Riemann integrable, there is some partition P of $[a, b]$ such that $U(P, f) - L(P, f) < \varepsilon/c$. But then, $U(P, cf) - L(P, cf) = cU(P, f) - cL(P, f) < c \cdot \frac{\varepsilon}{c} = \varepsilon$. Thus f satisfies the Cauchy criterion and is Riemann integrable. Furthermore,

$$\begin{aligned} \int_a^b cf(x)\,dx = \overline{\int_a^b} cf(x)\,dx &= \inf\{U(P, cf)\colon P \text{ partitions } [a, b]\} \\ &= \inf\{c \cdot U(P, f)\colon P \text{ partitions } [a, b]\} \\ &= c \cdot \inf\{U(P, f)\colon P \text{ partitions } [a, b]\} = c\overline{\int_a^b} f(x)\,dx = c\int_a^b f(x)\,dx. \end{aligned}$$

For the case $c < 0$, note that we have

$$\begin{aligned} M_i^{cf} &= \sup\{cf(t)\colon t \in [x_{i-1}, x_i]\} = c \cdot \inf\{f(t)\colon t \in [x_{i-1}, x_i]\} = cm_i^f, \\ m_i^{cf} &= \inf\{cf(t)\colon t \in [x_{i-1}, x_i]\} = c \cdot \sup\{f(t)\colon t \in [x_{i-1}, x_i]\} = cM_i^f \end{aligned}$$

for any partition $P = \langle x_0, x_1, \ldots, x_n \rangle$ of $[a, b]$ and any $i \leq n$. This implies $U(P, cf) = cL(P, f)$ and $L(P, cf) = cU(P, f)$.

Arguing as in the $c > 0$ case if $\varepsilon > 0$, then there is some partition P of $[a, b]$ such that $U(P, f) - L(P, f) < \varepsilon/-c$. But then, $U(P, cf) - L(P, cf) = cL(P, f) - cU(P, f) = -c(U(P, f) - L(P, f)) < -c \cdot \frac{\varepsilon}{-c} = \varepsilon$. Thus, f satisfies the Cauchy criterion and is Riemann integrable. Furthermore,

$$\begin{aligned} \int_a^b cf(x)\,dx = \overline{\int_a^b} cf(x)\,dx &= \sup\{U(P, cf)\colon P \text{ partitions } [a, b]\} \\ &= \sup\{c \cdot L(P, f)\colon P \text{ partitions } [a, b]\} \\ &= c \cdot \inf\{L(P, f)\colon P \text{ partitions } [a, b]\} = c\underline{\int_a^b} f(x)\,dx = c\int_a^b f(x)\,dx. \end{aligned}$$

Finally, the $c=0$ case is easy: In this case, cf is the constant function 0, which is clearly Riemann integrable with integral 0.

For (2), observe that for any partition $P=\langle x_0, x_1, \ldots, x_n\rangle$ of $[a, b]$,

$$\begin{aligned}
M_i^{f+g} &= \sup\{f(t)+g(t)\colon t\in[x_{i-1}, x_i]\} \\
&\leq \sup\{f(t)\colon t\in[x_{i-1}, x_i]\} + \sup\{g(t)\colon t\in[x_{i-1}, x_i]\} = M_i^f + M_i^g, \\
m_i^{f+g} &= \inf\{f(t)+g(t)\colon t\in[x_{i-1}, x_i]\} \\
&\geq \inf\{f(t)\colon t\in[x_{i-1}, x_i]\} + \sup\{g(t)\colon t\in[x_{i-1}, x_i]\} = m_i^f + m_i^g.
\end{aligned}$$

It follows that

$$\begin{aligned}
U(P, f+g) &= \sum_{i=1}^{n} M_i^{f+g}\Delta x_i \leq \sum_{i=1}^{n}(M_i^f + M_i^g)\Delta x_i \\
&= \sum_{i=1}^{n} M_i^f\Delta x_i + \sum_{i=1}^{n} M_i^g\Delta x_i = U(P, f) + U(P, g), \text{ and} \\
L(P, f+g) &= \sum_{i=1}^{n} m_i^{f+g}\Delta x_i \geq \sum_{i=1}^{n}(m_i^f + m_i^g)\Delta x_i \\
&= \sum_{i=1}^{n} m_i^f\Delta x_i + \sum_{i=1}^{n} m_i^g\Delta x_i = L(P, f) + L(P, g).
\end{aligned}$$

Now, let $\varepsilon > 0$. Because $f(x)$ and $g(x)$ are Riemann integrable, there are partitions P_f and P_g of $[a, b]$ with $U(P_f, f) - L(P_f, f) < \varepsilon/2$ and $U(P_g, f) - L(P_g, f) < \varepsilon/2$. Let $Q = \langle x_0, x_1, \ldots, x_n\rangle$ be a common refinement of P_f and P_g, so that $U(Q, f) - L(Q, f) < \varepsilon/2$ and $U(Q, g) - L(Q, g) < \varepsilon/2$. But then,

$$U(Q, f+g) - L(Q, f+g) \leq U(Q, f) + U(Q, g) - L(Q, f) - L(Q, g) < \tfrac{\varepsilon}{2} + \tfrac{\varepsilon}{2} = \varepsilon.$$

Thus, $f+g$ satisfies the Cauchy criterion and is Riemann integrable. Furthermore, note that

$$L(Q, f) + L(Q, g) \leq \int_a^b f(x)\,dx + \int_a^b g(x)\,dx \leq U(Q, f) + U(Q, g).$$

Because we also have

$$\begin{aligned}
L(Q, f) + L(Q, g) \leq L(Q, f+g) &\leq \int_a^b f(x) + g(x)\,dx \\
&\leq U(Q, f+g) \leq U(Q, f) + U(Q, g)
\end{aligned}$$

and $(U(Q,f) + U(Q,g)) - (L(Q,f) - L(Q,g)) < \varepsilon$, this implies

$$\left| \int_a^b f(x) + g(x)\, dx - \left(\int_a^b f(x)\, dx + \int_a^b g(x)\, dx \right) \right| < \varepsilon.$$

As ε was arbitrary, $\int_a^b f(x) + g(x)\, dx = \int_a^b f(x)\, dx + \int_a^b g(x)\, dx$.

For (3), first note that because f and g are both bounded on $[a, b]$, there is a constant $B > 0$ such that $|f(x)| < B$ and $|g(x)| < B$ for all $x \in [a, b]$. It follows that for any $s, t \in [a, b]$,

$$\begin{aligned}
|f(s)g(s) - f(t)g(t)| &= |f(s)g(s) - f(s)g(t) + f(s)g(t) - f(t)g(t)| \\
&= |f(s)(g(s) - g(t)) - g(t)(f(s) - f(t))| \\
&\leq |f(s)| \cdot |g(s) - g(t)| + |g(t)| \cdot |f(s) - f(t)| \\
&\leq B\left(|g(s) - g(t)| + |f(s) - f(t)|\right).
\end{aligned}$$

Applying Proposition 6.13, for any partition $P = \langle x_0, x_1, \ldots, x_n \rangle$ of $[a, b]$

$$\begin{aligned}
M_i^{fg} - m_i^{fg} &= \sup\left\{f(s)g(s) - f(t)g(t) \colon s, t \in [x_{i-1}, x_i]\right\} \\
&\leq \sup\left\{B\left(|g(s) - g(t)| + |f(s) - f(t)|\right) \colon s, t \in [x_{i-1}, x_i]\right\} \\
&\leq B \sup\left\{|g(s) - g(t)| \colon s, t \in [x_{i-1}, x_i]\right\} \\
&\qquad + B \sup\left\{|f(s) - f(t) \colon s, t \in [x_{i-1}, x_i]\right\} \\
&= B(M_i^g - m_i^g) + B(M_i^f - m_i^f).
\end{aligned}$$

It follows that

$$\begin{aligned}
U(P, fg) - L(P, fg) &= \sum_{i=1}^{n} (M_i^{fg} - m_i^{fg}) \Delta x_i \\
&\leq \sum_{i=1}^{n} \left(B(M_i^g - m_i^g) + B(M_i^f - m_i^f)\right) \Delta x_i \\
&= B\left(U(P,f) - L(P,f)\right) + B\left(U(P,g) - L(P,g)\right).
\end{aligned}$$

Now, let $\varepsilon > 0$. Because $f(x)$ and $g(x)$ are Riemann integrable, there are partitions P_f and P_g of $[a, b]$ such that $U(P_f, f) - L(P_f, f) < \varepsilon/2B$ and $U(P_g, f) - L(P_g, f) < \varepsilon/2B$. If $Q = \langle x_0, x_1, \ldots, x_n \rangle$ is a common refinement of P_f and P_g, then $U(Q, f) - L(Q, f) < \varepsilon/2B$ and $U(Q, g) - L(Q, g) < \varepsilon/2B$. But then,

$$\begin{aligned}
U(Q, fg) - L(Q, fg) &\leq B\left(U(Q,f) - L(Q,f)\right) + B\left(U(Q,g) - L(Q,g)\right) \\
&< B\tfrac{\varepsilon}{2B} + B\tfrac{\varepsilon}{2B} = \varepsilon.
\end{aligned}$$

Thus, fg satisfies the Cauchy criterion and is Riemann integrable.

For (4), observe that for any $s, t \in [a, b]$,

$$\left|\frac{1}{f(s)} - \frac{1}{f(t)}\right| = \left|\frac{f(t) - f(s)}{f(s)f(t)}\right| = \frac{|f(t) - f(s)|}{|f(s)||f(t)|} \le \frac{1}{c^2}|f(s) - f(t)|.$$

Applying Proposition 6.13, for any partition $P = \langle x_0, x_1, \ldots, x_n\rangle$ of $[a, b]$

$$\begin{aligned} M_i^{1/f} - m_i^{1/f} &= \sup\{1/f(s) - 1/f(t) : s, t \in [x_{i-1}, x_i]\} \\ &\le \sup\left\{\tfrac{1}{c^2}|f(s) - f(t)| : s, t \in [x_{i-1}, x_i]\right\} = \tfrac{1}{c^2}(M_i^f - m_i^f). \end{aligned}$$

It follows that

$$\begin{aligned} U(P, 1/f) - L(P, 1/f) &= \sum_{i=1}^{n}(M_i^{1/f} - m_i^{1/f})\Delta x_i \\ &\le \sum_{i=1}^{n}\frac{1}{c^2}(M_i^f - m_i^f)\Delta x_i = \frac{1}{c^2}\left(U(P, 1/f) - L(P, 1/f)\right). \end{aligned}$$

Now, let $\varepsilon > 0$. Because $f(x)$ is Riemann integrable, there is a partition P of $[a, b]$ such that $U(P,f) - L(P_f,f) < c^2\varepsilon$. But then,

$$U(P, 1/f) - L(P, 1/f) \le \tfrac{1}{c^2}\left(U(P,f) - L(P,f)\right) < \tfrac{1}{c^2}\cdot c^2\varepsilon = \varepsilon.$$

Thus, $1/f$ satisfies the Cauchy criterion and is Riemann integrable. □

We now consider compositions of functions. Perhaps surprisingly, the composition of two Riemann integrable functions is not necessarily Riemann integrable, as the following example demonstrates.

Example 6.20. Let $f : [0, 1] \to \mathbb{R}$ denote Thomae's function

$$f(x) = \begin{cases} 1/q & \text{if } x = p/q \text{ where } q > 0 \text{ and } p \text{ and } q \text{ are co-prime integers,} \\ 0 & \text{otherwise.} \end{cases}$$

This function is Riemann integrable. (This follows from the Lebesgue–Vitali Theorem, proved later. But it can also be shown directly from the definition of Riemann integrability, and doing so is one of the end-of-chapter exercises.) Define $g : [0, 1] \to \mathbb{R}$ by

$$g(x) = \begin{cases} 1 & \text{if } x \neq 0 \\ 0 & \text{if } x = 0. \end{cases}$$

This function is Riemann integrable as well, because it has only one point of discontinuity. However,

$$g \circ f(x) = \begin{cases} 1 & \text{if } x \in \mathbb{Q} \\ 0 & \text{if } x \notin \mathbb{Q}. \end{cases}$$

In other words, $g \circ f$ is the Dirichlet function, which is not Riemann integrable by Example 6.4.

Despite this example, the following theorem shows we can still prove something about compositions of functions: If we assume that the function on the left is not merely integrable, but continuous, then the composition is integrable. Like in the proof of Theorem 6.15 from the previous section, the proof works by dividing our rectangles into two groups, one in which we can control the width but not the height and one in which we can control the height but not the width.

Theorem 6.21. *Suppose $f : [a, b] \to \mathbb{R}$ is Riemann integrable. Suppose also that the image of f is contained in the interval $[c, d]$, and g is a continuous function $[c, d] \to \mathbb{R}$. Then, $g \circ f$ is Riemann integrable.*

Proof. First, observe that g is bounded because it is a continuous function with a compact domain. Because g is bounded, $\sup\{g \circ f(t) : t \in [a, b]\}$ and $\inf\{g \circ f(t) : t \in [a, b]\}$ both exist. Let $H = \sup\{g \circ f(t) : t \in [a, b]\} - \inf\{g \circ f(t) : t \in [a, b]\} + 1$.

Let $\varepsilon > 0$. Because g is a continuous function and $[c, d]$ is compact, g is uniformly continuous on $[c, d]$. Hence, there is some $\delta > 0$ such that if $x, y \in [c, d]$ and $|x - y| < \delta$, then $|g(x) - g(y)| < \varepsilon/2(b-a)$.

Because f is Riemann integrable, there is a partition $P = \langle x_0, x_1, \ldots, x_n\rangle$ of $[a, b]$ such that $U(P, f) - L(P, f) < \delta\varepsilon/2H$.

We claim that $U(P, g \circ f) - L(P, g \circ f) < \varepsilon$. To see this, divide the members of P into two groups: Let B denote the set of all those $i \leq n$ such that $M_i^f - m_i^f < \delta$, and let $A = \{1, 2, \ldots, n\} \setminus B$.

If $i \in B$, then we claim $M_i^{g\circ f} - m_i^{g\circ f} \leq \varepsilon/2(b-a)$. To see this, fix any $s, t \in [x_{i-1}, x_i]$. Observe that $|f(s) - f(t)| < \delta$ (because $i \in B$), and this implies $|g \circ f(s) - g \circ f(t)| < \varepsilon/2(b-a)$. Because s and t were arbitrary members of $[x_{i-1}, x_i]$, this implies $M_i^{g\circ f} - m_i^{g\circ f} \leq \varepsilon/2(b-a)$ by Proposition 6.13.

If $i \in A$, then we cannot put any bounds on $M_i^{g\circ f} - m_i^{g\circ f}$ except the trivial bound $M_i^{g\circ f} - m_i^{g\circ f} \leq H$. However, observe that $\delta \leq M_i^f - m_i^f$ for all $i \in A$, by the definition of A, and this implies

$$\begin{aligned} \sum_{i \in A} \delta \Delta x_i &\leq \sum_{i \in A} (M_i^f - m_i^f) \Delta x_i \\ &\leq \sum_{i \leq n} (M_i^f - m_i^f) \Delta x_i = U(P, f) - L(P, f) < \frac{\delta\varepsilon}{2H}. \end{aligned}$$

Dividing through by δ, we get $\sum_{i\in A} \Delta x_i < {}^{\varepsilon}/{}_{2H}$. Therefore,

$$\begin{aligned}
&U(P, g\circ f) - L(P, g\circ f)\\
&= \sum_{i=1}^{n} M_i^{g\circ f}\Delta x_i - \sum_{i=1}^{n} m_i^{g\circ f}\Delta x_i = \sum_{i=1}^{n}(M_i^{g\circ f} - m_i^{g\circ f})\Delta x_i\\
&= \sum_{i\in A}(M_i^{g\circ f} - m_i^{g\circ f})\Delta x_i + \sum_{i\in B}(M_i^{g\circ f} - m_i^{g\circ f})\Delta x_i\\
&< \sum_{i\in A} H\Delta x_i + \sum_{i\in B}\frac{\varepsilon}{2(b-a)}\Delta x_i = H\sum_{i\in A}\Delta x_i + \frac{\varepsilon}{2(b-a)}\sum_{i\in B}\Delta x_i\\
&\leq H\frac{\varepsilon}{2H} + \frac{\varepsilon}{2(b-a)}(b-a) = \varepsilon.
\end{aligned}$$

Thus, f satisfies the Cauchy criterion and is Riemann integrable. □

Theorem 6.21 raises the question of what happens when we switch the order of composition. Is Riemann integrability preserved for compositions of the form (integrable) ∘ (continuous) since we just showed that it is for (continuous) ∘ (integrable)? Surprisingly, the answer is no!

Example 6.22. In Example 6.42, we will see that there is a closed set $F \subseteq \mathbb{R}$, called a "fat Cantor set," with the property that the function

$$f(x) = \begin{cases} 1 & \text{if } x \in F \\ 0 & \text{if } x \notin F \end{cases}$$

is not Riemann integrable. Assuming for now that such a set F exists, let us prove that there is a function of the form (integrable) ∘ (continuous) that is not Riemann integrable. For our continuous function, define

$$g(x) = \inf\{|x-y| : y \in F\}$$

for all $x \in [0, 1]$. In other words, $g(x)$ is the distance from x to F. Using the triangle inequality, one may show that $|g(x) - g(y)| \leq |x - y|$ for all $x, y \in [0, 1]$, and this implies g is continuous (take $\delta = \varepsilon$ in our original ε-δ definition of continuity). Note that because F is closed, $x \in F$ if and only if $g(x) = 0$. For our integrable function, define $h : [0, 1] \to \mathbb{R}$ by

$$h(x) = \begin{cases} 1 & \text{if } x = 1 \\ 0 & \text{if } x \in [0, 1). \end{cases}$$

This function is integrable by Theorem 6.16 because it has only one point of discontinuity. But $h \circ g = f$, which is not Riemann integrable.

Theorem 6.23. *Suppose f is a Riemann integrable function $[a, b] \to \mathbb{R}$. Then, $|f(x)|$ is also Riemann integrable, and $\left|\int_a^b f(x)\,dx\right| \leq \int_a^b |f(x)|\,dx$.*

Proof. The function $\phi(x) = |x|$ is continuous. Thus by Theorem 6.21, the composite function $\phi \circ f$ is Riemann integrable.

Generally speaking, if g and h are Riemann integrable functions with $g(x) \leq h(x)$ for all x, then $\int_a^b g(x)\,dx \leq \int_a^b h(x)\,dx$. (This is not difficult to see, and proving it is an end-of-chapter exercise.) Using this fact, let us consider two cases. If $\int_a^b f(x)\,dx > 0$, then

$$\left|\int_a^b f(x)\,dx\right| = \int_a^b f(x)\,dx \leq \int_a^b |f(x)|\,dx$$

because $f(x) \leq |f(x)|$ for all x. If instead $\int_a^b f(x)\,dx < 0$, then

$$\left|\int_a^b f(x)\,dx\right| = -\int_a^b f(x)\,dx = \int_a^b -f(x)\,dx \leq \int_a^b |f(x)|\,dx. \qquad \square$$

The fundamental theorems of calculus

In this section, we prove two of the most important theorems of real analysis, namely, the fundamental theorems of calculus.

The first of these two theorems states roughly that if $f(x)$ is continuous, then it is the derivative of the function $F(x) = \int_a^x f(t)\,dt$. In other words, integration and differentiation are inverse operations, provided that we stick to functions whose derivatives are continuous.

Definition 6.24. Suppose f is a Riemann integrable function $[a, b] \to \mathbb{R}$. An *antiderivative* for f on $[a, b]$ is a continuous function $F : [a, b] \to \mathbb{R}$ such that F is differentiable at every $x \in (a, b)$, and $F'(x) = f(x)$.

Let us henceforth set the convention that $\int_a^a f(x)\,dx = 0$ for any $a \in \mathbb{R}$ and any function f defined at a.

Theorem 6.25 (The first fundamental theorem of calculus). *Suppose f is a Riemann integrable function $[a, b] \to \mathbb{R}$. Define $F : [a, b] \to \mathbb{R}$ by setting*

$$F(x) = \int_a^x f(t)\,dt$$

for all $x \in [a, b]$. Then, F is continuous on $[a, b]$. Furthermore, if f is continuous at some $x \in (a, b)$, then F is differentiable at x and $F'(x) = f(x)$. In particular, if f is continuous on (a, b), then F is an antiderivative for f, and $\int_a^b f(t)\,dt = F(b) = F(b) - F(a)$.

Proof. Because f is Riemann integrable on $[a, b]$, it is bounded on $[a, b]$. Fix $B > 0$ such that $|f(x)| < B$ for all $x \in [a, b]$.

Let $\varepsilon > 0$, and let $\delta = \varepsilon/B$. If $a \leq x < y \leq b$ and $|x - y| < \delta$, then

$$\begin{aligned}
|F(x) - F(y)| &= \left|\int_a^x f(t)\,dt - \int_a^y f(t)\,dt\right| \\
&= \left|\int_a^x f(t)\,dt - \left(\int_a^x f(t)\,dt + \int_x^y f(t)\,dt\right)\right| \\
&= \left|\int_x^y f(t)\,dt\right| \leq \int_x^y |f(t)|\,dt \leq \int_x^y B\,dt \\
&= B|x - y| \leq B\frac{\varepsilon}{B} = \varepsilon.
\end{aligned}$$

A similar argument (reversing the roles of x and y) shows that if $a \leq y < x \leq b$ and $|x - y| < \delta$, then $|F(x) - F(y)| < \varepsilon$. As ε was arbitrary, this shows f is (uniformly) continuous on $[a, b]$.

To prove the "furthermore" part of the theorem, fix some $x \in (a, b)$ such that f is continuous at x. Let $c = f(x)$. Let $\varepsilon > 0$, and pick $\delta > 0$ small enough that if $|x - t| < \delta$, then $|f(x) - f(t)| = |f(t) - c| < \varepsilon$. If $x < y$ and $|x - y| < \delta$, then

$$\begin{aligned}
\left|\frac{F(y) - F(x)}{y - x} - c\right| &= \frac{1}{y - x}\left|F(y) - F(x) - c(y - x)\right| \\
&= \frac{1}{y - x}\left|\int_a^y f(t)\,dt - \int_a^x f(t)\,dt - \int_x^y c\,dt\right| \\
&= \frac{1}{y - x}\left|\int_x^y f(t)\,dt - \int_x^y c\,dt\right| \\
&= \frac{1}{y - x}\left|\int_x^y f(t) - c\,dt\right| \\
&\leq \frac{1}{y - x}\int_x^y |f(t) - c|\,dt \\
&\leq \frac{1}{y - x}\int_x^y \varepsilon\,dt = \frac{1}{y - x}(y - x)\varepsilon = \varepsilon.
\end{aligned}$$

Similarly (reversing the roles of x and y in the computation), if $y < x$ and $|x - y| < \delta$, then $\left|\frac{F(y)-F(x)}{y-x} - c\right| < \varepsilon$. Thus for every $\varepsilon > 0$, there is some $\delta > 0$ such that for any y, if $|x - y| < \delta$, then $\left|\frac{F(y)-F(x)}{y-x} - c\right| < \varepsilon$. In other words, $c = \lim_{y \to x} \frac{F(y)-F(x)}{y-x} = F'(x)$.

The "in particular" part of the theorem follows directly from the "furthermore" part. □

The "furthermore" part of this proof essentially just formalizes the idea that if y is very close to x and f is continuous, then $\frac{F(y)-F(x)}{y-x}$ is very close to $f(x)$, as shown in the next illustration.

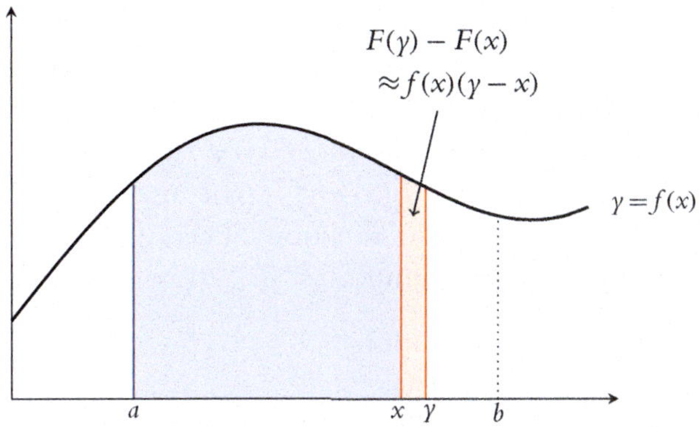

One important consequence of Theorem 6.25 is that if f is continuous, then f has an antiderivative, namely, the function $F(x) = \int_a^x f(t)\,dt$. The following few examples and observations show that the continuity assumption cannot be ignored: Not every Riemann integrable function has an antiderivative.

Example 6.26. Consider the function

$$f(x) = \begin{cases} 1 & \text{if } x = 2 \\ 2 & \text{if } x \neq 2. \end{cases}$$

Then, $F(x) = \int_a^x f(t)\,dt = x$, and $F'(x) = 1$, for all x.

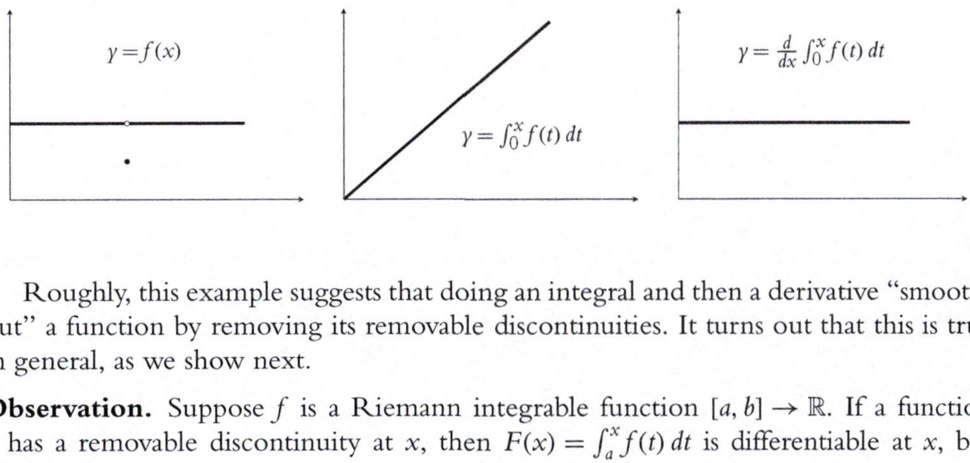

Roughly, this example suggests that doing an integral and then a derivative "smooths out" a function by removing its removable discontinuities. It turns out that this is true in general, as we show next.

Observation. Suppose f is a Riemann integrable function $[a,b] \to \mathbb{R}$. If a function f has a removable discontinuity at x, then $F(x) = \int_a^x f(t)\,dt$ is differentiable at x, but $F'(x) \neq f(x)$.

Proof. Suppose a Riemann integrable function f has a removable discontinuity at a point z. (Recall that this means $\lim_{t\to z} f(t)$ exists, but $\lim_{t\to z} f(t) \neq f(z)$.) Define g to be equal to f everywhere but at z, with $g(z) = \lim_{t\to z} f(t)$. Then, g is integrable, and $F(x) = \int_a^x f(t)\,dt = \int_a^x g(t)\,dt$ for all x. Furthermore, g is continuous at z, so by Theorem 6.25, $F'(z) = g(z) \neq f(z)$.

For a more extreme illustration of this observation, consider Thomae's function. It has a removable discontinuity at every rational number and is Riemann integrable (by an end-of-chapter exercise). But if f denotes Thomae's function, then $F(x) = \int_0^x f(t)\,dt$ is the constant function 0, and in particular $F'(x) = 0 \neq f(x)$ whenever $x \in \mathbb{Q}$.

Thus if f is not continuous, then $F(x) = \int_a^x f(t)\,dt$ may not be an antiderivative for f. The next example shows that the function $F(x) = \int_a^x f(t)\,dt$ can also fail to be differentiable for a discontinuous function f.

Example 6.27. Consider the step function

$$f(x) = \begin{cases} 0 & \text{if } x \leq 2 \\ 1 & \text{if } x > 2. \end{cases}$$

This function is Riemann integrable on $[0, 4]$ because it has just one point of discontinuity. However, f cannot be the derivative of any function defined on $[0, 4]$. This is because derivatives always satisfy the conclusion of the Intermediate Value Theorem, but of course f does not.

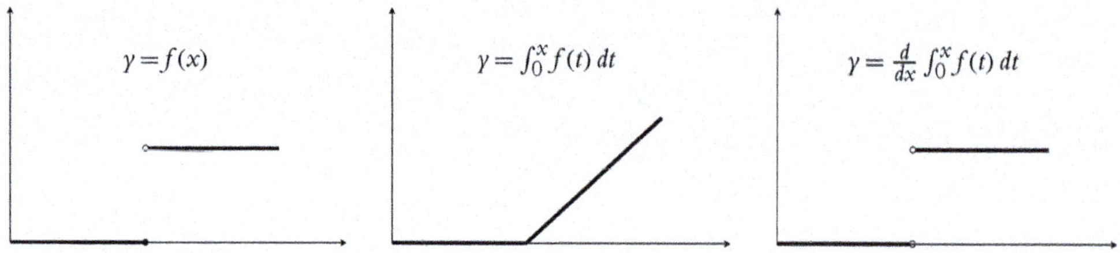

The function $F(x) = \int_0^x f(t)\,dt$ is of course still well-defined because f is Riemann integrable. But F fails to be differentiable at $x = 2$, where f has its jump discontinuity.

Just like our last example, this example also illustrates a more general phenomenon:

Observation. Suppose f is a Riemann integrable function $[a, b] \to \mathbb{R}$. If f has a jump discontinuity at x, then f has no antiderivative, and the function $F(x) = \int_a^x f(t)\,dt$ is not differentiable at x.

Proof. By Corollary 5.27 in Chap. 5, if F is a differentiable function, then its derivative F' does not have any discontinuities of the first kind. This implies that if f has a jump discontinuity, then f does not have an antiderivative. To finish proving the observation, we need to show that if f has a jump discontinuity at x, then $F(x) = \int_a^x f(t)\,dt$ is not differentiable at x.

We claim that, by the same argument as in the proof of the "furthermore" part of Theorem 6.25, $\lim_{t\to x^+} \frac{F(x)-F(t)}{x-t} = \lim_{t\to x^+} f(t)$.

To see this, let $c = \lim_{t\to x^+} f(t)$. Let $\varepsilon > 0$, and pick $\delta > 0$ small enough that if $t \in (x, x+\delta)$, then $|f(x) - f(t)| = |f(t) - c| < \varepsilon$. If $x < y$ and $|x - y| < \delta$, then

$$\begin{aligned}\left|\frac{F(y)-F(x)}{y-x} - c\right| &= \frac{1}{y-x}\left|F(y) - F(x) - c(y-x)\right| \\ &= \frac{1}{y-x}\left|\int_a^y f(t)\,dt - \int_a^x f(t)\,dt - \int_x^y c\,dt\right| \\ &= \frac{1}{y-x}\left|\int_x^y f(t)\,dt - \int_x^y c\,dt\right| \\ &= \frac{1}{y-x}\left|\int_x^y f(t) - c\,dt\right| \\ &\le \frac{1}{y-x}\int_x^y |f(t) - c|\,dt \\ &\le \frac{1}{y-x}\int_x^y \varepsilon\,dt = \frac{1}{y-x}(y-x)\varepsilon = \varepsilon.\end{aligned}$$

As ε was arbitrary, this shows $\lim_{t\to x^+} \frac{F(x)-F(t)}{x-t} = \lim_{t\to x^+} f(t)$. A similar computation, but with $y < x$, shows $\lim_{t\to x^-} \frac{F(x)-F(t)}{x-t} = \lim_{t\to x^-} f(t)$.

Recall that f having a jump discontinuity at x means that $\lim_{t\to x^+} f(t) \neq \lim_{t\to x^-} f(t)$. By the previous paragraph, this implies $\lim_{t\to x^+} \frac{F(x)-F(t)}{x-t} \neq \lim_{t\to x^-} \frac{F(x)-F(t)}{x-t}$, which means F is not differentiable at x. ☕

For a more extreme example illustrating this observation, Exercise 12 in Chap. 4 shows that there is an increasing function $f(x)$ that has a jump discontinuity at x whenever $x \in \mathbb{Q}$. This function is Riemann integrable by Theorem 6.15 from the previous section. But the function $F(x) = \int_a^x f(t)\,dt$ is not differentiable at any rational number, by the previous observation.

These examples show that not every Riemann integrable function has an antiderivative. We turn now to the second fundamental theorem of calculus, which roughly speaking says that when an integrable function does have an antiderivative, then it can be used to compute the integral.

Lemma 6.28. *Suppose F and G are both antiderivatives of the same function $f : [a, b] \to \mathbb{R}$. Then, there is a constant $c \in \mathbb{R}$ such that $F(x) = G(x) + c$.*

Proof. Let $H(x) = F(x) - G(x)$. Then, H is differentiable on (a, b), and $H'(x) = F'(x) - G'(x) = f(x) - f(x) = 0$ for all $x \in (a, b)$. By Corollary 5.23 from Chap. 5, this implies $H(x)$ is a constant function. □

Corollary 6.29. *Suppose f is a continuous function $[a, b] \to \mathbb{R}$. If F is an antiderivative for f on $[a, b]$, then $\int_a^b f(x)\,dx = F(b) - F(a)$.*

Proof. Define $G : [a, b] \to \mathbb{R}$ by setting $G(x) = \int_a^x f(t)\,dt$ for all $x \in [a, b]$. By Theorem 6.25, G is an antiderivative for f on $[a, b]$, and $G(b) - G(a) = \int_a^b f(x)\,dx$. But Lemma 6.28 implies there is some constant c with $F(x) = G(x) + c$, hence $F(b) - F(a) = G(b) - G(a) = \int_a^b f(x)\,dx$. □

This much follows from the first fundamental theorem. The second fundamental theorem says something similar to Corollary 6.29, but with a much weaker hypothesis. What Corollary 6.29 says is true of continuous functions the second fundamental theorem asserts for every Riemann integrable that has an antiderivative.

Theorem 6.30 (The second fundamental theorem of calculus). *Suppose f is a Riemann integrable function $[a, b] \to \mathbb{R}$, and that f has an antiderivative. If F is an antiderivative for f, then*

$$\int_a^b f(x)\,dx = F(b) - F(a).$$

Proof. Let $\varepsilon > 0$, and let $P = \langle x_0, x_1, \ldots, x_n \rangle$ be a partition of $[a, b]$ such that $U(P, f) - L(P, f) < \varepsilon$. For each $i \leq n$, there is by the Mean Value Theorem some $t_i \in (x_{i-1}, x_i)$ such that

$$f(t_i) = F'(t_i) = \frac{1}{x_i - x_{i-1}}\Big(F(x_i) - F(x_{i-1})\Big) = \frac{1}{\Delta x_i}\Big(F(x_i) - F(x_{i-1})\Big)$$

or, equivalently, $f(t_i)\Delta x_i = F(x_i) - F(x_{i-1})$. Therefore,

$$\begin{aligned} L(P, f) &= \sum_{i=1}^{n} m_i \Delta x_i \leq \sum_{i=1}^{n} f(t_i)\Delta x_i = \sum_{i=1}^{n} F(x_i) - F(x_{i-1}) \\ &= F(b) - F(a), \quad \text{and} \\ U(P, f) &= \sum_{i=1}^{n} M_i \Delta x_i \geq \sum_{i=1}^{n} f(t_i)\Delta x_i = \sum_{i=1}^{n} F(x_i) - F(x_{i-1}) \\ &= F(b) - F(a). \end{aligned}$$

Consequently, $L(P, f) \leq F(b) - F(a) \leq U(P, f)$. But $U(P, f) - L(P, f) < \varepsilon$, and $L(P, f) \leq \int_a^b f(x)\,dx \leq U(P, f)$ as well, so this implies

$$\left| \big(F(b) - F(a)\big) - \int_a^b f(x)\,dx \right| < \varepsilon.$$

As ε was arbitrary, $\displaystyle\int_a^b f(x)\,dx = F(b) - F(a)$. □

Example 6.31. In Example 5.24 of Chap. 5, we saw a function $f(x)$ that is differentiable, but whose derivative $f'(x)$ has a discontinuity (of the second kind). So $f'(x)$ is an example of a Riemann integrable function that has an antiderivative but is not continuous. In particular, the second fundamental theorem of calculus applies to $f'(x)$, but Corollary 6.29 does not.

Theorem 6.30 requires not just that f has an antiderivative, but also integrability. This raises the question of whether a function can have an antiderivative without being Riemann integrable. The answer is yes!

Example 6.32. Consider the function

$$F(x) = \begin{cases} x^2 \sin\left(\frac{1}{x^2}\right) & \text{if } x \neq 0 \\ 0 & \text{if } x = 0. \end{cases}$$

This function is differentiable on $\mathbb{R}$. (The proof that it is differentiable at 0 is essentially the same as in Example 5.24.) Its derivative is

$$F'(x) = \begin{cases} 2x \sin\left(\frac{1}{x^2}\right) - \frac{2}{x}\cos\left(\frac{1}{x^2}\right) & \text{if } x \neq 0 \\ 0 & \text{if } x = 0. \end{cases}$$

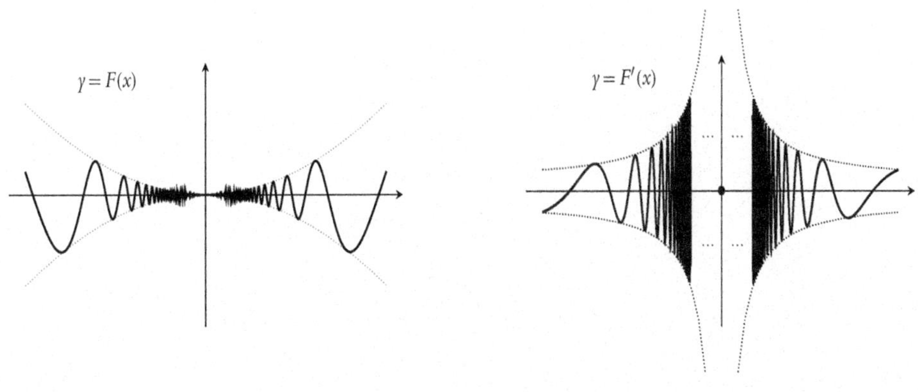

The function $F'(x)$ is unbounded on $[-1, 1]$, hence not Riemann integrable, despite the fact that it has an antiderivative on $[-1, 1]$.

One might object to this example on the grounds that F' fails to be Riemann integrable only on a technicality. After all, some unbounded functions can be integrated via improper integration, even though they are not technically integrable. Could some trick like that work here?

At the end of this chapter, we will see an example of a differentiable function whose derivative fails to be integrable in a more fundamental, unfixable way: a function $F(x)$, known as *Volterra's function*, such that $F'(x)$ exists and is bounded, but is not Riemann integrable.

Such examples show, in short, that derivatives are not always integrable. In other words, the old motto often given to beginning calculus students, "differentiation and integration are inverse operations," is not really true. This motto works well enough for very tame functions (polynomials, exponentials, etc.), but fails in general. Derivatives are not always integrable (like Example 6.32), and integrals are not always differentiable (like Example 6.27). Example 6.32 shows that there are differentiable functions F, with derivative $F'(x) = f(x)$, such that the equality $\int_a^b f(x)\,dx = F(b) - F(a)$ cannot be true, because the left-hand side is not even well-defined.

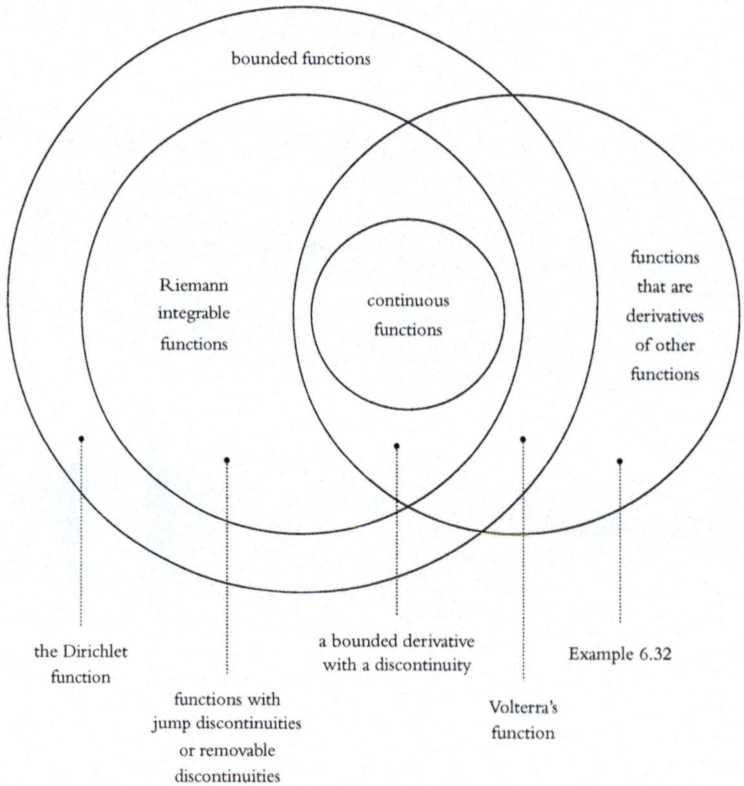

However, in fairness to the old motto "differentiation and integration are inverse operations," let us point out that this motto can be rescued to some extent. It is almost true. In the latter half of this chapter, we will prove that if f is a Riemann integrable function, then for *almost all* points $x \in [a, b]$, the function $F(x) = \int_a^x f(t)\,dt$ is differentiable at x and $F'(x) = f(x)$. What precisely the phrase "almost all" means in this sentence is the topic of the next section.

Lebesgue null sets

Definition 6.33. For every interval $I \subseteq \mathbb{R}$, let $\lambda(I)$ denote the *length* of I: i.e., $\lambda(I) = 0$ if $I = \emptyset$, $\lambda(I) = \infty$ if I is unbounded, and $\lambda(I) = \sup(I) - \inf(I)$, otherwise. (In particular, the length of an open interval (a, b) is $b - a$.)

Recall that a set $\mathcal{I}$ of intervals *covers* a set $X \subseteq \mathbb{R}$ if $X \subseteq \bigcup \mathcal{I}$. The *total length* of a finite set $\{I_1, I_2, \ldots, I_n\}$ of intervals is defined to be $\sum_{j=1}^{n} \lambda(I_j)$. (We do not subtract anything to account for overlaps.) Similarly, the total length of a countably infinite set of intervals $\{I_n \colon n \in \mathbb{N}\}$ is $\sum_{n=1}^{\infty} \lambda(I_n)$. Note that this is a sum of nonnegative extended real numbers, so the order of the summation does not matter, and the total length is always either a nonnegative real number or ∞.

A subset N of $\mathbb{R}$ is a *Lebesgue null set* (or simply a *null set*) if for any $\varepsilon > 0$, there is a countable set of open intervals that covers N and has total length at most ε. Roughly, this means N has total "size" effectively 0.

Proposition 6.34. *If $M \subseteq N \subseteq \mathbb{R}$ and N is null, then M is also null.*

Proof. Any set of open intervals covering N also covers M. □

Proposition 6.35. *Every countable subset of $\mathbb{R}$ is Lebesgue null.*

Proof. Let X be a countable subset of $\mathbb{R}$, and fix an enumeration of X, say $X = \{x_n \colon n \in \mathbb{N}\}$. Given $\varepsilon > 0$, note that $\left\{(x_n - \frac{\varepsilon}{2^{n+1}}, x_n + \frac{\varepsilon}{2^{n+1}}) \colon n \in \mathbb{N}\right\}$ is a countable set of open intervals that covers X and has total length ε. □

Example 6.36. The Cantor set K is an uncountable Lebesgue null set. Recall that K is a countable intersection $\bigcap_{n\in\mathbb{N}} K_n$, where each of the sets K_n consists of 2^n closed intervals having total length $\left(\frac{2}{3}\right)^n$. Given $\varepsilon > 0$, fix n big enough that $\left(\frac{2}{3}\right)^n < \varepsilon$, and let $I_1, I_2, \ldots, I_{2^n}$ denote the 2^n intervals comprising K_n. Each I_k can be "fattened up" a bit to an open interval J_k: Specifically, let $\delta = \varepsilon - \left(\frac{2}{3}\right)^n$, and for each $k \leq 2^n$ let $J_k = \left(\inf(I_k) - \frac{\delta}{2^{n+1}}, \sup(I_k) + \frac{\delta}{2^{n+1}}\right)$. Then, $\{J_1, J_2, \ldots, J_{2^n}\}$ is a set of open intervals covering K with total length ε.

Theorem 6.37. *Suppose $\{N_n \colon n \in \mathbb{N}\}$ is a collection of Lebesgue null subsets of $\mathbb{R}$. Then, $N = \bigcup_{n\in\mathbb{N}} N_n$ is also Lebesgue null. In other words, a countable union of null sets is a null set.*

Proof. Let $\varepsilon > 0$. For each $n \in \mathbb{N}$, let $\mathcal{I}_n$ be a countable set of open intervals that covers N_n and has total length at most $\varepsilon/2^n$. Then, $\bigcup \mathcal{I}_n$ is a countable set of open intervals that covers $N = \bigcup_{n\in\mathbb{N}} N_n$ and has total length at most $\sum_{n=1}^{\infty} \frac{\varepsilon}{2^n} = \varepsilon$. □

Roughly, we may think of a Lebesgue null set of reals as being negligibly small. This intuition is reflected in the following terminology.

Definition 6.38. Let f and g be functions $[a, b] \to \mathbb{R}$.

- We say that f is *continuous almost everywhere* on $[a, b]$ (or just continuous almost everywhere) if $\{x \in [a, b] : f \text{ is not continuous at } x\}$ is Lebesgue null.
- We say f and g are *equal almost everywhere* if $\{x \in D : f(x) \neq g(x)\}$ is Lebesgue null.
- We say that g is an *antiderivative of* f *almost everywhere* if $[a, b] \setminus \{x \in [a, b] : g \text{ is differentiable at } x \text{ and } g'(x) = f(x)\}$ is null.

Generally speaking, a statement is said to hold "almost everywhere" on $[a, b]$ if the set of all x for which the statement does not hold is Lebesgue null. ☕

Of course, we have not yet proved that there are any nonnull sets, and Definition 6.38 would be rather silly if there were not. We show next (in Theorem 6.40) that, indeed, not all sets are null. Note that by Proposition 6.35, and the fact that $\mathbb{Q}$ is countable, our proof of this theorem must somehow make use of the Axiom of Completeness.

Lemma 6.39. *If $\mathcal{I} = \{I_1, I_2, \ldots, I_n\}$ is a finite set of open intervals covering a closed bounded interval J, then the total length of $\mathcal{I}$ is at least $\lambda(J)$.*

Proof. The proof is by induction on n. The base case $n = 1$ is obvious. Now, suppose that the lemma holds true for sets of n intervals. Let $\{I_1, I_2, \ldots, I_n, I_{n+1}\}$ be a set of $n+1$ open intervals covering $J = [a, b]$. There is some $j \leq n + 1$ such that $b \in I_j$. But I_j is an open interval, so $I_j = (c, d)$ for some $c, d \in \mathbb{R}$ with $c < b < d$. Removing I_j from our set results in a set $\{I_1, \ldots, I_{j-1}, I_{j+1}, \ldots, I_n\}$ of n open intervals that covers $[a, c]$. By the inductive hypothesis, the total length of this new set of intervals is at least $c - a$. Thus, the total length of the original set of intervals is at least $c - a + \lambda(I_j) = c - a + d - c = d - a > b - a$. □

Theorem 6.40. *If $a < b$, then any set of open intervals that covers $[a, b]$ has total length at least $b - a$. In particular, $[a, b]$ is not a null set.*

Proof. Suppose $\mathcal{I} = \{I_n : n \in \mathbb{N}\}$ is a countable set of open intervals covering $[a, b]$. Because $[a, b]$ is compact, and $\mathcal{I}$ is an open cover of $[a, b]$, it has a finite subcover, i.e., there is some $N \in \mathbb{N}$ such that $\{I_1, I_2, \ldots, I_N\}$ covers $[a, b]$. By Lemma 6.39, the total length of $\{I_1, I_2, \ldots, I_N\}$ is at least $b - a$. This implies the total length of $\mathcal{I}$ is at least $b - a$. □

Corollary 6.41. *If I is an interval in $\mathbb{R}$, then any set of open intervals covering I has total length at least $\lambda(I)$. In particular, nondegenerate intervals are not null.*

Proof. Suppose I is a nondegenerate bounded interval (of the form $[a, b]$, (a, b), $(a, b]$, or $[a, b)$). Given $\varepsilon > 0$, let $\delta = \min\{\frac{b-a}{2}, \varepsilon\}$. The interval $\left[a + \frac{\delta}{2}, b - \frac{\delta}{2}\right]$ is contained in I. By Theorem 6.40, since $\left[a - \frac{\delta}{2}, b + \frac{\delta}{2}\right]$ cannot be covered with a set of open intervals of total length less than $\lambda(I) - \varepsilon$, therefore, neither can I. As this holds for any $\varepsilon > 0$, any set of open intervals covering I has total length at least $\lambda(I)$. If I is unbounded, then it contains arbitrarily long bounded intervals, so the previous sentence implies that any set of open intervals covering I has total length $\infty = \lambda(I)$. □

As the following examples show, there are also nonnull subsets of $\mathbb{R}$ that do not contain any intervals.

Example 6.42. The *fat Cantor set* of measure α, also known as a *Smith–Volterra–Cantor set*, is a variant of the middle-thirds Cantor set from Chap. 2. It is defined as the intersection of a decreasing sequence $K_0, K_1, K_2, K_3, \ldots$ of closed subsets of $[0, 1]$, which are described as follows.

First, fix some α with $0 < \alpha < 1$, and let $\langle \varepsilon_n : n \in \mathbb{N} \rangle$ be a sequence of positive reals with $\sum_{n=1}^{\infty} \varepsilon_n = 1 - \alpha$. (For example, we may take $\varepsilon_n = {}^{(1-\alpha)}/_{2^n}$ for all n.) Let $K_0 = [0, 1]$. For each $n \geq 0$, K_n consists of 2^n disjoint closed intervals, and we obtain K_{n+1} from K_n by removing an open interval of length ${}^1/_{(2^n \varepsilon_n)}$ from the center of each interval of F_n, so that the sum of the lengths of the intervals removed from K_n to obtain K_{n+1} is ε_n. In the end, let $K = \bigcap_{n \in \mathbb{N}} K_n$. The sum of the lengths of all the intervals removed from $[0, 1]$ in the formation of K is $\sum_{n=0}^{\infty} \varepsilon_n = 1 - \alpha$.

We claim that K is not Lebesgue null. In fact, we claim that any countable set of open intervals covering K has total length at least α.

To see this, suppose instead, aiming for a contradiction, that $\{I_n : n \in \mathbb{N}\}$ is a countable set of open intervals covering K with total length $L < \alpha$. Let $\mathcal{J}$ denote the set of intervals removed from $[0, 1]$ during the construction of K. But then, $\mathcal{J} \cup \{I_n : n \in \mathbb{N}\}$ is a countable set of open intervals covering $[0, 1]$, with total length $L + 1 - \alpha < 1$. This contradicts Theorem 6.40.

Example 6.43. The set of irrational numbers $\mathbb{R} \setminus \mathbb{Q}$ is not Lebesgue null. To see this, note that $\mathbb{Q}$ is a null set by Proposition 6.35, and the union of two null sets is null by Theorem 6.37. Thus, if $\mathbb{R} \setminus \mathbb{Q}$ were Lebesgue null, $(\mathbb{R} \setminus \mathbb{Q}) \cup \mathbb{Q} = \mathbb{R}$ would be, too. But then every subset of $\mathbb{R}$ would be null by Proposition 6.34, contradicting Theorem 6.40.

Lebesgue outer measure

Definition 6.44. Given any $A \subseteq \mathbb{R}$, the *Lebesgue outer measure* of A, denoted $\lambda^*(A)$, is defined as the infimum of all $L \geq 0$ such that A can be covered by a countable set of open intervals with total length at most L.

Roughly, the outer measure of a set A is a quantification of its size, as approximated by supersets of A consisting of open intervals.

Proposition 6.45. *A set $A \subseteq \mathbb{R}$ is Lebesgue null if and only if $\lambda^*(A) = 0$.*

The following observation generalizes Proposition 6.34:

Proposition 6.46. *Given $A, B \subseteq \mathbb{R}$, if $A \subseteq B$, then $\lambda^*(A) \leq \lambda^*(B)$.*

Proof. If $A \subseteq B$, then any set of intervals covering B also covers A. Hence, the infimum in the definition of $\lambda^*(A)$ is taken over a possibly larger set of numbers than the infimum in the definition of $\lambda^*(B)$. ☕

Theorem 6.47. *For any interval $I \subseteq \mathbb{R}$, $\lambda^*(I) = \lambda(I)$.*

Proof. It follows from Corollary 6.41 that $\lambda^*(I) \geq \lambda(I)$. For the reverse inequality, it suffices to show that for any $\varepsilon > 0$, there is a countable set of open intervals covering I with total length at most $\lambda(I) + \varepsilon$. But this is easy: If $a = \inf I$ and $b = \sup I$, then $\{(a - \frac{\varepsilon}{2}, b + \frac{\varepsilon}{2})\}$ works. □

Example 6.48. If K denotes the fat Cantor set of measure α from Example 6.42, then $\lambda^*(K) = \alpha$. In fact, we showed in Example 6.42 that no set of open intervals with total length less than α can cover K, and it follows immediately that $\lambda^*(K) \geq \alpha$.

For the reverse inequality, it suffices to show that for any $\varepsilon > 0$, there is a countable set of open intervals covering K with total length at most $\alpha + \varepsilon$. To do this, first fix some N large enough that $\sum_{n=1}^{N} \varepsilon_n > 1 - \alpha - \varepsilon$. (Here, we are using the terminology from Example 6.42.) The set K_N consists of 2^N closed intervals with total length $L = 1 - \sum_{n=1}^{N} \varepsilon_n < \alpha + \varepsilon$. Fattening up each of these intervals slightly, we can get 2^N open intervals with total length $\alpha + \varepsilon$ covering these 2^N closed intervals, thus also covering K. ☕

To end this section, we prove a theorem that demonstrates how the concept of Lebesgue outer measure can relate to Riemann integration.

Lemma 6.49. *For any $A \subseteq \mathbb{R}$, any countable set of (not necessarily open) intervals that covers A has total length at least $\lambda^*(A)$.*

Proof. We prove the lemma first for finite sets of intervals. Suppose, aiming for a contradiction, $\{I_n : n \leq N\}$ is a set of intervals covering A with total length $L < \lambda^*(A)$. Fix ε with $0 < \varepsilon < \lambda^*(A) - L$. For each $n \leq N$, let $a_n = \inf I_n$, $b_n = \sup I_n$, and $J_n = (a_n - \frac{\varepsilon}{2N}, b_n + \frac{\varepsilon}{2N})$. Then, $\{J_n : n \leq N\}$ is a finite set of intervals covering A with total length $L + \varepsilon < \lambda^*(A)$, contradicting the definition of $\lambda^*(A)$. The proof is similar for a countably infinite set of intervals: Just define $J_n = (a_n - \frac{\varepsilon}{2^{n+1}}, b_n + \frac{\varepsilon}{2^{n+1}})$. □

Theorem 6.50. *Suppose f and g are Riemann integrable functions $[a, b] \to \mathbb{R}$. If $f(x) < g(x)$ for all $x \in [a, b]$, then $\int_a^b f(x)\,dx < \int_a^b g(x)\,dx$.*

Proof. Let $h(x) = g(x) - f(x)$. By Theorem 6.19, h is Riemann integrable, and $\int_a^b h(x)\,dx = \int_a^b g(x)\,dx - \int_a^b f(x)\,dx$. So, to prove the theorem, it suffices to show $\int_a^b h(x)\,dx > 0$.

For each $n \in \mathbb{N}$, let $X_n = \{x \in [a,b]: h(x) > 1/n\}$. Because $h(x) > 0$ for all $x \in [a,b]$, we have $\bigcup_{n\in\mathbb{N}} X_n = [a,b]$. If each of the X_n were a Lebesgue null set, then their union $[a,b]$ would also be a null set by Theorem 6.37. But, by Theorem 6.40, $[a,b]$ is not null. Therefore, there is some $N \in \mathbb{N}$ such that X_N is not a null set, or in other words, $\lambda^*(X_N) > 0$.

Let $P = \langle x_0, x_1, \ldots, x_n \rangle$ be a partition of $[a,b]$. Let A denote the set of all those $i \leq n$ such that $[x_{i-1}, x_i] \cap X_N \neq \emptyset$, and let $B = \{1, 2, \ldots, n\} \setminus A$.

For any $i \in A$, there is some $t_i \in [x_{i-1}, x_i] \cap X_N$. Thus,

$$M_i = \sup\{h(t): t \in [x_{i-1}, x_i]\} \geq h(t_i) > 1/N.$$

Furthermore, because $\{[x_{i-1}, x_i]: i \in A\}$ is a set of intervals covering X_N, Lemma 6.49 implies $\sum_{i\in A} \Delta x_i \geq \lambda^*(X_N)$. Hence,

$$U(P,f) = \sum_{i=1}^{n} M_i \Delta x_i \geq \sum_{i\in A} \frac{1}{N} \Delta x_i = \frac{1}{N} \sum_{i\in A} \Delta x_i \geq \frac{\lambda^*(X_N)}{N}.$$

As P was arbitrary, this shows $\int_a^b h(x)\,dx = \overline{\int_a^b} h(x)\,dx \geq \frac{\lambda^*(X_N)}{N} > 0$. □

The Lebesgue–Vitali theorem

The following theorem provides an exact characterization of the Riemann integrable functions on an interval $[a,b]$. It was first proved independently by Henri Lebesgue and Giuseppe Vitali in 1907.

Theorem 6.51 (The Lebesgue–Vitali Theorem). *Let $[a,b]$ be an interval in $\mathbb{R}$, and let f be a bounded function $[a,b] \to \mathbb{R}$. Then, f is Riemann integrable on $[a,b]$ if and only if f is continuous almost everywhere on $[a,b]$.*

Proof. Fix a bounded function $f: [a,b] \to \mathbb{R}$. Let

$$D = \{x \in [a,b]: f \text{ is discontinuous at } x\}.$$

For each $n \in \mathbb{N}$, define $D_{1/n} = \{x \in [a,b]: \omega_f(x) \geq 1/n\}$. Recalling from Chap. 4 that f is continuous at x if and only if $\omega_f(x) = 0$, we have

$$D = \{x \in [a,b]: \omega_f(x) > 0\} = \bigcup_{n\in\mathbb{N}} D_{1/n}.$$

We aim to prove f is Riemann integrable if and only if D is a null set.

For the "only if" direction of the proof, let us show the contrapositive: If D is not a null set, then f is not Riemann integrable.

So, suppose D is not a null set. As $D = \bigcup_{n\in\mathbb{N}} D_{1/n}$, there is some $n \in \mathbb{N}$ such that $D_{1/n}$ is not null since otherwise D would be null by Theorem 6.37. Fix some $N \in \mathbb{N}$ such that $D_{1/N}$ is not a null set, i.e., such that $\lambda^*(D_{1/N}) > 0$.

Let $P = \langle x_0, x_1, \ldots, x_n \rangle$ be a partition of $[a, b]$. Let A denote the set of all those $i \leq n$ such that $(x_{i-1}, x_i) \cap D_{1/N} \neq \emptyset$, and let $B = \{1, 2, \ldots, n\} \setminus A$.

For any $i \in A$, there is some $t_i \in (x_{i-1}, x_i) \cap D_{1/N}$. Thus,

$$M_i - m_i = \omega_f([x_{i-1}, x_i]) \geq \omega_f((x_{i-1}, x_i)) \geq \omega_f(t_i) \geq 1/N.$$

Furthermore, note that $\{[x_{i-1}, x_i] : i \in A\} \cup \{\{x_i\} : i \leq n\}$ is a set of intervals covering $D_{1/N}$, with total length $\sum_{i\in A} \Delta x_i$. By Lemma 6.49, this implies $\sum_{i\in A} \Delta x_i \geq \lambda^*(D_{1/N})$. Hence,

$$\begin{aligned} U(P, f) - L(P, f) &= \sum_{i=1}^{n} M_i \Delta x_i - \sum_{i=1}^{n} m_i \Delta x_i \geq \sum_{i\in A} (M_i - m_i) \Delta x_i \\ &\geq \sum_{i\in A} \frac{1}{N} \Delta x_i = \frac{1}{N} \sum_{i\in A} \Delta x_i \geq \frac{\lambda^*(D_{1/N})}{N}. \end{aligned}$$

As P was an arbitrary partition of $[a, b]$, this shows that f does not satisfy the Cauchy criterion. Therefore, f is not Riemann integrable.

For the "if" direction, suppose that D is a null set. By Proposition 6.34, it follows that $D_{1/n}$ is a null set for every $n \in \mathbb{N}$.

Because f is bounded, $\sup\{f(t) : t \in [a, b]\}$ and $\inf\{f(t) : t \in [a, b]\}$, both exist. Let $H = \sup\{f(t) : t \in [a, b]\} - \inf\{f(t) : t \in [a, b]\} + 1$.

Let $\varepsilon > 0$. Pick N large enough that $1/N \leq \varepsilon/2(b-a)$. Because $D_{1/N}$ is Lebesgue null, there is a countable set $\mathcal{I} = \{I_n : n \in \mathbb{N}\}$ of open intervals covering $D_{1/N}$ with total length less than $\varepsilon/2H$. Recall from Chap. 4 that for any $c > 0$, $\{x \in [a, b] : \omega_f(x) \geq c\}$ is closed. In particular, $D_{1/N}$ is closed, and therefore compact. It follows that $\mathcal{I}$ has a finite subcover, i.e., there is some $m \in \mathbb{N}$ such that $\{I_1, I_2, \ldots, I_m\}$ covers $D_{1/N}$. And of course the total length of $\{I_1, I_2, \ldots, I_m\}$ is at most the total length of $\mathcal{I}$, less than $\varepsilon/2H$.

Without loss of generality, we may (and do) assume that the intervals $I_1, I_2, \ldots, I_m$, have disjoint closures. (To see why we may assume this, suppose instead that some of the intervals have overlapping closures. Pick some such I_j and I_k, and note that we may combine them into a single interval $I = \mathrm{Int}(\overline{I_j \cup I_k})$ with $\lambda(I) \leq \lambda(I_j) + \lambda(I_k)$. We may then delete I_j and I_k from our set of intervals, and add I in instead. Repeating this process finitely many times results in a set of [possibly fewer] open intervals, with the same total length, any two of which have disjoint closures.)

The set $[a, b] \setminus \bigcup_{j\leq m} I_j$ is closed and bounded, hence compact. Furthermore, $\omega_f(x) \leq 1/N < \varepsilon/2(b-a)$ for every point x in this set. By Theorem 4.25 from Chap. 4, this implies there is some $\delta > 0$ such that if $x, y \in [a, b] \setminus \bigcup_{j \le m} I_j$ and $|x - y| < \delta$, then $|f(x) - f(y)| < \varepsilon/2(b-a)$.

Let $P = \langle x_0, x_1, \ldots, x_n \rangle$ be a partition of $[a, b]$ such that

(1) $\Delta x_i < \delta$ for all $i \le n$, and

(2) For any $j \le m$, if the endpoints of I_j are in $[a, b]$, then they are in P.

We claim that $U(P,f) - L(P,f) < \varepsilon$.

To see this, divide the members of P into two groups: Let A denote the set of all those $i \le n$ such that there is some $j \le m$ with x_{i-1} and x_i both in $\overline{I_j}$, and let $B = \{1, 2, \ldots, n\} \setminus A$.

If $i \in B$, then $[x_{i-1}, x_i]$ is disjoint from all the I_j. (This follows from the definition of A, plus the fact that the closures of the I_j are nonoverlapping.) So, if $s, t \in [x_{i-1}, x_i]$, then $s, t \in [a, b] \setminus \bigcup_{j \le m} I_j$, and $|s - t| < \delta$ (by our choice of P). Therefore, $|f(s) - f(t)| < \varepsilon/2(b-a)$ (by our choice of δ). Hence,

$$M_i - m_i = \sup\{f(s) - f(t) : s, t \in [x_{i-1}, x_i]\} \le \varepsilon/2(b-a).$$

If $i \in A$, then we cannot put any bounds on $M_i - m_i$ except the trivial bound $M_i - m_i \le H$. However, observe that if $i \in A$, then there is some $j \le m$ such that $[x_{i-1}, x_i] \subseteq I_j$. Hence, $\sum_{i \in A} \Delta x_i \le \sum_{j \le m} \lambda(I_j) < \frac{\varepsilon}{2H}$.

Putting these pieces together, we get

$$\begin{aligned}
U(P,f) - L(P,f) &= \sum_{i=1}^{n} M_i \Delta x_i - \sum_{i=1}^{n} m_i \Delta x_i = \sum_{i=1}^{n} (M_i - m_i)\Delta x_i \\
&= \sum_{i \in A} (M_i - m_i)\Delta x_i + \sum_{i \in B} (M_i - m_i)\Delta x_i \\
&\le \sum_{i \in A} H \Delta x_i + \sum_{i \in B} \frac{\varepsilon}{2(b-a)} \Delta x_i \\
&= H \sum_{i \in A} \Delta x_i + \frac{\varepsilon}{2(b-a)} \sum_{i \in B} \Delta x_i \\
&\le H \frac{\varepsilon}{2H} + \frac{\varepsilon}{2(b-a)}(b-a) = \varepsilon.
\end{aligned}$$

Thus, f satisfies the Cauchy criterion, and is Riemann integrable. □

As an important corollary to this theorem, we see that the old motto, that "differentiation and integration are inverse operations," is almost half true. More specifically, the half of the motto that states differentiation undoes integration is almost true, in the following sense.

Corollary 6.52. *Let f be a Riemann integrable function $[a, b] \to \mathbb{R}$, and define $F(x) = \int_a^x f(t)\,dt$ for all $x \in [a, b]$. Then F is an antiderivative for f almost everywhere.*

Proof. Suppose f is a Riemann integrable function $[a, b] \to \mathbb{R}$. By the Lebesgue–Vitali theorem, f is continuous almost everywhere. And if f is continuous at x, the first fundamental theorem of calculus states that F is differentiable at x and $F'(x) = f(x)$. □

The other half of the motto, the half stating that integration undoes differentiation, is addressed in the following section. It turns out that this half of the motto cannot be rescued as easily.

Volterra's function

In this section, we show there is a differentiable function F, called *Volterra's function*, such that its derivative F' is bounded but not Riemann integrable. It is named for Vito Volterra, who discovered (a slightly different version of) this function in 1881. This function shows that it is overly simplistic to claim that Darboux–Riemann integration is an "inverse" of differentiation. This is true for certain classes of functions, but not generally true.

To start, recall the function

$$f(x) = \begin{cases} x^2 \sin\left(\frac{1}{x}\right) & \text{if } x \neq 0 \\ 0 & \text{if } x = 0 \end{cases}$$

from Chap. 5. This function is differentiable, with derivative

$$f'(x) = \begin{cases} 2x \sin\left(\frac{1}{x}\right) - \cos\left(\frac{1}{x}\right) & \text{if } x \neq 0 \\ 0 & \text{if } x = 0. \end{cases}$$

Note that f' is bounded on the interval $[0, 1]$ by some number $B \approx 1$. (A bound of 3 is easy to prove, though it is far from optimal.) Also note that f' has a discontinuity (of the second kind) at $x = 0$. More specifically, it is not too difficult to check that $\omega_{f'}(0) = 2$.

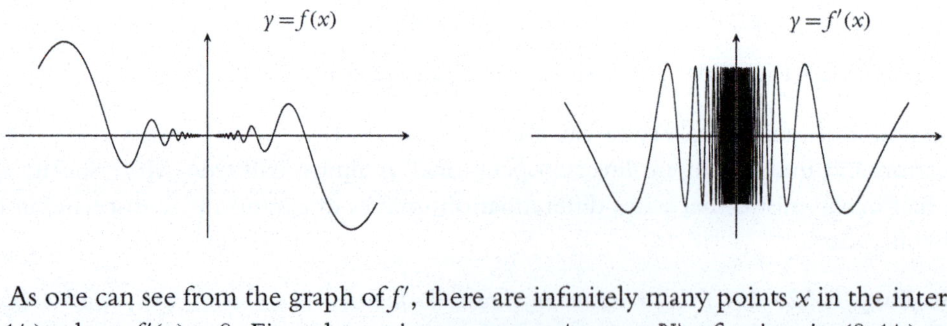

As one can see from the graph of f', there are infinitely many points x in the interval $(0, 1/2)$ where $f'(x) = 0$. Fix a decreasing sequence $\langle x_n \colon n \in \mathbb{N} \rangle$ of points in $(0, 1/2)$ such that $\lim_{n\to\infty} x_n = 0$ and $f'(x_n) = 0$ for all n. For each $n \in \mathbb{N}$, define a function $g^n_{[0,1]} : \mathbb{R} \to \mathbb{R}$ by

$$g^n_{[0,1]}(x) = \begin{cases} f(x) & \text{if } 0 \le x \le x_n, \\ f(x_n) & \text{if } x_n \le x \le 1 - x_n, \\ f(1-x) & \text{if } 1 - x_n \le x \le 1, \\ 0 & \text{if } x \notin [0,1]. \end{cases}$$

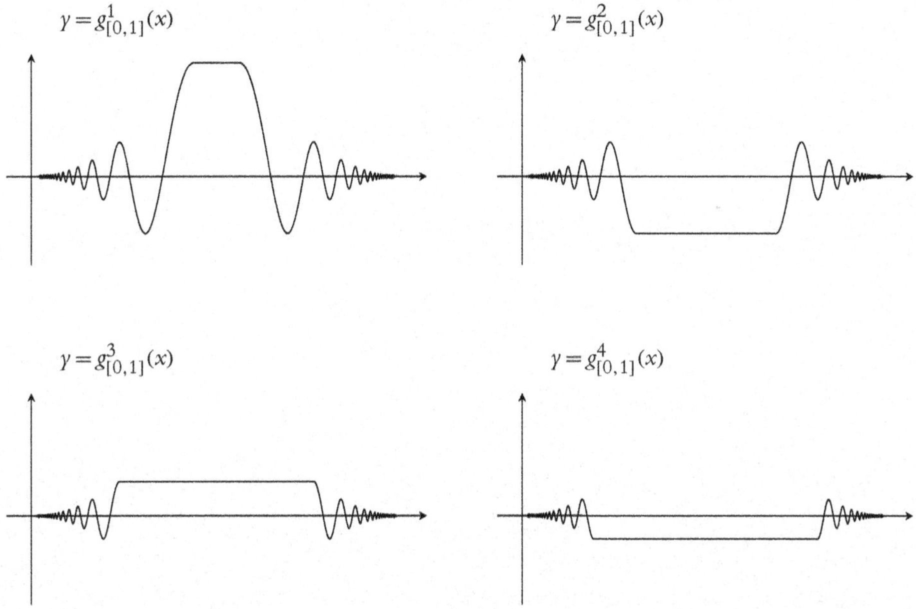

Note that, by our choice of the x_n, these functions are all differentiable:

$$(g^n_{[0,1]})'(x) = \begin{cases} f'(x) & \text{if } 0 \le x \le x_n, \\ 0 & \text{if } x_n \le x \le 1 - x_n, \\ -f'(1-x) & \text{if } 1 - x_n \le x \le 1, \\ 0 & \text{if } x \notin [0,1]. \end{cases}$$

The derivative $(g^n_{[0,1]})'(x)$ is bounded by B (just like f'), and it has discontinuities at 0 and 1. More precisely, $\omega_{(g^n_{[0,1]})'}(0) = \omega_{(g^n_{[0,1]})'}(1) = 2$, and if $x \neq 0, 1$ then $(g^n_{[0,1]})'$ is continuous at x.

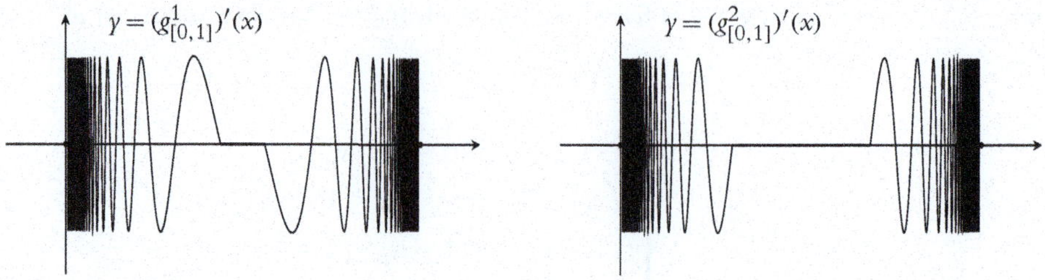

For each $n \in \mathbb{N}$, there is some slope $m_n > 0$ such that

$$\left| \frac{g^n_{[0,1]}(y) - g^n_{[0,1]}(x)}{y - x} \right| < m_n$$

for any $x \in [0, 1]$ and $y \notin [0, 1]$. In other words, the slope of the secant line connecting $(x, g^n_{[0,1]}(x))$ and $(y, g^n_{[0,1]}(y))$ is between $-m_n$ and m_n.

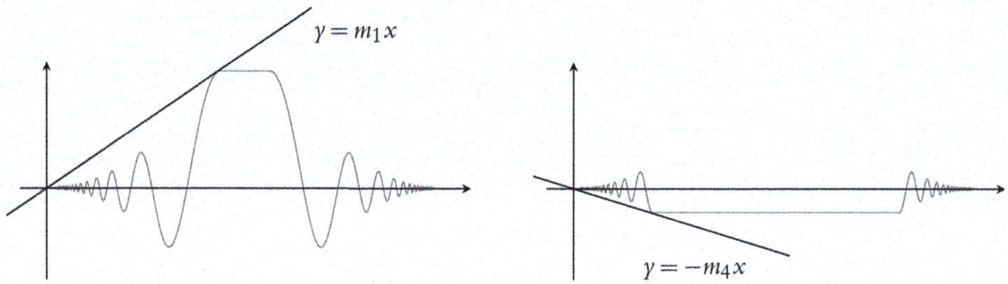

In fact, there is some $y_n \in (0, x_n]$ such that $m_n = \left|\frac{g^n_{[0.1]}(y_n)}{y_n}\right| = \left|\frac{g^1_{[0.1]}(y_n)}{y_n}\right|$. But $\lim_{n\to\infty} y_n = 0$ (because $\lim_{n\to\infty} x_n = 0$), and it follows that

$$\lim_{n\to\infty} m_n = \lim_{n\to\infty} \frac{g^1_{[0.1]}(y_n)}{y_n} = \lim_{t\to 0} \frac{g^1_{[0.1]}(t)}{t} = (g^1_{[0.1]})'(0) = 0.$$

For every $\ell, r \in \mathbb{R}$ with $\ell < r$, and for every $n \in \mathbb{N}$, define a function $g^n_{[\ell,r]} : \mathbb{R} \to \mathbb{R}$ by rescaling and shifting $g^n_{[0,1]}$ as follows:

$$g^n_{[\ell,r]}(x) = (r - \ell)\, g^n_{[0,1]}\left(\tfrac{x-\ell}{r-\ell}\right).$$

In other words, the graph of $g^n_{[\ell,r]}$ looks just like the graph of $g^n_{[0,1]}$, but scaled and shifted so that the nonzero part of the function sits inside $[\ell, r]$. This function $g^n_{[\ell,r]}$ is a differentiable function $\mathbb{R} \to \mathbb{R}$, and (applying the chain rule) $(g^n_{[\ell,r]})'(x) = (g^n_{[0,1]})'\left(\frac{x-\ell}{r-\ell}\right)$. Note that the scaling factor of $r - \ell$ disappears in the derivative $(g^n_{[\ell,r]})'$. In particular, $(g^n_{[\ell,r]})'$ is bounded by B, and $\omega_{(g^n_{[\ell,r]})'}(\ell) = \omega_{(g^n_{[\ell,r]})'}(r) = 2$.

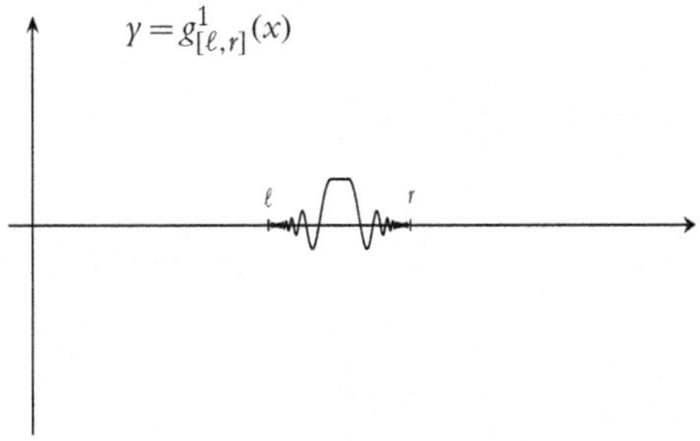

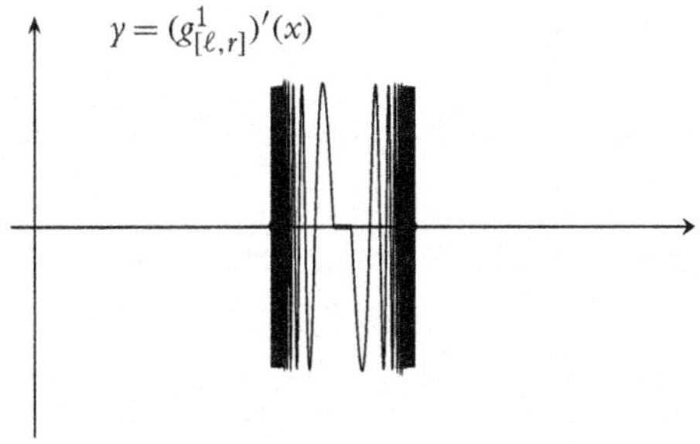

Note that the defining property of m_n is not changed by scaling or translation: Thus, for each $n \in \mathbb{N}$ and each $\ell, r \in \mathbb{R}$ with $\ell < r$,

$$\left| \frac{g^n_{[\ell,r]}(y) - g^n_{[\ell,r]}(x)}{y - x} \right| < m_n$$

for any $x \in [\ell, r]$ and $y \notin [\ell, r]$. In other words, the slope of the secant line connecting $(x, g^n_{[\ell,r]}(x))$ and $(y, g^n_{[\ell,r]}(y))$ is between $-m_n$ and m_n, just as with $g^n_{[0,1]}$.

To define our function F, let K denote the "fat Cantor set" of measure 1/2, as described in Example 6.42. Recall that $\lambda^*(K) = 1/2$ by Example 6.48.

The Volterra function $F : [0, 1] \to \mathbb{R}$ is defined as follows:

$$F(x) = \begin{cases} g^n_{[\ell,r]}(x) & \text{if } x \in (\ell, r), \text{ and } (\ell, r) \text{ is one of the } 2^{n-1} \text{ open} \\ & \quad \text{intervals removed from } K_{n-1} \text{ to obtain } K_n. \\ 0 & \text{if } x \in K. \end{cases}$$

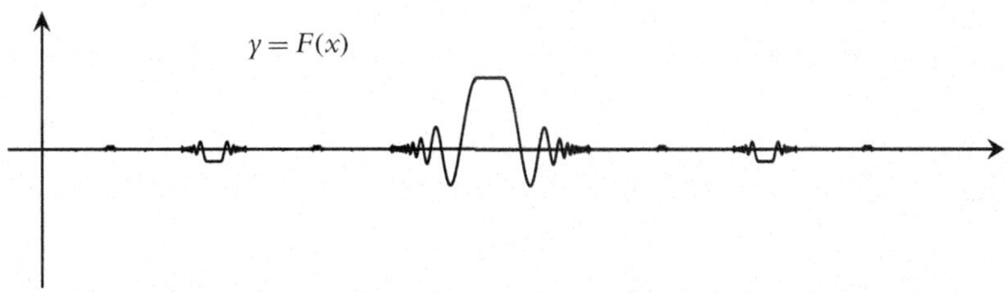

We claim that this function is differentiable, and

$$F'(x) = \begin{cases} (g^n_{[\ell,r]})'(x) & \text{if } x \in (\ell, r), \text{ and } (\ell, r) \text{ is one of the } 2^{n-1} \text{ open} \\ & \quad \text{intervals removed from } K_{n-1} \text{ to obtained } K_n. \\ 0 & \text{if } x \in K. \end{cases}$$

Indeed, it follows from the definition of F that $F'(x) = (g_{[\ell,r]})'(x)$ whenever $x \in (\ell, r)$ and (ℓ, r) is one of the 2^{n-1} open intervals removed from K_{n-1} to obtained K_n. We must show $F'(x) = 0$ whenever $x \in K$.

Fix $x \in K$. If $x = \ell$ for one of the intervals (ℓ, r) removed at stage n of the construction of K, then $\lim_{t\to x^+} \frac{F(t)-F(x)}{t-x} = \lim_{t\to \ell^+} \frac{g^n_{[\ell,r]}(t)-g^n_{[\ell,r]}(x)}{t-x} = (g^n_{[\ell,r]})'(\ell) = 0$. If not (i.e., if x is not the left-hand endpoint of any such interval), then for any $N \in \mathbb{N}$, there is some $\delta > 0$ such that the interval $(x, x+\delta)$ does not intersect any interval (ℓ, r) removed before stage N of the construction of K. It follows that if $t \in (x, \delta)$, then $\left|\frac{F(t)-F(x)}{t-x}\right| \leq m_n$ for some $n \geq N$. Because this holds for any N and because $\lim_{n\to\infty} m_n = 0$, this implies $\lim_{t\to x^+} \frac{F(t)-F(x)}{t-x} = 0$.

In either case, we see that $\lim_{t\to x^+} \frac{F(t)-F(x)}{t-x} = 0$ for any $x \in K$. A similar argument shows that $\lim_{t\to x^-} \frac{F(t)-F(x)}{t-x} = 0$ for any $x \in K$. Hence, $F'(x) = \lim_{t\to x} \frac{F(t)-F(x)}{t-x} = 0$ for any $x \in K$.

Thus, F is differentiable, and F' is as just described. Note that F' is bounded below by -1 and above by 1 because each of the functions $g^n_{[\ell,r]}$ is. It remains to show that F' is not Riemann integrable.

If (ℓ, r) is one of the 2^{n-1} open intervals removed from K_{n-1} to obtained K_n, then $F'(x) = (g_{[\ell,r]})'(x)$ for all $x \in (\ell, r)$, and therefore $\omega_{F'}(\ell) = \omega_F(r) = 2$. We claim F' is discontinuous at every point of K.

To see this, fix $x \in K$ and let $\delta > 0$. There is some $N \in \mathbb{N}$ such that each of the 2^N intervals comprising K_N has length $<\delta$. Of course, x is in one of these intervals, say I. Letting (ℓ, r) denote the open interval removed from I to obtain K_{N+1}, we have $(\ell, r) \subseteq (x-\delta, x+\delta)$. But $\omega_{F'}(\ell) = \omega_{F'}(r) = 2$ by the previous paragraph, and so $\omega_{F'}(B_\delta(x)) \geq 2$. It follows that $\omega_{F'}(x) \geq 2$. In particular, F' is not continuous at x.

Because $\lambda^*(K) = 1/2$, the set of all points where F' is discontinuous is not a Lebesgue null set. By the Lebesgue–Vitali theorem, this implies F' is not Riemann integrable. ☕

Exercises

1. Suppose f and g are both Riemann integrable functions $[a, b] \to \mathbb{R}$. Prove that if $f(x) \leq g(x)$ for all $x \in [a, b]$, then $\int_a^b f(x)\,dx \leq \int_a^b g(x)\,dx$.
2. Suppose $f : [a, b] \to \mathbb{R}$ is continuous, and $\int_a^b |f(x)|\,dx = 0$. Prove that $f(x) = 0$ for all $x \in [a, b]$.
3. Recall that Thomae's function, or the "stars over Babylon" function, $f : [0, 1] \to \mathbb{R}$ is defined as follows:

$$f(x) = \begin{cases} 1/q & \text{if } x = p/q \text{ where } q > 0 \text{ and } p \text{ and } q \text{ are co-prime,} \\ 0 & \text{otherwise.} \end{cases}$$

Show from the definition of Riemann integrability (without using the Lebesgue–Vitali Theorem) that this function is Riemann integrable.

4. Let F denote a "fat Cantor set" of measure $\alpha > 0$, as described in Example 6.42. Show that the function

$$f(x) = \begin{cases} 1 & \text{if } x \in F \\ 0 & \text{if } x \notin F \end{cases}$$

is not Riemann integrable.

5. Find an example of a function $f : [a, b] \to \mathbb{R}$ such that $f(x)$ is not Riemann integrable, but the function $g(x) = (f(x))^2$ is Riemann integrable. On the other hand, show that if $(f(x))^3$ is Riemann integrable, then $f(x)$ is, too.

6. Suppose $f : [a, b] \to \mathbb{R}$ is Riemann integrable, and $g : [a, b] \to \mathbb{R}$ is equal to f at all but finitely many points. Prove that g is Riemann integrable, and $\int_a^b g(x)\,dx = \int_a^b f(x)\,dx$.

7. Prove the following statement, which describes the technique known as *integration by parts*: Suppose f and g are Riemann integrable functions $[a, b] \to \mathbb{R}$ and that F and G are antiderivatives for f and g, respectively. Then,

$$\int_a^b F(x)g(x)\,dx = F(b)G(b) - F(a)G(a) - \int_a^b f(x)G(x).$$

8. Prove the following statement, which describes the technique known as *u-substitution*: Suppose ϕ is a differentiable function $[a, b] \to \mathbb{R}$ and that ϕ' is Riemann integrable. If $f : \mathbb{R} \to \mathbb{R}$ is continuous, then

$$\int_a^b f(\phi(x)) \cdot \phi'(x)\,dx = \int_{\phi(a)}^{\phi(b)} f(u)\,du.$$

9. Suppose f is a function $[a, b] \to \mathbb{R}$, and is Riemann integrable on $[c, b]$ for all $c \in (a, b)$. Define $\int_a^b f(x)\,dx = \lim_{c \to a^+} \int_c^b f(x)\,dx$ if this limit exists. (This is called the *improper* integral of f from a to b.) Show that if f is Riemann integrable on $[a, b]$, then this definition does not conflict with the usual definition of $\int_a^b f(x)\,dx$. (That is, the improper integral equals the regular integral whenever both are well-defined.)

10. Suppose f is a function $[a, \infty) \to \mathbb{R}$ and is Riemann integrable on $[a, b]$ for all $b > a$. Define $\int_a^\infty f(x)\,dx = \lim_{b \to \infty} \int_a^b f(x)\,dx$ if this limit exists. Show that if $f : [1, \infty) \to \mathbb{R}$ is nonincreasing, then $\int_a^\infty f(x)\,dx$ exists and is finite if and only if $\sum_{n=1}^\infty f(n)$ converges.

11. Suppose f is a function $[a, \infty) \to \mathbb{R}$ and is Riemann integrable on $[a, b]$ for all $b > a$. Suppose further that there is a strictly positive $g : [a, \infty) \to \mathbb{R}$ such that $\int_a^\infty g(x)\,dx$ exists and is finite, and such that $|f(x)| \le g(x)$ for all $x \in [a, \infty)$. Prove that $\int_a^\infty f(x)\,dx$ exists and $\int_a^\infty f(x)\,dx \le \int_a^\infty |g(x)|\,dx$.

12. Suppose f is a Riemann integrable function on $[a, b]$. Let B denote the image of $[a, b]$ under f, and suppose g is a continuous, bounded function $B \to \mathbb{R}$. Prove that $g \circ f$ is Riemann integrable. (Note that we are not assuming g is defined on an interval containing B, which means that Theorem 6.21 does not apply.)

13. If the following statement is true prove it, and if it is false, then provide a counterexample: If two functions $f, g : [a, b] \to \mathbb{R}$ are equal almost everywhere, and if f is Riemann integrable, then g must be too.

14. Suppose $\mathcal{I}$ is an uncountable collection of open intervals in $\mathbb{R}$. Prove that there is a countable subset of $\mathcal{I}$ with infinite total length.

15. Prove that if $\mathcal{I}$ is a finite set of open intervals covering $[a, b] \cap \mathbb{Q}$, then the total length of $\mathcal{I}$ is at least $b - a$. Thus, for any $\varepsilon > 0$, we can cover $[a, b] \cap \mathbb{Q}$ with countably many intervals of total length $< \varepsilon$, but we cannot do it with finitely many intervals.

16. Show that if two continuous functions $\mathbb{R} \to \mathbb{R}$ are equal almost everywhere, then they are equal.

17. Prove that if two Riemann integrable functions $[a, b] \to \mathbb{R}$ are equal almost everywhere, then their integrals are equal.

18. Give an example of a function that is Riemann integrable but has uncountably many points of discontinuity.

19. Let $\varepsilon > 0$. Give an example of a set $A \subseteq \mathbb{R}$ such that $\lambda^*(A \cap I) > 0$ for every non-empty open interval I, but also $\lambda^*(A) < \varepsilon$.

20. Give an example of a Lebesgue null set N and a meager set M such that $M \cup N = \mathbb{R}$.

21. Give an example of a nonconstant, continuous function $f : [0, 1] \to \mathbb{R}$ such that the constant function $g(x) = 0$ is an almost everywhere antiderivative for f.

22. In our definition of Volterra's function F, we set $F(x) = g^n_{[\ell,r]}(x)$ whenever $x \in (\ell, r)$ and (ℓ, r) is one of the 2^{n-1} open intervals removed from K_{n-1} to obtain K_n. One might wonder whether it is necessary to use $g^n_{[\ell,r]}(x)$ here, rather than just $g^1_{[0,1]}$ every time. Show that the modified Volterra function

$$F(x) = \begin{cases} g^1_{[\ell,r]}(x) & \text{if } x \in (\ell, r) \text{, and } (\ell, r) \text{ is one of the intervals} \\ & \quad \text{removed at some point in constructing of } K. \\ 0 & \text{if } x \in K \end{cases}$$

is not differentiable at every point of $[0, 1]$.

CHAPTER 7

Subsets of $\mathbb{R}$

In this chapter, we look at three important classes of subsets of $\mathbb{R}$: the Borel sets, the sets with the property of Baire, and the Lebesgue measurable sets. Each of these classes consists of a somehow "well-behaved" collection of sets. Our goal is simply to describe these three classes of sets and to prove a few of their many interesting features. We also will show how the Axiom of Choice can be used to obtain a subset of $\mathbb{R}$ that is not in any of our three well-behaved classes of sets: a set that is not Borel, not Lebesgue measurable, and does not have the property of Baire.

σ-algebras and Borel sets

We begin with the abstract notion of a σ-algebra, which gives three basic features that a "well-behaved" collection of subsets of some set X should possess.

Definition 7.1. Let X be a non-empty set. A collection $\mathcal{A}$ of subsets of X is a *σ-algebra on X* if

1. $\emptyset \in \mathcal{A}$ and $X \in \mathcal{A}$,
2. if $A \in \mathcal{A}$, then $X \setminus A \in \mathcal{A}$, and
3. if $\mathcal{F}$ is a countable subset of $\mathcal{A}$, then $\bigcup \mathcal{F} \in \mathcal{A}$.

Properties (2) and (3) are summarized by saying that $\mathcal{A}$ is *closed under complementation* and *closed under countable unions*, respectively. ☕

Example 7.2. If X is a non-empty set, then $\mathcal{A} = \mathcal{P}(X)$ (the power set of X) is a σ-algebra on X. This is the largest σ-algebra on X, in the sense that if $\mathcal{B}$ is any other σ-algebra on X, then $\mathcal{B} \subseteq \mathcal{A}$.

$\mathcal{A} = \{\emptyset, X\}$ is also a σ-algebra on X. This is the smallest σ-algebra on X in the sense that if $\mathcal{B}$ is any other σ-algebra on X, then $\mathcal{B} \supseteq \mathcal{A}$. ☕

Example 7.3. The open subsets of $\mathbb{R}$ do not form a σ-algebra on $\mathbb{R}$. This class of sets contains $\emptyset$ and $\mathbb{R}$ and is closed under countable unions, but it is not closed under complementation.

Let $\mathcal{A} = \{A \subseteq \mathbb{R}:\ A \text{ is finite}\} \cup \{A \subseteq \mathbb{R}:\ \mathbb{R} \setminus A \text{ is finite}\}$. This is also not a σ-algebra on $\mathbb{R}$. It contains $\emptyset$ and $\mathbb{R}$ and is closed under complementation, but it is not closed under countable unions. ☕

In addition to being closed under complementation and countable unions, σ-algebras are closed under several other operations as well.

Foundations of Real Analysis
https://doi.org/10.1016/B978-0-44-348527-5.00011-1

Proposition 7.4. *Let X be a set and let $\mathcal{A}$ be a σ-algebra on X.*

1. *$\mathcal{A}$ is closed under finite unions, i.e., if $A_1, A_2, \ldots, A_n \in \mathcal{A}$, then $\bigcup_{i \le n} A_i \in \mathcal{A}$.*
2. *$\mathcal{A}$ is closed under finite intersections, i.e., if $A_1, A_2, \ldots, A_n \in \mathcal{A}$, then $\bigcap_{i \le n} A_i \in \mathcal{A}$.*
3. *$\mathcal{A}$ is closed under countable intersections, i.e., if $\mathcal{F}$ is a countable subset of $\mathcal{A}$, then $\bigcap \mathcal{F} \in \mathcal{A}$.*
4. *$\mathcal{A}$ is closed under relative complements, i.e., if $A, B \in \mathcal{A}$, then $A \setminus B \in \mathcal{A}$.*

Proof. Part (1) of the proposition follows immediately from part (3) of Definition 7.1: If $A_1, A_2, \ldots, A_n \in \mathcal{A}$, then $\{A_1, A_2, \ldots, A_n\}$ is a countable subset of $\mathcal{A}$, and hence $\bigcup\{A_1, A_2, \ldots, A_n\} \in \mathcal{A}$.

For part (3) of the proposition, suppose $\mathcal{F}$ is a countable subset of $\mathcal{A}$. Observe that

$$\bigcap \mathcal{F} = X \setminus \bigcup \{X \setminus A \colon A \in \mathcal{F}\}.$$

(This equality is known as De Morgan's law.) Because $\mathcal{A}$ is a σ-algebra, $X \setminus A \in \mathcal{F}$ for every $A \in \mathcal{F}$. Therefore, $\{X \setminus A \colon A \in \mathcal{F}\}$ is a countable subset of $\mathcal{A}$, which means its union $\bigcup\{X \setminus A \colon A \in \mathcal{F}\}$ is in $\mathcal{A}$, which means the complement of this union $\bigcap \mathcal{F} = X \setminus \bigcup\{X \setminus A \colon A \in \mathcal{F}\}$ is in $\mathcal{A}$.

In the same way that part (1) of the proposition follows from part (3) of Definition 7.1, part (2) of the proposition follows from part (3).

For part (4), suppose $A, B \in \mathcal{A}$. Then, $X \setminus B \in \mathcal{A}$ and, by part (2) of the proposition, $A \setminus B = A \cap (X \setminus B) \in \mathcal{A}$. □

Proposition 7.5. *Let $\mathcal{C}$ be a collection of subsets of a set X. Then, there is a σ-algebra on X, denoted $\langle\langle \mathcal{C} \rangle\rangle$, such that $\mathcal{C} \subseteq \langle\langle \mathcal{C} \rangle\rangle$, and if $\mathcal{A}$ is any σ-algebra on X with $\mathcal{C} \subseteq \mathcal{A}$, then $\langle\langle \mathcal{C} \rangle\rangle \subseteq \mathcal{A}$. In other words, $\langle\langle \mathcal{C} \rangle\rangle$ is the smallest σ-algebra on X that contains $\mathcal{C}$.*

Proof. Consider the collection of all σ-algebras on X that contain $\mathcal{C}$. Note that this collection is non-empty because there is at least one σ-algebra on X containing $\mathcal{C}$, namely $\mathcal{P}(X)$ (the power set of X), which is the largest σ-algebra on X. Let

$$\langle\langle \mathcal{C} \rangle\rangle = \bigcap \{\mathcal{A} \colon \mathcal{A} \text{ is a } \sigma\text{-algebra on } X \text{ containing } \mathcal{C}\}.$$

To finish the proof, we must show that this is a σ-algebra, that it contains $\mathcal{C}$, and that it is the smallest σ-algebra on X containing $\mathcal{C}$. For convenience, let $\Sigma = \{\mathcal{A} \colon \mathcal{A} \text{ is a } \sigma\text{-algebra on } X \text{ containing } \mathcal{C}\}$, so that $\langle\langle \mathcal{C} \rangle\rangle = \bigcap \Sigma$.

To see that $\langle\langle \mathcal{C} \rangle\rangle$ is a σ-algebra on X, first note $\emptyset, X \in \mathcal{A}$ for every $\mathcal{A} \in \Sigma$, and this implies $\emptyset, X \in \langle\langle \mathcal{C} \rangle\rangle$. Secondly, suppose $A \in \langle\langle \mathcal{C} \rangle\rangle$. Then, $A \in \mathcal{A}$ for all $\mathcal{A} \in \Sigma$ and since each $\mathcal{A} \in \Sigma$ is a σ-algebra on X, this implies $X \setminus A \in \mathcal{A}$ for all $\mathcal{A} \in \Sigma$. Consequently, $X \setminus A \in \langle\langle \mathcal{C} \rangle\rangle$. Finally, suppose $\mathcal{F}$ is a countable subset of $\langle\langle \mathcal{C} \rangle\rangle$. But then, $\mathcal{F}$ is a countable subset of $\mathcal{A}$ for every $\mathcal{A} \in \Sigma$ and, since each $\mathcal{A} \in \Sigma$ is a σ-algebra, this implies $\bigcup \mathcal{F} \in \mathcal{A}$ for all $\mathcal{A} \in \Sigma$. Consequently, $\bigcup \mathcal{F} \in \langle\langle \mathcal{C} \rangle\rangle$.

Hence $\langle\langle \mathcal{C} \rangle\rangle$ is a σ-algebra. If $\mathcal{A}$ is any σ-algebra on X with $\mathcal{C} \subseteq \mathcal{A}$, then $\mathcal{A} \in \Sigma$, and therefore $\mathcal{A} \supseteq \bigcap \Sigma = \langle\langle \mathcal{C} \rangle\rangle$. □

Definition 7.6. Let X be a set and let $\mathcal{C}$ be a collection of subsets of X. The *σ-algebra generated by $\mathcal{C}$* is the σ-algebra

$$\langle\langle \mathcal{C} \rangle\rangle = \bigcap \left\{ \mathcal{A} \colon \mathcal{A} \text{ is a } \sigma\text{-algebra on } X \text{ containing } \mathcal{C} \right\},$$

the least σ-algebra on X containing $\mathcal{C}$, meaning that if $\mathcal{A}$ is any σ-algebra on X with $\mathcal{C} \subseteq \mathcal{A}$, then $\langle\langle \mathcal{C} \rangle\rangle \subseteq \mathcal{A}$.

Example 7.7. In Example 7.3, we observed that the finite and cofinite subsets of $\mathbb{R}$ do not form a σ-algebra on $\mathbb{R}$. More generally, if X is infinite, then

$$\mathcal{C} = \{A \subseteq X \colon\ A \text{ is finite}\} \cup \{A \subseteq X \colon\ X \setminus A \text{ is finite}\}$$

is not a σ-algebra on X since it is not closed under countable unions. (Proving this is an end-of-chapter exercise.) The σ-algebra generated by $\mathcal{C}$ is

$$\langle\langle \mathcal{C} \rangle\rangle = \{A \subseteq X \colon\ A \text{ is countable}\} \cup \{A \subseteq X \colon\ X \setminus A \text{ is countable}\}.$$

Definition 7.8. The *Borel σ-algebra* on a metric space X is the σ-algebra generated by the collection of open subsets of X. A subset of X is called a *Borel set* if it is a member of this σ-algebra. Henceforth, let $\mathcal{B}$ denote the collection of Borel subsets of $\mathbb{R}$.

Example 7.9. Every open subset of $\mathbb{R}$ is a Borel set, by definition. Every closed set is also Borel because it is the complement of an open set.

Recall that a countable union of closed sets is called an F_σ set. Every F_σ subset of $\mathbb{R}$ is Borel because closed sets are Borel and $\mathcal{B}$ is closed under taking countable unions. This shows that $\mathcal{B}$ contains sets that are neither open nor closed. For example, $\mathbb{Q}$ is an F_σ set because $\mathbb{Q} = \bigcup \left\{ \{q\} \colon q \in \mathbb{Q} \right\}$. Hence, $\mathbb{Q}$ is a nonopen, nonclosed Borel set.

Recall that a countable intersection of open sets is called a G_δ set. Every G_δ subset of $\mathbb{R}$ is Borel because open sets are Borel and $\mathcal{B}$ is closed under taking countable intersections. For example, $\mathbb{R} \setminus \mathbb{Q}$ is an G_δ set because $\mathbb{R} \setminus \mathbb{Q} = \bigcap \left\{ \mathbb{R} \setminus \{q\} \colon q \in \mathbb{Q} \right\}$. Hence, $\mathbb{R} \setminus \mathbb{Q}$ is Borel.

We may go further. Countable intersections of F_σ sets are called $F_{\sigma\delta}$, and countable unions of G_δ sets are called $G_{\delta\sigma}$. Countable intersections of $G_{\delta\sigma}$ sets are called $G_{\delta\sigma\delta}$, and countable unions of $F_{\sigma\delta}$ sets are called $F_{\sigma\delta\sigma}$, etc. This notation becomes cumbersome rather quickly, but there is an alternative naming scheme for these classes of sets: They are called $\boldsymbol{\Sigma}_n^0$ or $\boldsymbol{\Pi}_n^0$, as shown in the following diagram.

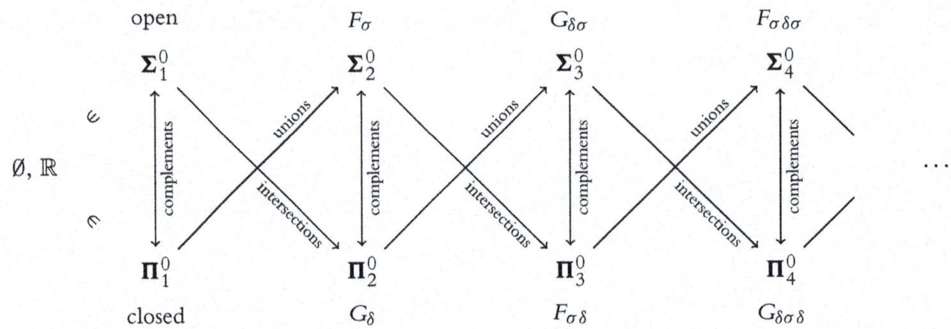

This diagram gives a glimpse of the first few stages of what is known as the *Borel hierarchy*. While we do not prove these facts here, it is worth pointing out that:

- If $\alpha < \beta$, then every $\mathbf{\Sigma}^0_\alpha$ or $\mathbf{\Pi}^0_\alpha$ set is both $\mathbf{\Sigma}^0_\beta$ and $\mathbf{\Pi}^0_\beta$. In other words, these set classes grow larger moving to the right in the diagram.
- If $\alpha < \beta$, then there are $\mathbf{\Sigma}^0_\beta$ sets and $\mathbf{\Pi}^0_\beta$ sets that are neither $\mathbf{\Sigma}^0_\alpha$ nor $\mathbf{\Pi}^0_\alpha$. In other words, these set classes grow strictly larger moving to the right in the diagram. At no point along the way do we stop getting new sets.

In fact, this diagram does not really give the full picture and shows only a small part of the complexity of the Borel sets. One can see this using the (unproved) facts just stated, as follows. Each interval of the form $(n, n+1)$ is like a small copy of $\mathbb{R}$, and on the interval $(n, n+1)$ there is a set A_n, i.e., $\mathbf{\Sigma}^0_n$ but not $\mathbf{\Sigma}^0_m$ for any $m < n$. But then, the set $A = \bigcup_{n\in\mathbb{N}} A_n$ is a Borel subset of $\mathbb{R}$ (a countable union of Borel sets), and for any $n \in \mathbb{N}$, it is not in the class $\mathbf{\Sigma}^0_n$ because the piece of A on the interval $(n, n+1)$ is not $\mathbf{\Sigma}^0_n$. Such a set A is in the class named $\mathbf{\Sigma}^0_\omega$. There are further classes beyond this one, reaching up through all the countable ordinals. Each new level of the uncountably long Borel hierarchy contains new sets not in any lower level. A fuller description of the Borel hierarchy, and proofs of the facts stated above, are the beginnings of the field known as *descriptive set theory*. For now, let us just say that the Borel sets form a very large class, including every subset of $\mathbb{R}$ discussed in this book up to now, and every subset of $\mathbb{R}$ whose definition can be easily written. The Borel sets are the "nicely definable" subsets of $\mathbb{R}$.

The property of Baire

Recall that the *symmetric difference* of two sets A and B, denoted $A\triangle B$, is the set of all things in A or B but not both as pictured:

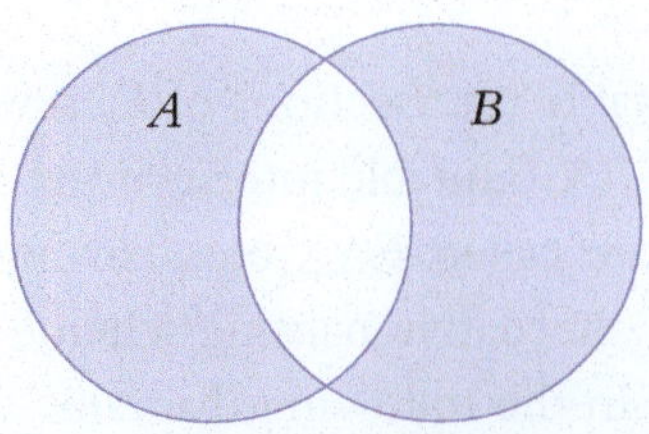

$$A \triangle B = (A \cup B) \setminus (A \cap B) = (A \setminus B) \cup (B \setminus A).$$

Some basic rules concerning the symmetric difference, which are left for proof in the end-of-chapter exercises, are:

- $A \triangle (A \triangle B) = B$,
- if $A, B \subseteq X$ then $(X \setminus A) \triangle (X \setminus B) = A \triangle B$,
- $A \triangle (B \triangle C) \subseteq (A \setminus B) \cup C$,
- $A \setminus (A \triangle B) \subseteq B$.

Definition 7.10. A subset A of a metric space X has the *property of Baire* if there is an open $U \subseteq X$ and a meager $M \subseteq X$ such that $A = U \triangle M$.

Lemma 7.11. *If X is a metric space and U is an open subset of X, then $\overline{U} \setminus U$ is a closed nowhere dense subset of X.*

Proof. Recall that $\overline{U} \setminus U$ being nowhere dense means that for every non-empty open $V \subseteq X$, there is some non-empty open $W \subseteq V$ such that $W \cap (\overline{U} \setminus U) = \emptyset$. Let V be a non-empty open subset of X. If $V \cap U = \emptyset$, then $X \setminus V$ is a closed set containing U, which means $\overline{U} \subseteq X \setminus V$, or equivalently, $V \cap \overline{U} = \emptyset$. Taking $W = V$, we have $W \cap (\overline{U} \setminus U) \subseteq W \cap \overline{U} = \emptyset$. If $V \cap U \neq \emptyset$, then let $W = V \cap U$. Then $W \subseteq U$, which implies $W \cap (\overline{U} \setminus U) = \emptyset$. □

Theorem 7.12. *If X is a metric space, then*

$$\mathcal{R} = \{A \subseteq X \colon A \text{ has the property of Baire}\}$$

is a σ-algebra on X.

Proof. First, note that $\emptyset$ and X are both open subsets of X, and $\emptyset$ is a meager subset of X. It follows that $X = X \triangle \emptyset$ and $\emptyset = \emptyset \triangle \emptyset$ both have the property of Baire.

Second, fix $A \in \mathcal{R}$. This means that $A = U \triangle M$ for some open $U \subseteq X$ and meager $M \subseteq X$. Let $V = X \setminus \overline{U}$ and $N = V \triangle (X \setminus A)$. Then,

$$V \triangle N = V \triangle (V \triangle (X \setminus A)) = X \setminus A.$$

(This equality uses the first of our three basic rules of the symmetric difference previously listed.) Clearly, V is an open set, so to show $X \setminus A = V \triangle N$ has the property of Baire, we need to show N is meager. To see this, observe that, applying the second and third of our three basic rules of the symmetric difference listed,

$$N = V \triangle (X \setminus A) = (X \setminus \overline{U}) \triangle \big(X \setminus (U \triangle M)\big) = \overline{U} \triangle (U \triangle M) \subseteq (\overline{U} \setminus U) \cup M.$$

The preceding lemma shows $\overline{U} \setminus U$ is nowhere dense, hence meager, and M is meager by assumption. Because countable unions of meager sets are meager, $(\overline{U} \setminus U) \cup M$ is meager, and because subsets of meager sets are meager, $N \subseteq (\overline{U} \setminus U) \cup M$ is meager. Hence, $X \setminus A \in \mathcal{R}$. As A was an arbitrary member of $\mathcal{R}$, this shows $\mathcal{R}$ is closed under complementation.

Finally, suppose $\mathcal{F} = \{A_n \colon n \in \mathbb{N}\}$ is a countable subset of $\mathcal{R}$, and let $A = \bigcup \mathcal{F}$. For each $n \in \mathbb{N}$, there is an open set U_n and a meager set M_n such that $A_n = U_n \Delta M_n$. Let $U = \bigcup_{n=1}^{\infty} U_n$, and let $M = U \Delta A$. Then, U is open and, by our first basic rule of symmetric differences,

$$U \Delta M = U \Delta (U \Delta A) = A.$$

To show A has the property of Baire, we need to show M is meager. If $x \in M = U \Delta A$, then either $x \in U \setminus A$ or $x \in A \setminus U$.

If $x \in U \setminus A = \left(\bigcup_{n=1}^{\infty} U_n\right) \setminus A$, then there is some $n \in \mathbb{N}$ such that $x \in U_n \setminus A = U_n \setminus \left(\bigcup_{i=1}^{\infty} A_i\right) \subseteq U_n \setminus A_n \subseteq M_n$. (This last inclusion uses our fourth basic rule about symmetric differences.)

If $x \in A \setminus U = \left(\bigcup_{n=1}^{\infty} A_n\right) \setminus U$, then there is some $n \in \mathbb{N}$ such that $x \in A_n \setminus U = A_n \setminus \left(\bigcup_{i=1}^{\infty} U_i\right) \subseteq A_n \setminus U_n \subseteq M_n$. (Again, his last inclusion uses our fourth basic rule about symmetric differences.)

In either case, we see that if $x \in M$, then $x \in \bigcup_{n=1}^{\infty} M_n$. Hence, $M \subseteq \bigcup_{n=1}^{\infty} M_n$. Because countable unions of meager sets are meager and subsets of meager sets are meager, this implies M is meager. Hence, $A = U \Delta M$ has the property of Baire, and this shows $\mathcal{R}$ is closed under countable unions. □

Corollary 7.13. *Every Borel set has the property of Baire.*

Proof. Let X be a metric space. Every open $U \subseteq X$ has the property of Baire because $\emptyset$ is a meager set and $U = U \Delta \emptyset$. Therefore,

$$\mathcal{R} = \{A \subseteq X \colon A \text{ has the property of Baire}\}$$

is a σ-algebra on X containing all the open subsets of X. This means that $\mathcal{R}$ contains all the Borel sets. □

Recall the infinite-length Banach–Mazur games discussed in Chap. 3. The game $\mathsf{BM}(X, A)$ is played by two players, called Hit and Miss, who take turns building an infinite decreasing sequence of open subsets of X. In the end, Hit wins if the intersection of these sets contains a point of A, and otherwise Miss wins.

$$\begin{array}{lccccccccccccc}
\textsc{Hit} & U_1 & & & & U_2 & & & & U_3 & & \cdots & & U_n & & & & U_{n+1} & & \cdots \\
 & & \supseteq & & \supseteq & & \supseteq & & \supseteq & & & & \supseteq & & \supseteq & & \supseteq & & & \\
\textsc{Miss} & & & V_1 & & & & V_2 & & & & V_{n-1} & & & & V_n & & & &
\end{array}$$

We proved in Chap. 3 that if X is a separable metric space (like $\mathbb{R}$), then Miss has a winning tactic in $\mathsf{BM}(X, A)$ if and only if A is meager, and Hit has a winning tactic if and only if $U \setminus A$ is meager for some non-empty open $U \subseteq X$. In the remainder of this section, we explore the close connection between the property of Baire and the existence of a winning tactic for Hit or Miss in Banach–Mazur games.

Definition 7.14. Let X be a metric space and let $A \subseteq X$. The Banach–Mazur game $\mathsf{BM}(X, A)$ is *determined* if one of the two players Hit or Miss has a winning tactic.

Proposition 7.15. *Suppose X is a separable metric space. If $A \subseteq X$ has the property of Baire, then the Banach–Mazur game* $\mathsf{BM}(X, A)$ *is determined.*

Proof. Suppose $A \subseteq \mathbb{R}$ has the property of Baire, i.e., $A = U \Delta M$ for some open $U \subseteq \mathbb{R}$ and meager $M \subseteq \mathbb{R}$. If $U = \emptyset$, then $A = \emptyset \Delta M = M$, so A is meager. By Theorem 3.74 in Chap. 3, this implies Miss has a winning tactic in $\mathsf{BM}(\mathbb{R}, A)$. On the other hand, if $U \neq \emptyset$, then $U \setminus A \subseteq M$ (by our fourth basic rule for symmetric differences), so $U \setminus A$ is meager. By Theorem 3.77 in Chap. 3, this implies Hit has a winning tactic in $\mathsf{BM}(\mathbb{R}, A)$. □

Corollary 7.16. *Suppose X is a separable metric space. If $A \subseteq X$ is Borel, then the Banach–Mazur game* $\mathsf{BM}(X, A)$ *is determined.*

Proof. This follows from the previous proposition plus Corollary 7.13. □

Example 7.17. There is a subset of $\mathbb{R}$ that does not have the property of Baire, but the game $\mathsf{BM}(X, A)$ is determined. In other words, the converse of the previous proposition is not true: Sets with the property of Baire are always determined, but determined sets do not necessarily have the property of Baire.

To see this, let us assume, for the moment, that there is a set $A \subseteq \mathbb{R}$ that does not have the property of Baire. (We have not yet proved that such a set exists, but we will do so in the final section of this chapter.) If both the sets $A^- = A \cap (-\infty, 0]$ and $A^+ = A \cap [0, \infty)$ had the property of Baire, then so would their union $A^- \cup A^+ = A$. Thus, either A^- or A^+ does not have the property of Baire; without loss of generality, let us assume A^- does not have the property of Baire. Define $B = A^- \cup [1, 2]$.

If B were to have the property of Baire, then so would the relative complement $B \setminus (0, \infty) = A^-$. Thus, B does not have the property of Baire. Nevertheless, Hit has a winning tactic in $\mathsf{BM}(\mathbb{R}, B)$. This follows from Theorem 3.77 in Chap. 3, because there is a non-empty open set $U = (1, 2)$ such that $U \setminus B$ is meager (or in this case, empty).

Roughly, this example shows that Hɪᴛ may have a winning tactic in $\mathsf{BM}(\mathbb{R}, B)$ because of how B looks in one place (between 1 and 2, for example), while B fails to have the property of Baire because of how it looks in another place (on the negative reals, for example). In order to get a set $A \subseteq \mathbb{R}$ for which $\mathsf{BM}(\mathbb{R}, B)$ is not determined, we wish to find a set A that fails to have the property of Baire "everywhere" on some non-empty open $U \subseteq \mathbb{R}$. This is, more or less, the content of the following lemma.

Lemma 7.18. *Let X be a separable metric space. Let $A \subseteq X$ and suppose A does not have the property of Baire. Then, there is some non-empty open $V \subseteq X$ such that $A \cap W$ fails to have the property of Baire for every non-empty open $W \subseteq V$.*

Proof. We prove the contrapositive: If the property stated in the conclusion of the lemma is false, then A has the property of Baire. So suppose the conclusion of the lemma is false, i.e., we suppose that for every non-empty open $V \subseteq X$, there is some non-empty open $W_V \subseteq V$ such that $A \cap W_V$ has the property of Baire. For each such W_V, we have $A \cap W_V = U_V \Delta M_V$ for some open $U_V \subseteq X$ and meager $M_V \subseteq X$. Define a function τ by setting

$$\tau(V) = \begin{cases} U_V & \text{if } U_V \neq \emptyset, \text{ and} \\ W_V & \text{if } U_V = \emptyset. \end{cases}$$

In the first case, $\tau(V) \setminus A = U_V \setminus A \subseteq M_V$, so $\tau(V) \setminus A$ is meager. In the second case, $\tau(V) \cap A = W_V \cap A = M_V$ is meager. Thus, if $U = \tau(V)$ for some V (if U is in the image of τ), either $\tau(V) \setminus A$ or $\tau(V) \cap A$ is meager.

Note that τ is a function $\mathcal{O}^+(X) \to \mathcal{O}^+(X)$. (Recall from Chap. 3 that $\mathcal{O}^+(X)$ denotes the set of all non-empty open subsets of X.) By Lemma 3.73 from Chap. 3, there is a collection $\mathcal{U}$ of disjoint open subsets of X such that $\overline{\mathcal{U}} = X$, and for each $U \in \mathcal{U}$ there is some $V \in \mathcal{O}^+(X)$ with $\tau(V) = U$.

Divide the members of $\mathcal{U}$ into two groups as follows:

$$\mathcal{U}_1 = \{U \in \mathcal{U} \colon U \setminus A \text{ is meager}\},$$
$$\mathcal{U}_2 = \{U \in \mathcal{U} \colon U \cap A \text{ is meager}\}.$$

By the comments following the definition of τ, $\mathcal{U} = \mathcal{U}_1 \cup \mathcal{U}_2$. Define $O = \bigcup \mathcal{U}_1$ and $M = O \Delta A$. Then, O is an open set, and $O \Delta M = O \Delta (O \Delta A) = A$. To show that A has the property of Baire, we need to show M is meager.

To see that M is meager, first note that $\mathcal{U}$ is countable by Exercise 18 in Chap. 3. This means $\mathcal{U}_1$ and $\mathcal{U}_2$ are both countable as well. Let

$$M_1 = \{U \setminus A \colon U \in \mathcal{U}_1\},$$
$$M_2 = \{U \cap A \colon U \in \mathcal{U}_2\},$$
$$M_3 = X \setminus \bigcup \mathcal{U}.$$

Because $U \setminus A$ is meager for every $U \in \mathcal{U}_1$, M_1 is a countable union of meager sets, and M_1 is meager. Because $U \cap A$ is meager for every $U \in \mathcal{U}_2$, M_2 is a countable union of meager sets, and so M_2 is meager. Because $\overline{\bigcup \mathcal{U}} = X$, $M_3 = \overline{\bigcup \mathcal{U}} \setminus \bigcup \mathcal{U}$ and $\bigcup \mathcal{U}$ is open in X, M_3 is meager by Lemma 7.11. Hence, $M_1 \cup M_2 \cup M_3$ is meager.

Because $\mathcal{U}_1$ and $\mathcal{U}_2$ have no sets in common and the members of $\mathcal{U} = \mathcal{U}_1 \cup \mathcal{U}_2$ are pairwise disjoint, $\bigcup \mathcal{U}_1$ and $\bigcup \mathcal{U}_2$ are disjoint open sets. Hence, $x \notin O = \bigcup \mathcal{U}_1$ if and only if either $x \in \bigcup \mathcal{U}_2$ or $x \notin \bigcup \mathcal{U}_1 \cup \bigcup \mathcal{U}_2 = \bigcup \mathcal{U}$. Therefore, $X \setminus O = \bigcup \mathcal{U}_2 \cup (X \setminus \bigcup \mathcal{U})$. Consequently,

$$\begin{aligned} O \setminus A &= \left(\bigcup \mathcal{U}_1\right) \setminus A = \bigcup \{U \setminus A \colon U \in \mathcal{U}_1\} = M_1, \text{ and} \\ A \setminus O &= A \cap \left(\bigcup \mathcal{U}_2 \cup (X \setminus \bigcup \mathcal{U})\right) \subseteq \left(A \cap \bigcup \mathcal{U}_2\right) \cup (X \setminus \bigcup \mathcal{U}) \\ &= \bigcup \{A \cap U \colon U \in \mathcal{U}_2\} \cup (X \setminus \bigcup \mathcal{U}) = M_2 \cup M_3. \end{aligned}$$

Thus, $M = O \Delta A = (O \setminus A) \cup (A \setminus O) \subseteq M_1 \cup M_2 \cup M_3$, which implies M is meager. Therefore, $A = O \Delta M$ has the property of Baire as claimed. □

Theorem 7.19. *Let X be a complete separable metric space. If there is a subset A of X without the property of Baire, then there is a subset B of A such that* $\mathsf{BM}(X, B)$ *is not determined.*

Proof. Suppose $A \subseteq X$ does not have the property of Baire. By the preceding lemma, there is a non-empty open $V \subseteq X$ such that $A \cap W$ fails to have the property of Baire for every non-empty open $W \subseteq V$. Let $B = V \cap A$.

Because $A \cap W$ fails to have the property of Baire for every non-empty open $W \subseteq V$, $B = A \cap V$ does not have the property of Baire. In particular, B is not meager. By Theorem 3.74 from Chap. 3, this implies Miss does not have a winning tactic in $\mathsf{BM}(X, B)$.

Let U be a non-empty open subset of X. We consider two cases. For the first case, suppose $U \setminus \overline{V} \neq \emptyset$. Then, $U \setminus \overline{V}$ is a non-empty open set, so its complement is not dense in X, and this implies via Theorems 3.63 and 3.65 from Chap. 3 that $U \setminus \overline{V}$ is not meager. Because $B \subseteq V \subseteq \overline{V}$, which means $U \setminus B \supseteq U \setminus \overline{V}$, this implies $U \setminus B$ is not meager. For the second case, suppose $U \setminus \overline{V} = \emptyset$. Then, $U \subseteq V \cup (\overline{V} \setminus V)$, and because $\overline{V} \setminus V$ is nowhere dense (by Lemma 7.11) and V is non-empty and open, this implies $U \cap V \neq \emptyset$. Setting $W = U \cap V$ in our description of V, we see that $A \cap U \cap V$ does not have the property of Baire. But U does have the property of Baire, and $A \cap U \cap V = B \cap U = U \setminus (U \setminus B)$. Because the sets with the property of Baire for a σ-algebra (closed under taking relative complements), this means $U \setminus B$ does not have the property of Baire. Thus, in either case, $U \setminus B$ does not have the property of Baire. By Theorem 3.77 from Chap. 3, this implies Hit does not have a winning tactic in $\mathsf{BM}(X, B)$. □

Lebesgue measurable sets

Recall from the previous chapter that, given any $A \subseteq \mathbb{R}$, the *Lebesgue outer measure* of A, denoted $\lambda^*(A)$, is defined as the infimum of all $L \geq 0$ such that A can be covered by a countable set of open intervals with a total length of at most L. In this section, we dig deeper into the notion of Lebesgue outer measure and use it to define our third important σ-algebra of subsets of $\mathbb{R}$, the σ-algebra of Lebesgue measurable sets.

Concerning the outer measure of subsets of $\mathbb{R}$, we already proved that

- If $A, B \subseteq \mathbb{R}$ and $A \subseteq B$, then $\lambda^*(A) \leq \lambda^*(B)$.
- If I is an interval in $\mathbb{R}$, then $\lambda^*(I) = \lambda(I)$.

The following theorem is simple but important. The assertion it makes is known as the *countable subadditivity* of the Lebesgue outer measure.

Theorem 7.20. *If $A_1, A_2, \ldots, A_n \subseteq \mathbb{R}$, then $\lambda^*\left(\bigcup_{i=1}^{n} A_i\right) \leq \sum_{i=1}^{n} \lambda^*(A_i)$. Likewise, if $A_1, A_2, \cdots \subseteq \mathbb{R}$, then $\lambda^*\left(\bigcup_{i=1}^{\infty} A_i\right) \leq \sum_{i=1}^{\infty} \lambda^*(A_i)$.*

Proof. The first assertion follows from the second assertion by taking $A_i = \emptyset$ for all $i > n$. So, let us prove the second assertion.

Let $\varepsilon > 0$. For each $n \in \mathbb{N}$, there is a countable set $\mathcal{I}_n$ of open intervals covering A_n with total length $L_n < \lambda^*(A_n) + \frac{\varepsilon}{2^n}$. But then, $\bigcup_{n\in\mathbb{N}} \mathcal{I}_n$ is a countable set of open intervals covering $\bigcup_{i=1}^{\infty} A_i$ with total length at most $\sum_{n=1}^{\infty} L_n < \varepsilon + \sum_{n=1}^{\infty} \lambda^*(A_n)$. As ε was arbitrary, it follows that $\lambda^*\left(\bigcup_{n=1}^{\infty} A_n\right) \leq \sum_{n=1}^{\infty} \lambda^*(A_n)$. □

Definition 7.21. A set $A \subseteq \mathbb{R}$ is *Lebesgue measurable* if for any $\varepsilon > 0$, there is a closed $F \subseteq \mathbb{R}$ and an open $U \subseteq \mathbb{R}$ such that $F \subseteq A \subseteq U$ and $\lambda^*(U \setminus F) < \varepsilon$. Let $\mathcal{L} = \{A \subseteq \mathbb{R} \colon A \text{ is Lebesgue measurable}\}$.

Do not confuse "Lebesgue measurable" with "has a well-defined outer measure." The latter property applies to every subset of $\mathbb{R}$. But as we shall see in the final section of this chapter, not every subset of $\mathbb{R}$ is in $\mathcal{L}$. We may think of $\mathcal{L}$ as a well-behaved class of subsets of $\mathbb{R}$. For sets in this class, the Lebesgue outer measure satisfies certain nice properties. For example, we prove that if A and B are disjoint subsets of $\mathbb{R}$ and $A, B \in \mathcal{L}$, then $\lambda^*(A \cup B) = \lambda^*(A) + \lambda^*(B)$. This is a nice property of λ^* for sets in $\mathcal{L}$, but is not necessarily true if $A, B \notin \mathcal{L}$.

Theorem 7.22. *Every interval is Lebesgue measurable.*

Proof. Suppose $a < b$ and I is an interval from a to b (i.e., one of (a, b) or $(a, b]$ or $[a, b)$ or $[a, b]$). Fix $\varepsilon > 0$, and let $\delta > 0$ be less than both $b - a$ and ε. Let $F = \left[a + \frac{\delta}{4}, b - \frac{\delta}{4}\right]$, and let $U = \left(a - \frac{\delta}{4}, b + \frac{\delta}{4}\right)$. Then, F is closed, U is open, and $U \setminus F = \left(a - \frac{\delta}{4}, a + \frac{\delta}{4}\right) \cup \left(b - \frac{\delta}{4}, b + \frac{\delta}{4}\right)$. By Theorem 7.20, and Theorem 6.47 from Chap. 6, $\lambda^*(U \setminus F) \leq \frac{\delta}{2} + \frac{\delta}{2} = \delta < \varepsilon$.

The proof for unbounded intervals is similar. For example, for an interval from a to ∞, we can take $U = \left(a - \frac{\delta}{2}, \infty\right)$ and $F = \left[a + \frac{\delta}{2}, \infty\right)$. For the doubly unbounded interval $\mathbb{R}$, we can take $U = F = \mathbb{R}$. □

Theorem 7.23. *Lebesgue null sets are Lebesgue measurable.*

Proof. Let $A \subseteq \mathbb{R}$ be Lebesgue null, and fix $\varepsilon > 0$. By the definition of a null set, there is a countable set $\mathcal{I}$ of open intervals covering A with total length $L < \varepsilon$. Let $F = \emptyset$ and $U = \bigcup \mathcal{I}$. Then, F is closed, U is open, and $\lambda^*(U \setminus F) = \lambda^*(U) \leq L < \varepsilon$. □

Theorem 7.24. *If $A \subseteq \mathbb{R}$ is Lebesgue measurable, then so is $\mathbb{R} \setminus A$. In other words, $\mathcal{L}$ is closed under the operation of taking complements.*

Proof. Let $\varepsilon > 0$. Assuming $A \in \mathcal{L}$, there is a closed set F and an open set U such that $F \subseteq A \subseteq U$ and $\lambda^*(U \setminus F) < \varepsilon$. But then, $U' = \mathbb{R} \setminus U$ is closed, $F' = \mathbb{R} \setminus F$ is open, $U' \subseteq \mathbb{R} \setminus A \subseteq F'$, and $\lambda^*(F' \setminus U') < \varepsilon$ (because $F' \setminus U' = U \setminus F$). This shows $\mathbb{R} \setminus A \in \mathcal{L}$ as well. □

Lemma 7.25. *If A and B are Lebesgue measurable, then so is $A \cap B$.*

Proof. Suppose $A, B \subseteq \mathbb{R}$ are Lebesgue measurable, and let $\varepsilon > 0$. There are some closed sets $F_A, F_B \subseteq \mathbb{R}$ and some open sets $U_A, U_B \subseteq \mathbb{R}$, such that $F_A \subseteq A \subseteq U_A$ and $F_B \subseteq B \subseteq U_B$, and $\lambda^*(U_A \setminus F_A) < \varepsilon/2$ and $\lambda^*(U_B \setminus F_B) < \varepsilon/2$. Let $U = U_A \cap U_B$ and $F = F_A \cap F_B$. Then, U is open, F is closed, and $U \setminus F \subseteq (U_A \setminus F_A) \cup (U_B \setminus F_B)$ (see the following picture), which implies by Theorem 7.20 that $\lambda^*(U \setminus F) \leq \lambda^*(U_A \setminus F_A) + \lambda^*(U_B \setminus F_B) < \varepsilon$. □

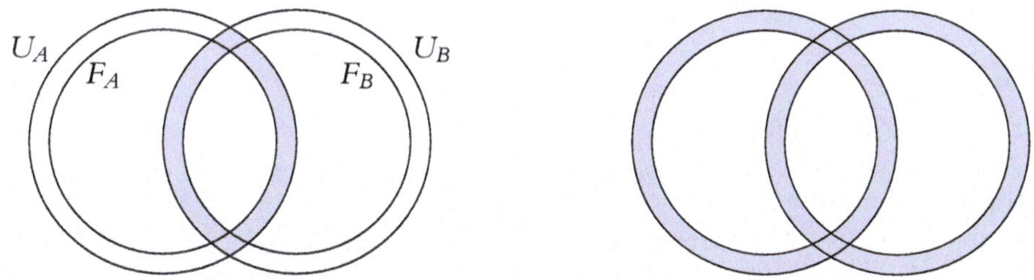

$(U_A \cap U_B) \setminus (F_A \cap F_B) \quad \subseteq \quad (U_A \setminus F_A) \cup (U_B \setminus F_B)$

Proposition 7.26. *If $A_1, A_2, \ldots, A_n$ are Lebesgue measurable, then so is $\bigcap_{i=1}^{n} A_i$. In other words, $\mathcal{L}$ is closed under taking finite intersections.*

Proof. This follows from Lemma 7.25 via induction. □

Proposition 7.27. *If A and B are Lebesgue measurable, then so is $B \setminus A$. In other words, $\mathcal{L}$ is closed under taking relative complements.*

Proof. This follows immediately from Theorem 7.24 and Lemma 7.25, because $B \setminus A = B \cap (\mathbb{R} \setminus A)$. □

Proposition 7.28. *If $A_1, A_2, \ldots, A_n$ are Lebesgue measurable, then so is $\bigcup_{i=1}^{n} A_i$. In other words, $\mathcal{L}$ is closed under taking finite unions.*

Proof. This follows from Theorem 7.24 and Proposition 7.26 via De Morgan's laws. In more detail: As $\mathbb{R}$ is Lebesgue measurable, $\mathbb{R} \setminus A_i \in \mathcal{L}$ for every i, which implies $\bigcap_{i=1}^{n}(\mathbb{R} \setminus A_i) \in \mathcal{L}$, so $\bigcup_{i=1}^{n} A_i = \mathbb{R} \setminus \bigcap_{i=1}^{n}(\mathbb{R} \setminus A_i) \in \mathcal{L}$. □

We have proved that $\emptyset, \mathbb{R} \in \mathcal{L}$ (they are intervals) and that $\mathcal{L}$ is closed under complementation, and under countable unions and intersections. Our goal is to prove that in fact $\mathcal{L}$ is closed under countable unions and intersections, i.e., that $\mathcal{L}$ is a σ-algebra. We also wish to prove that the outer measure λ^* is *countably additive* on $\mathcal{L}$, which means that if $A_1, A_2, A_3, \cdots \in \mathcal{L}$ and these sets are pairwise disjoint, then $\lambda^*(\bigcup_{i=1}^{\infty} A_i) = \sum_{i=1}^{\infty} \lambda^*(A_i)$. These things are more difficult to prove, and we first require a few lemmas.

Lemma 7.29. *Suppose F and G are disjoint compact subsets of a metric space. Then, there is some $\delta > 0$ such that the distance from any point in F to any point in G is at least δ.*

Proof. Because G is closed, there is for every $x \in F$ some $\delta_x > 0$ with $B_{\delta_x}(x) \cap G = \emptyset$. Observe that $\left\{B_{\frac{1}{2}\delta_x}(x) : x \in F\right\}$ is an open cover of F. Because F is compact, there are finitely many points $x_1, x_2, \ldots, x_n \in F$ such that $F \subseteq \bigcup_{i=1}^{n} B_{\frac{1}{2}\delta_{x_i}}(x_i)$. Let $\delta = \min\{\frac{1}{2}\delta_{x_1}, \frac{1}{2}\delta_{x_2}, \ldots, \frac{1}{2}\delta_{x_n}\}$. If $x \in F$, then there is some $i \leq n$ such that $x \in \frac{1}{2}\delta_{x_i}$. If y is any point with $d(x, y) < \delta$,

$$d(y, x_i) \leq d(y, x) + d(x, x_i) < \tfrac{1}{2}\delta + \tfrac{1}{2}\delta_{x_i} \leq \delta_{x_i}.$$

But, $B_{\delta_{x_i}}(x_i) \cap G = \emptyset$, so this shows that if $d(x, y) < \delta$, then $y \notin G$. As x was an arbitrary point of F, the lemma follows. □

Example 7.30. Note that the conclusion of this lemma is false if we only assume F and G are closed (but not necessarily compact). For example, we could take F and G to be subsets of the plane $\mathbb{R}^2$: Take F to be the graph of $y = 1/x$ and G the Y-axis. These sets are both closed in $\mathbb{R}^2$, but the distance between them is vanishingly small. ☕

Lemma 7.31. *Let $\mathcal{I}$ be a countable set of open intervals with total length L, and let $\varepsilon, \delta > 0$. There is a countable set $\mathcal{J}$ of open intervals with total length at most $L + \varepsilon$ such that $\bigcup \mathcal{I} = \bigcup \mathcal{J}$, and every interval in $\mathcal{J}$ has length less than δ.*

Proof. Fix $\delta > 0$. To begin, let us assume $\mathcal{I}$ consists of just one interval, i.e., $\mathcal{I} = \{(a, b)\}$, where $b - a = L$. Let $\eta > 0$, and fix $N \in \mathbb{N}$ such that $L/N < \delta$. It is not hard to see that we can write (a, b) as a union of $N + 1$ evenly spaced open intervals, slightly overlapping, each of length L/N.

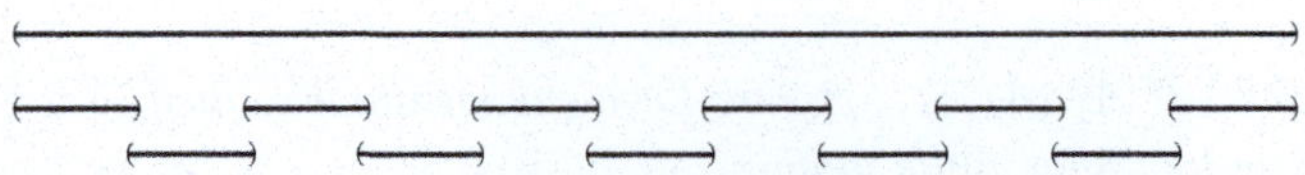

This is a finite set $\mathcal{J}$ of open intervals with total length $(N+1)\frac{L}{N} = L + \frac{1}{N} < L + \eta$, such that $\bigcup \mathcal{J} = (a, b)$, and every interval in $\mathcal{J}$ has length $<\delta$.

Now, consider the general case, where $\mathcal{I} = \{I_n : n \in \mathbb{N}\}$ consists of countably many intervals and has total length L. (Perhaps $\mathcal{I}$ is finite, of size k, in which case we could have $I_n = I_k$ for all $n \geq k$.) By the previous paragraph, there is for each n some finite collection $\mathcal{J}_n$ of open intervals with total length at most $\lambda^*(I_n) + \varepsilon/2^n$ such that $\bigcup \mathcal{J}_n = I_n$, and every interval in $\mathcal{J}_n$ has length at most δ. Then, $\mathcal{J} = \bigcup \mathcal{J}_n$ is a countable set of open intervals with total length at most $L + \varepsilon$ such that $\bigcup \mathcal{I} = \bigcup \mathcal{J}$, and every interval in $\mathcal{J}$ has length $<\delta$. □

Lemma 7.32. *If $F_1, F_2, \ldots, F_n$ are compact, pairwise disjoint subsets of $\mathbb{R}$, then $\lambda^*(\bigcup_{i\leq n} F_i) = \sum_{i\leq n} \lambda^*(F_i)$.*

Proof. Note that $\lambda^*(\bigcup_{i\leq n} F_i) \leq \sum_{i\leq n} \lambda^*(F_i)$ by Theorem 7.20. We need to prove the reverse inequality.

For each pair of numbers $i, j \leq n$ with $i \neq j$, the Lemma 7.29 says that there is some $\delta_{i,j} > 0$ such that the distance from any point in F_i to any point in F_j is at least $\delta_{i,j}$. Let $\delta = \min\{\delta_{i,j} : i, j \leq n\}$.

Let $\varepsilon > 0$, and let $\mathcal{I} = \{I_n : n \in \mathbb{N}\}$ be a countable set of open intervals with total length L covering $\bigcup_{i=1}^n F_i$. By the previous lemma, there is a countable set $\mathcal{J}$ of open intervals with total length at most $L + \varepsilon$ such that $\bigcup \mathcal{I} = \bigcup \mathcal{J}$, and every interval in $\mathcal{J}$ has length $<\delta$. For each $i \leq n$, let $\mathcal{K}_i = \{J \in \mathcal{J} : J \cap F_i \neq \emptyset\}$, which is a countable set of open intervals covering F_i, and therefore whose total length is at least $\lambda^*(F_i)$. But, if $i \neq j$, then $\mathcal{K}_i$ and $\mathcal{K}_j$ are disjoint, by our choice of δ and the fact that each of the intervals in $\mathcal{J}$ has a length $<\delta$. Thus, the total length of $\mathcal{J}$ is at least as large as the sum of the total lengths of the $\mathcal{K}_i$, i.e., the total length of $\mathcal{J}$ is at least $\sum_{i=1}^n \lambda^*(F_i)$. Because the total length of $\mathcal{J}$ is at most $L + \varepsilon$, this shows $L \geq \sum_{i=1}^n \lambda^*(F_i) - \varepsilon$. As ε was arbitrary, $L \geq \sum_{i=1}^n \lambda^*(F_i)$. As $\mathcal{I}$ was an arbitrary covering of $\bigcup_{i=1}^n F_i$ with open intervals, it follows that $\lambda^*(\bigcup_{i=1}^n F_i) \geq \sum_{i=1}^n \lambda^*(F_i)$. □

Lemma 7.33. *If $U \subseteq \mathbb{R}$ is open, then there is a countable set $\mathcal{I}$ of pairwise disjoint open intervals such that $U = \bigcup \mathcal{I}$.*

Proof. Define an equivalence relation $\sim$ on U by setting $x \sim y$ if and only if $[x, y] \subseteq U$. (It is clear that this relation is reflexive, symmetric, and transitive, i.e., it is in fact an equivalence relation.) Let $\mathcal{I}$ be the set of all equivalence classes for this relation.

We need to show that every member of $\mathcal{I}$ is an open interval. So, let I be an equivalence class for $\sim$. If $a, b \in I$, then $a \sim b$, which means $[a, b] \subseteq U$, so $[a, x] \subseteq U$ for all $x \in (a, b)$, which means $a \sim x$ for all $x \in (a, b)$, so $(a, b) \subseteq I$. This shows I is convex, hence it is an interval. To see that I is open, fix $a \in I$. Then, $a \in U$ and, as U is open, there is some $\varepsilon > 0$ such that $(a - \varepsilon, a + \varepsilon) \subseteq U$. But then, $a \sim x$ for all $x \in (a - \varepsilon, a + \varepsilon)$, so $(a - \varepsilon, a + \varepsilon) \subseteq I$. As a was an arbitrary point of I, this shows I is open.

Thus, $\mathcal{I}$ is a collection of pairwise disjoint open intervals. Clearly, $U = \bigcup \mathcal{I}$, and $\mathcal{I}$ is countable because every set of pairwise disjoint open intervals in $\mathbb{R}$ is countable. (This last fact follows from Exercise 18 in Chap. 3 or simply from the observation that each $I \in \mathcal{I}$ contains a rational number q_I, and the map $I \mapsto q_I$ is an injection $\mathcal{I} \to \mathbb{Q}$.) $\square$

Lemma 7.34. *Suppose U is a bounded open subset of $\mathbb{R}$, and $\varepsilon > 0$. There is a closed $F \subseteq U$ such that $\lambda^*(U) - \lambda^*(F) < \varepsilon$.*

Proof. If U is a finite union of open intervals, then this is easy to see: If $U = I_1 \cup I_2 \cup \cdots \cup I_n$, then, for each $i \leq n$, let $F_i \subseteq I_i$ be a closed interval with length $\lambda(F_i) > \lambda(I_i) - \varepsilon/n$, and take $F = \bigcup_{i \leq n} F_i$.

So, suppose U is not a finite union of intervals. Because U is bounded, there is some $B > 0$ with $U \subseteq [-B, B]$. Because U is open, the previous lemma shows there is a countable set $\{I_n \colon n \in \mathbb{N}\}$ of pairwise disjoint open intervals such that $U = \bigcup_{n=1}^{\infty} I_n$. Fix $\varepsilon > 0$, and for each $n \in \mathbb{N}$ let $F_n \subseteq I_n$ be a closed interval with $\lambda(F_n) > \lambda(I_n) - \varepsilon/2^{n+1}$.

For any $N \in \mathbb{N}$, observe that $[-B, B] \supseteq \bigcup_{n \leq N} F_n$, and so

$$2B = \lambda^*([-B, B]) \geq \lambda^*\left(\bigcup_{n=1}^{N} F_n\right) = \sum_{n=1}^{N} \lambda^*(F_n),$$

where the last equality follows from Lemma 7.32. Because this is true for every N, the infinite series $\sum_{n=1}^{\infty} \lambda^*(F_n)$ converges (to some number $\leq 2B$). Fix N large enough that $\sum_{n=1}^{\infty} \lambda^*(F_n) < \varepsilon/2 + \sum_{n=1}^{N} \lambda^*(F_n)$. Then,

$$\lambda^*(U) \leq \sum_{n=1}^{\infty} \lambda^*(I_n) \leq \tfrac{\varepsilon}{2} + \sum_{n=1}^{\infty} \lambda^*(F_n) < \varepsilon + \sum_{n=1}^{N} \lambda^*(F_n).$$

Because $\lambda^*\left(\bigcup_{n=1}^{N} F_n\right) = \sum_{n=1}^{N} \lambda^*(F_n)$ by Lemma 7.32, the closed set $F = \bigcup_{n=1}^{N} F_n$ satisfies the conclusion of the lemma. $\square$

Lemma 7.35. *Suppose U is a bounded open subset of $\mathbb{R}$ and F is a closed subset of U. Then, $\lambda^*(U \setminus F) = \lambda^*(U) - \lambda^*(F)$.*

Proof. Theorem 7.20 states that $\lambda^*(U) \leq \lambda^*(U \setminus F) + \lambda^*(F)$, and therefore $\lambda^*(U) - \lambda^*(F) \leq \lambda^*(U \setminus F)$.

For the reverse inequality, let $\varepsilon > 0$. Observe that $U \setminus F$ is open, and applying the previous lemma, let K be a closed subset of $U \setminus F$ such that $\lambda^*(U \setminus F) - \lambda^*(K) < \varepsilon$. By Lemma 7.32, $\lambda^*(F) + \lambda^*(K) = \lambda^*(F \cup K) \leq \lambda^*(U)$. Adding the two preceding inequalities, we get $\lambda^*(U \setminus F) + \lambda^*(F) < \lambda^*(U) + \varepsilon$, or equivalently, $\lambda^*(U \setminus F) < \lambda^*(U) - \lambda^*(F) + \varepsilon$. As ε was arbitrary, this shows $\lambda^*(U \setminus F) \leq \lambda^*(U) - \lambda^*(F)$. $\square$

Lemma 7.36. *Every bounded open subset of $\mathbb{R}$ is Lebesgue measurable.*

Proof. Suppose $U \subseteq \mathbb{R}$ is bounded and open. By the previous two lemmas, there is for any $\varepsilon > 0$ some closed set F such that $F \subseteq U \subseteq U$, and $\lambda^*(U \setminus F) = \lambda^*(U) - \lambda^*(F) < \varepsilon$. Thus, U is Lebesgue measurable. $\square$

Lemma 7.37. *Every compact subset of $\mathbb{R}$ is Lebesgue measurable.*

Proof. Suppose $K \subseteq \mathbb{R}$ is closed and bounded. Let $a = \inf K$ and $b = \sup K$, and let $U = [a, b] \setminus K$. Then U is a bounded open set, and $K = [a, b] \setminus U$. From this, Theorem 7.22, Proposition 7.27, and Lemma 7.36, it follows that K is also Lebesgue measurable. □

Lemma 7.38. *Suppose $A_1, A_2, A_3, \ldots$ is an infinite sequence of pairwise disjoint Lebesgue measurable sets, and $\bigcup_{i=1}^{\infty} A_i$ is bounded. Then, $\bigcup_{i=1}^{\infty} A_i$ is Lebesgue measurable, and $\lambda^*\left(\bigcup_{i=1}^{\infty} A_i\right) = \sum_{i=1}^{\infty} \lambda^*(A_i)$.*

Proof. By Theorem 7.20, $\lambda^*\left(\bigcup_{i=1}^{\infty} A_i\right) \leq \sum_{i=1}^{\infty} \lambda^*(A_i)$. We must prove the reverse inequality and that $\bigcup_{i=1}^{\infty} A_i$ is Lebesgue measurable.

Fix $\varepsilon > 0$. For each $i \in \mathbb{N}$, there is a closed F_i and an open U_i such that $F_i \subseteq A_i \subseteq U_i$ and $\lambda^*(U_i \setminus F_i) = \lambda^*(U_i) - \lambda^*(F_i) < \varepsilon/2^{n+1}$.

As $\bigcup_{i=1}^{\infty} A_i$ is bounded, there is some $B > 0$ with $\bigcup_{i=1}^{\infty} A_i \subseteq [-B, B]$. Thus, $\lambda^*\left(\bigcup_{i=1}^{\infty} A_i\right) \leq \lambda^*([\mathcal{B}, B]) = 2B$, and in particular $\lambda^*\left(\bigcup_{i=1}^{\infty} A_i\right)$ is finite. For each $n \in \mathbb{N}$,

$$\sum_{i=1}^{n} \lambda^*(F_i) = \lambda^*\left(\bigcup_{i=1}^{n} F_i\right) \leq \lambda^*\left(\bigcup_{i=1}^{n} A_i\right) \leq \lambda^*\left(\bigcup_{i=1}^{\infty} A_i\right).$$

Thus, $\sum_{i=1}^{\infty} \lambda^*(F_i)$ converges (to at most $\lambda^*\left(\bigcup_{i=1}^{\infty} A_i\right) \leq 2B$). This implies $\sum_{i=1}^{\infty} \lambda^*(U_i)$ also converges, and $\sum_{i=1}^{\infty} \lambda^*(U_i) < \frac{\varepsilon}{2} + \sum_{i=1}^{\infty} \lambda^*(F_i)$.

Fix $N \in \mathbb{N}$ large enough that $\frac{\varepsilon}{2} + \sum_{i=1}^{N} \lambda^*(F_i) > \sum_{i=1}^{\infty} \lambda^*(F_i)$. Let $U = \bigcup_{i=1}^{\infty} U_i$, and let $F = \bigcup_{i=1}^{N} F_i$. Then, F is closed, U is open, and $F \subseteq \bigcup_{i=1}^{N} A_i \subseteq \bigcup_{i=1}^{N} U_i \subseteq U$. Furthermore,

$$\begin{aligned} \lambda^*(F) = \lambda^*\left(\bigcup_{i=1}^{N} F_i\right) &= \sum_{i=1}^{N} \lambda^*(F_i) > \left(\sum_{i=1}^{\infty} \lambda^*(F_i)\right) - \frac{\varepsilon}{2} \\ &\geq \left(\sum_{i=1}^{\infty} \lambda^*(U_i)\right) - \varepsilon \geq \lambda^*(U) - \varepsilon. \end{aligned}$$

(The last inequality uses Theorem 7.20, which says $\lambda^*(U) \leq \sum_{i=1}^{\infty} \lambda^*(U_i)$.) So, $\lambda^*(U \setminus F) = \lambda^*(U) - \lambda^*(F) < \varepsilon$. This shows $\bigcup_{i=1}^{\infty} A_i$ is measurable.

Furthermore, $\lambda^*(F) \leq \lambda^*\left(\bigcup_{i=1}^{\infty} A_i\right)$ (because $F \subseteq \bigcup_{i=1}^{\infty} A_i$), and by replacing each U_i with A_i in the previous computation, we see that $\lambda^*(F) \geq \left(\sum_{i=1}^{\infty} \lambda^*(A_i)\right) - \varepsilon$. Therefore, $\sum_{i=1}^{\infty} \lambda^*(A_i) \leq \lambda^*(F) + \varepsilon \leq \lambda^*\left(\bigcup_{i=1}^{\infty} A_i\right) + \varepsilon$. As ε was arbitrary, this shows $\sum_{i=1}^{\infty} \lambda^*(A_i) \leq \lambda^*\left(\bigcup_{i=1}^{\infty} A_i\right)$. □

Lemma 7.39. *Suppose $A \subseteq \mathbb{R}$, and suppose $A \cap [-n, n] \in \mathcal{L}$ for every $n \in \mathbb{N}$. Then, $A \in \mathcal{L}$ and $\lambda^*(A) = \lim_{n\to\infty} \lambda^*(A \cap [-n, n])$.*

Proof. For each $n \in \mathbb{N}$, let $A_n = (A \cap [-n-1, n+1]) \setminus (A \cap [-n, n]) = A \cap ([-n-1, n+1] \setminus [-n, n])$. Note that A_n is Lebesgue measurable by Proposition 7.27. For each n, let F_n be a closed set and U_n an open set such that $F_n \subseteq A_n \subseteq U_n$ and $\lambda^*(U_n \setminus F_n) < \varepsilon/2^n$.

Let $F = \bigcup_{n=1}^{\infty} F_n$, and let $U = \bigcup_{n=1}^{\infty} U_n$. For every $x \in \mathbb{R}$, the ball $(x-1, x+1)$ meets at most 2 of the F_n. If x is not in either of these two sets, then, because they are closed, we may find an even smaller ball $(x-\varepsilon, x+\varepsilon)$ avoiding both sets, hence avoiding $\bigcup_{n=1}^{\infty} F_n$. This shows that if $x \notin \bigcup_{n=1}^{\infty} F_n$, then $x \notin \overline{\bigcup_{n=1}^{\infty} F_n}$. Therefore, $\bigcup_{n\in\mathbb{N}} F_n = \overline{\bigcup_{n\in\mathbb{N}} F_n}$, which means $F = \bigcup_{n=1}^{\infty} F_n$ is closed. Furthermore U is open, $F \subseteq A \subseteq U$, and

$$\lambda^*(U \setminus F) \leq \lambda^*\left(\bigcup_{n=1}^{\infty} U_n \setminus F_n\right) \leq \sum_{n=1}^{\infty} \lambda^*(U_n \setminus F_n) < \varepsilon.$$

Thus, A is Lebesgue measurable. Furthermore,

$$\begin{aligned}\lambda^*(A) &\leq \lambda^*\left(\bigcup_{n=1}^{\infty} U_n\right) \leq \sum_{n=1}^{\infty} \lambda^*(U_n) \leq \varepsilon + \sum_{n=1}^{\infty} \lambda^*(A_n) \\ &= \varepsilon + \lim_{N\to\infty} \sum_{n=1}^{N} \lambda^*(A_n) = \varepsilon + \lim_{N\to\infty} \lambda^*(A \cap [-N, N]).\end{aligned}$$

The last equality here uses the fact that $\sum_{n=1}^{N} \lambda^*(A_n) = \lambda^*\left(\bigcup_{n=1}^{N} A_n\right) = \lambda^*\left(A \cap [-N, N]\right)$, which follows from the previous lemma by setting $A_n = \emptyset$ for all $n > N$. As ε was arbitrary, $\lambda^*(A) \leq \lim_{n\to\infty} \lambda^*\left(A \cap [-n, n]\right)$. For the reverse inequality, simply note that $\lambda^*\left(A \cap [-n, n]\right) \leq \lambda^*(A)$ for every n because $A \cap [-n, n] \subseteq A$. □

Theorem 7.40. *If $A_1, A_2, A_3, \ldots$ are Lebesgue measurable sets, then so is $\bigcup_{i=1}^{\infty} A_i$. In other words, $\mathcal{L}$ is closed under taking countable unions.*

Proof. Suppose $A_1, A_2, A_3, \ldots$ are all Lebesgue measurable sets, and let $A = \bigcup_{n\in\mathbb{N}} A_n$. For each $m, n \in \mathbb{N}$, define $B_n^m = [-m, m] \cap A_n \setminus \bigcup_{i<n} A_i$. By Propositions 7.26, 7.27, and 7.28, B_n^m is Lebesgue measurable for every $m, n \in \mathbb{N}$.

Furthermore, for any given $m \in \mathbb{N}$ the sets $B_1^m, B_2^m, B_3^m, \ldots$ are pairwise disjoint, and $\bigcup_{n\in\mathbb{N}} B_n^m = [-m, m] \cap A$. Lemma 7.38 implies $\bigcup_{n\in\mathbb{N}} B_n^m$ is Lebesgue measurable, so, $[-m, m] \cap A$ is Lebesgue measurable for every $m \in \mathbb{N}$. This implies A is Lebesgue measurable by Lemma 7.39. □

Theorem 7.41. *$\mathcal{L}$ is a σ-algebra.*

Proof. This follows from Theorems 7.22, 7.24, and 7.40. □

Corollary 7.42. *Every Borel set is Lebesgue measurable.*

Proof. First we claim that every open $U \subseteq \mathbb{R}$ is Lebesgue measurable. If U is bounded then this holds by Lemma 7.36. If U is unbounded, then $U = \bigcup_{n=1}^{\infty} \left(U \cap (-n, n)\right)$ is a countable union of bounded open sets, and so $U \in \mathcal{L}$ by Lemma 7.36 together with Theorem 7.40.

Thus, $\mathcal{L}$ is a σ-algebra on $\mathbb{R}$ containing all the open sets. It follows that every Borel set is in $\mathcal{L}$. □

Recall that our goal was not only to prove that $\mathcal{L}$ is a σ-algebra, but also that λ^* is countably additive on $\mathcal{L}$.

Theorem 7.43 (The countable additivity of Lebesgue measure). *Suppose $A_1, A_2, A_3, \ldots$ is an infinite sequence of pairwise disjoint Lebesgue measurable sets. Then, $\lambda^*\big(\bigcup_{n=1}^{\infty} A_i\big) = \sum_{n=1}^{\infty} \lambda^*(A_n)$.*

Proof. Theorem 7.20 tells us that $\lambda^*\big(\bigcup_{i=1}^{\infty} A_i\big) \leq \sum_{i=1}^{\infty} \lambda^*(A_i)$. We must prove the reverse inequality. We consider two cases.

For the first case, suppose $\sum_{i=1}^{\infty} \lambda^*(A_i)$ converges. Let $\varepsilon > 0$. There is some $N \in \mathbb{N}$ large enough that $\frac{\varepsilon}{2} + \sum_{i=1}^{N} \lambda^*(A_i) > \sum_{i=1}^{\infty} \lambda^*(A_i)$. Also, by Lemma 7.39, there is for each $i \leq N$ some m_i large enough that $\frac{\varepsilon}{2N} + \lambda^*([-m_i, m_i] \cap A_i) > \lambda^*(A_i)$. Let $m = \max\{m_i \colon i \leq N\}$, and observe that $\varepsilon + \sum_{i=1}^{N} \lambda^*([-m, m] \cap A_i) > \sum_{i=1}^{\infty} \lambda^*(A_i)$. By Lemma 7.38,

$$\textstyle\sum_{i=1}^{N} \lambda^*([-m, m] \cap A_i) = \lambda^*\big(\bigcup_{i=1}^{N} [-m, m] \cap A_i\big) \leq \lambda^*\big(\bigcup_{i=1}^{\infty} A_i\big).$$

Therefore, $\sum_{i=1}^{\infty} \lambda^*(A_i) \leq \varepsilon + \lambda^*\big(\bigcup_{i=1}^{\infty} A_i\big)$. As ε was arbitrary, this shows $\sum_{i=1}^{\infty} \lambda^*(A_i) \leq \lambda^*\big(\bigcup_{i=1}^{\infty} A_i\big)$.

For the second case, suppose $\sum_{i=1}^{\infty} \lambda^*(A_i)$ diverges. If $\lambda^*(A_n) = \infty$ for some particular $n \in \mathbb{N}$, then $\lambda^*\big(\bigcup_{i=1}^{\infty} A_i\big) = \infty$, and the conclusion of the theorem holds. So we must consider the case where $\lambda^*(A_i)$ is finite for all $i \in \mathbb{N}$, but $\sum_{i=1}^{\infty} \lambda^*(A_i)$ diverges.

Fix $M > 0$. Let $N \in \mathbb{N}$ be large enough that $\sum_{i=1}^{N} \lambda^*(A_i) > M + 1$. By Lemma 7.39, there is for each $i \leq N$ some m_i large enough that $\frac{1}{2N} + \lambda^*\big([-m_i, m_i] \cap A_i\big) > \lambda^*(A_i)$. Let $m = \max\{m_i \colon i \leq N\}$, and observe that $1 + \sum_{i=1}^{N} \lambda^*\big([-m, m] \cap A_i\big) > \sum_{i=1}^{N} \lambda^*(A_i)$. By Lemma 7.38,

$$\textstyle\sum_{i=1}^{N} \lambda^*\big([-m, m] \cap A_i\big) = \lambda^*\big(\bigcup_{i=1}^{N} [-m, m] \cap A_i\big) \leq \lambda^*\big(\bigcup_{i=1}^{\infty} A_i\big).$$

Therefore, $\lambda^*\big(\bigcup_{i=1}^{\infty} A_i\big) \geq \sum_{i=1}^{N} \lambda^*\big([-m, m] \cap A_i\big) \geq \big(\sum_{i=1}^{N} \lambda^*(A_i)\big) - 1 > M$. As M was arbitrary, $\lambda^*\big(\bigcup_{i=1}^{\infty} A_i\big) = \infty$. □

Corollary 7.44. *Suppose $A_1, A_2, A_3, \ldots, A_n$ are pairwise disjoint Lebesgue measurable sets. Then, $\lambda^*\big(\bigcup_{i=1}^{n} A_i\big) = \sum_{i=1}^{n} \lambda^*(A_i)$.*

Proof. This follows from Theorem 7.43 by setting $A_i = \emptyset$ for all $i > n$. □

Corollary 7.45. *If $A, B \in \mathcal{L}$ and $B \subseteq A$, then $\lambda^*(A \setminus B) = \lambda^*(A) - \lambda^*(B)$.*

Proof. It follows from Corollary 7.44 that $\lambda^*(A \setminus B) + \lambda^*(B) = \lambda^*(A)$. □

Corollary 7.46. *Suppose $A_1, A_2, A_3, \ldots$ is a sequence of Lebesgue measurable sets with $A_1 \subseteq A_2 \subseteq A_3 \subseteq \ldots$. Then, $\lambda^*\big(\bigcup_{n=1}^{\infty} A_n\big) = \lim_{n\to\infty} \lambda^*(A_n)$.*

Proof. For each $n \in \mathbb{N}$, define $B_n = A_n \setminus \bigcup_{i<n} A_i$. Then, $\bigcup_{n=1}^{\infty} A_n = \bigcup_{n=1}^{\infty} B_n$, and the previous two corollaries imply $\lambda^*(A_n) = \sum_{i\leq n} \lambda^*(B_i)$. Applying Theorem 7.43, $\lambda^*\big(\bigcup_{n=1}^{\infty} A_n\big) = \lambda^*\big(\bigcup_{n=1}^{\infty} B_n\big) = \sum_{n=1}^{\infty} \lambda^*(B_n) = \lim_{n\to\infty} \sum_{i=1}^{n} \lambda^*(B_i) = \lim_{n\to\infty} \lambda^*(A_n)$. □

The last three corollaries say more or less that the measures of sets in $\mathcal{L}$ add up as they should. Interpreting $\lambda^*(X)$ as the "size" of a set X, it seems obvious that if $A \cap B = \emptyset$, then we should have $\lambda^*(A) + \lambda * (B) = \lambda^*(A \cup B)$. This is true if $A, B \in \mathcal{L}$ by Corollary 7.44. But it is shown in the end-of-chapter exercises that it is not true for all $A, B \subseteq \mathbb{R}$.

Recall that sets with the property of Baire are, by definition, sets that differ from an open set by a meager set. We end this section by proving a similar characterization of the Lebesgue measurable sets: They are the sets that differ from a Borel set (or even just an F_σ or a G_δ set) by a null set.

Theorem 7.47. *Let $A \subseteq \mathbb{R}$. The following are equivalent:*

1. *A is Lebesgue measurable.*
2. *There is a Borel set B and a Lebesgue null set N such that $A = B \Delta N$.*
3. *There is a G_δ set G and a Lebesgue null set N such that $A = G \setminus N$.*
4. *There is an F_σ set F and a Lebesgue null set N such that $A = F \cup N$.*

Proof. To prove the theorem, we show that (3) and (4) both imply (2), (2) implies (1), and (1) implies both (3) and (4).

To see that (3) implies (2), suppose G is G_δ, N is null, and $A = G \setminus N$. But then, G is Borel, $N \cap G \subseteq N$ is null, and $A = G \setminus N = G \setminus (G \cap N) = G \Delta (G \cap N)$, so (2) holds.

The proof that (4) implies (2) is similar. Suppose F is F_σ, N is null, and $A = F \cup N$. Then, F is Borel, $N \setminus F \subseteq N$ is null, and $A = F \cup N = F \cup (N \setminus F) = F \Delta (N \setminus F)$, so (2) holds.

To see that (2) implies (1), if B is Borel and N is null, then $B \in \mathcal{L}$ by Corollary 7.42, $N \in \mathcal{L}$ by Theorem 7.23, and this implies $B \Delta N \in \mathcal{L}$ by Theorem 7.41.

To see that (1) implies (3) and (4), fix $A \subseteq \mathbb{R}$ and suppose that $A \in \mathcal{L}$. This means that for each $n \in \mathbb{N}$ there is a closed set F_n and an open set U_n such that $F_n \subseteq A \subseteq U_n$ and $\lambda^*(U_n \setminus F_n) < 1/n$. Let $G = \bigcap_{n=1}^{\infty} U_n$ and $F = \bigcup_{n=1}^{\infty} F_n$. It is clear that G is a G_δ set and F is an F_σ set, and we claim that G witnesses (3) and F witnesses (4).

To see that G witnesses (3), let $N = G \setminus A$. Because $U_n \supseteq A$ for all n, $G = \bigcap_{n=1}^{\infty} U_n \supseteq A$. Hence, $A = G \setminus N$. Furthermore, $F_n \subseteq A \subseteq G \subseteq U_n$ for every n, which implies $G \setminus A \subseteq U_n \setminus F_n$ and therefore $\lambda^*(G \setminus A) \leq \lambda^*(U_n \setminus F_n) < 1/n$ for every n. Thus, $\lambda^*(G \setminus A) = 0$, so N is a null set.

The proof that F witnesses (4) is similar. Let $N = A \setminus F$. Because $F_n \subseteq A$ for all n, $F = \bigcup_{n=1}^{\infty} F_n \subseteq A$, and so $A = F \cup N$. Furthermore, $F_n \subseteq F \subseteq A \subseteq U_n$ for every n, which implies $A \setminus F \subseteq U_n \setminus F_n$, and therefore $\lambda^*(A \setminus F) \leq \lambda^*(U_n \setminus F_n) < 1/n$ for every n. Thus $\lambda^*(A \setminus F) = 0$, meaning that N is a null set. □

The Axiom of Choice

In this section, we prove that not every subset of $\mathbb{R}$ is Lebesgue measurable, or has the property of Baire, and not every Banach–Mazur game is determined. This proof requires what is known as the *Axiom of Choice*, or *AC* for short.

Definition 7.48. The *Axiom of Choice* is the statement:

(AC) Suppose $\mathcal{C}$ is a set of non-empty sets. Then, there is a function f with domain $\mathcal{C}$ such that $f(X) \in X$ for every $X \in \mathcal{C}$.

In this context, the function f is called a *choice function* for $\mathcal{C}$ because it "chooses" simultaneously one element from each set in $\mathcal{C}$.

The AC is a set-theoretic principle that is independent of the more intuitive, and less controversial, axioms of set theory that we have been using until now, an axiom system denoted $\mathsf{ZF} + \mathsf{DC}$. The word "independent" has a precise meaning in this context: it means that it is impossible to prove or to disprove the AC from $\mathsf{ZF} + \mathsf{DC}$.

Along the same lines, Robert Solovay proved in 1968 that it is impossible to prove the existence of a subset of $\mathbb{R}$ that is not Lebesgue measurable, or that lacks the property of Baire, using only $\mathsf{ZF} + \mathsf{DC}$. In other words, we cannot prove the main theorems of the remainder of this chapter using only the axiom system $\mathsf{ZF} + \mathsf{DC}$ that we have relied upon so far. Our proofs of Theorems 7.61 and 7.65 that follow use the AC, and thanks to Solovay's work, we know that this is a necessary feature of these proofs.

Definition 7.49. Given a set X, a *partition* of X is a collection $\mathcal{P}$ of pairwise disjoint, non-empty subsets of X such that $\bigcup \mathcal{P} = X$.

If $\sim$ is an equivalence relation on a set X, then, for each $a \in X$,

$$[a] = \{x \in X : a \sim x\}.$$

This set is known as the *equivalence class of a*. Every $a \in X$ is in an equivalence class, namely $[a]$, and any two equivalence classes $[a]$ and $[b]$, either $[a] = [b]$ or $[a] \cap [b] = \emptyset$. Thus, the equivalence classes of $\sim$ form a partition of X.

Definition 7.50. If $\sim$ is an equivalence relation on X, then

$$\mathcal{P}_{\sim} = \{[a] : a \in X\}$$

is called the *partition induced by* $\sim$. If $\mathcal{P}$ is a partition of X, then the relation $\sim_{\mathcal{P}}$ defined by

$$a \sim_{\mathcal{P}} b \quad \Leftrightarrow \quad a \text{ and } b \text{ are elements of the same member of } \mathcal{P}$$

is called the *equivalence relation induced by* $\mathcal{P}$.

The following proposition shows that equivalence relations and partitions are two different ways of expressing essentially the same underlying mathematical idea.

Proposition 7.51. *Let X be a set. If $\sim$ is an equivalence relation on X, then $\mathcal{P}_\sim$ is a partition of X, and $\sim_{\mathcal{P}_\sim}$ is the same relation as $\sim$. If $\mathcal{P}$ is a partition of X, then $\sim_\mathcal{P}$ is an equivalence relation on X, and $\mathcal{P}_{\sim_\mathcal{P}} = \mathcal{P}$.*

Proof. If $\sim$ is an equivalence relation on X, then clearly $\mathcal{P}_\sim$ is a partition of X, and $a \sim_{\mathcal{P}_\sim} b$ if and only if $a \in [b]$ if and only if $a \sim b$, so $\sim_{\mathcal{P}_\sim}$ and $\sim$ are the same relation on X.

If $\mathcal{P}$ is a partition of X, then clearly $\sim_\mathcal{P}$ is an equivalence relation on X, and $A \in \mathcal{P}$ if and only if A is an equivalence class of $\sim_\mathcal{P}$ if and only if $A \in \mathcal{P}_{\sim_\mathcal{P}}$, so $\mathcal{P} = \mathcal{P}_{\sim_\mathcal{P}}$. □

Definition 7.52. Suppose X is a set, and $\mathcal{P}$ is a partition of X. A *selector* for $\mathcal{P}$ is a set $S \subseteq X$ such that $|S \cap A| = 1$ for every $A \in \mathcal{P}$.

Example 7.53. Let $\mathcal{P}$ denote the set of all vertical lines in $\mathbb{R}^2$. This is a partition of $\mathbb{R}^2$ into infinitely many pieces. A subset of $\mathbb{R}^2$ is a selector for $\mathcal{P}$ if and only if it is the graph of a function.

Fix $n \in \mathbb{N}$, and let $\mathcal{P}$ be the partition of $\mathbb{Z}$ induced by the equivalence relation $a \sim b$ if and only if $n \mid (a - b)$. (This equivalence relation is called *congruence modulo n* and is often written with $\equiv_n$ instead of $\sim$.) The partition of $\mathbb{Z}$ induced by this equivalence relation consists of exactly n infinite sets (known as *congruence classes* modulo n). A selector for this partition consists of exactly one member of each set, for example, $\{0, 1, 2, \dots, n-1\}$.

The equality relation on a set X induces the partition of X into singletons, $\{\{x\} : x \in X\}$. This partition has just one selector, X itself.

Let $\sim$ denote the relation on $\mathbb{R}$ defined by setting $x \sim y$ if and only if $|x - y| \in \mathbb{Z}$. This is an equivalence relation, and $[0, 1)$ is a selector for the partition induced by $\sim$. Other selectors for this partition include $[6, 7)$, or $[0, 1/2) \cup [3/2, 2)$, or $[0, 1/2) \cup [2 - 1/2, 2 - 1/4) \cup [3 - 1/4, 3 - 1/8) \cup \dots$.

If $\sim$ is an equivalence relation on a set X, then $S \subseteq X$ is a selector for $\mathcal{P}_\sim$ if and only if for every $a \in X$, there is exactly one $b \in S$ such that $a \sim b$. In other words, one may think of S as "choosing" precisely one member from every equivalence class of $\sim$. This suggests the following lemma.

Lemma 7.54. *The Axiom of Choice is equivalent to the statement that every partition of every set has a selector.*

Proof. For the forward implication, let $\mathcal{P}$ be a partition of some set X. By definition, $\mathcal{P}$ is a collection of non-empty sets. Using the AC, there is a choice function for $\mathcal{P}$, i.e., there is a function f with domain $\mathcal{P}$ such that $f(A) \in A$ for every $A \in \mathcal{P}$. Then, the image of this function $f[\mathcal{P}] = \{f(A) : A \in \mathcal{P}\}$ is a selector for $\mathcal{P}$.

For the reverse implication, let $\mathcal{C}$ be a collection of non-empty sets. If the members of $\mathcal{C}$ were disjoint, then $\mathcal{C}$ would be a partition of $\bigcup \mathcal{C}$, and we could define a choice function for $\mathcal{C}$ from a selector for this partition. Since the members of $\mathcal{C}$ may not be disjoint, we "disjointify" them as follows. Define

$$\mathcal{C}' = \{A \times \{A\} : A \in \mathcal{C}\}.$$

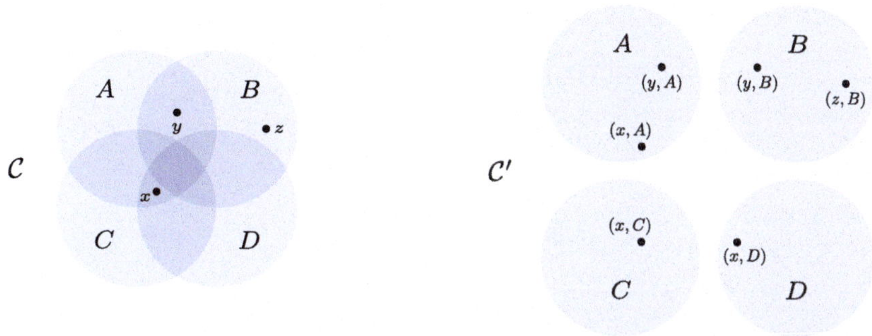

The members of $\mathcal{C}'$ are disjoint: If $A, B \in \mathcal{C}$ and $A \neq B$, then every member of $A \times \{A\}$ is of the form (x, A), while every member of $B \times \{B\}$ is of the form (y, B), and $A \neq B$ implies $(x, A) \neq (y, B)$, so $(A \times \{A\}) \cap (B \times \{B\}) = \emptyset$. Because the members of $\mathcal{C}'$ are disjoint, $\mathcal{C}'$ is a partition of $\bigcup \mathcal{C}'$. Let S be a selector for $\mathcal{C}'$. Then, define a function f with domain $\mathcal{C}$ by setting $f(A)$ equal to the first coordinate of the (unique) member of $(A \times \{A\}) \cap S$. It follows from our definition of $\mathcal{C}'$ that $f(A) \in A$ for every $A \in \mathcal{C}$. In other words, f is a choice function for $\mathcal{C}$. □

Vitali sets

Our proof that there is a subset of $\mathbb{R}$ that lacks the property of Baire and is not Lebesgue measurable is due to Giuseppe Vitali. It was first published in 1905, only a few years after Lebesgue published his work on measure theory.

Definition 7.55. Define a relation $\equiv$ on the interval $[0, 1)$ by setting

$$x \equiv y \qquad \text{if and only if} \qquad x - y \in \mathbb{Q}.$$

It is not difficult to see that this relation is an equivalence relation on $[0, 1)$. A *Vitali set* is a selector for $\mathcal{P}_{\equiv}$, the partition of $[0, 1)$ induced by this equivalence relation. ☕

Each equivalence class of $\equiv$ is countable: $[a] = \{a + q : a \in \mathbb{Q}\}$. Because $\mathbb{R}$ is the union of these equivalence classes, it follows that $\equiv$ has uncountably many equivalence classes. Hence, a selector for $\mathcal{P}_{\equiv}$, a Vitali set, must be uncountable.

In Example 7.53, we were able to explicitly define selectors for various partitions. It is an interesting exercise in futility to try to do so for the partition from Definition 7.55. No simple-looking definition that avoids the Axiom of Choice can describe a selector for this partition. However, Lemma 7.54 tells us, by an application of the Axiom of Choice, that Vitali sets must exist.

Definition 7.56. For all $x, y \in [0, 1)$, define

$$x \oplus y = \lfloor x + y \rceil = \begin{cases} x + y & \text{if } x + y < 1 \\ x + y - 1 & \text{if } x + y \geq 1. \end{cases}$$

If $x \in [0, 1)$ and $A \subseteq [0, 1)$, define

$$x \oplus A = \{x \oplus a \colon a \in A\}.$$

One may identify $[0, 1)$ with the unit circle in a natural way, via the map $x \mapsto (\cos(2\pi x), \sin(2\pi x))$. Under this identification, the operation "$x\oplus$" can be interpreted geometrically as an instruction to rotate by the angle corresponding to x. So, $x \oplus y$ is the rotation of y by this angle, and $x \oplus A$ is the rotation of the entire set A by this angle.

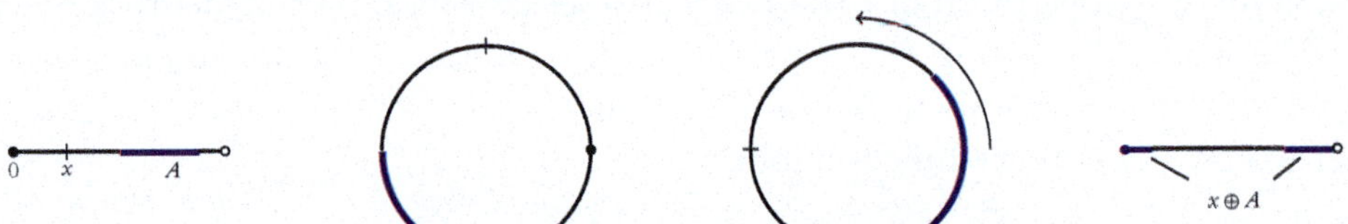

Alternatively, one may view $\oplus$ simply as an "addition with wraparound" operation, or addition modulo 1, on the interval $[0, 1)$.

The following facts about this operation (the first four of which can be summarized by saying that $([0, 1), \oplus)$ is an abelian group) follow easily from the definition:

- (associativity) $(a \oplus b) \oplus c = a \oplus (b \oplus c) = \lfloor a + b + c \rceil$ for all $a, b, c \in [0, 1)$. Because of this, we may (and do) simply write $a \oplus b \oplus c$.
- (commutativity) $a \oplus b = b \oplus a$ for all $a, b \in [0, 1)$.
- (inverses) $a \oplus (1 - a) = 0$ for all a. From this and the previous two bullet points, it follows that $a \oplus b \oplus (1 - a) = b$ for all $a, b \in [0, 1)$.
- (cancellation) For any $a, b, c \in [0, 1)$, $b = c$ if and only if $a \oplus b = a \oplus c$.
- if $q \in \mathbb{Q} \cap [0, 1)$, then $x \equiv x \oplus q$ for any $x \in [0, 1)$.

Lemma 7.57. *Suppose S is a Vitali set. If $q, r \in \mathbb{Q} \cap [0, 1)$ and $q \neq r$, then* $(q \oplus S) \cap (r \oplus S) = \emptyset$.

Proof. Let $\equiv$ denote the equivalence relation from Definition 7.55. Fix some $q, r \in \mathbb{Q} \cap [0, 1)$ with $q \neq r$ and, aiming for a contradiction, suppose that $z \in (q \oplus S) \cap (r \oplus S)$. In other words, $z = q \oplus x$ for some $x \in S$ and $z = r \oplus y$ for some $y \in S$. The fact that $z = q \oplus x$ implies $z \equiv x$, and the fact that $z = r \oplus y$ implies $z \equiv y$. Hence, $x \equiv y$ and, since $x, y \in S$ and S is a selector for $\mathcal{P}_{\equiv}$, this means $x = y$. On the other hand, using the facts about $\oplus$ listed previously and the fact that $1 - q \neq 1 - r$,

$$x = q \oplus x \oplus (1 - q) = z \oplus (1 - q) \neq z \oplus (1 - r) = r \oplus y \oplus (1 - r) = y.$$

But, we already showed $x = y$, so this is a contradiction. □

Lemma 7.58. *Suppose S is a Vitali set. Then, $[0, 1) = \bigcup_{q \in \mathbb{Q} \cap [0,1)} q \oplus S$.*

Proof. Again, let $\equiv$ denote the equivalence relation from Definition 7.55. Fix $z \in [0, 1)$. Because S is a selector for $\mathcal{P}_{\equiv}$, there is some $x \in S$ such that $z \equiv x$. This means $z - x = r \in \mathbb{Q}$. Because $x, z \in [0, 1)$, we have $|r| < 1$. If $0 \leq r < 1$, then $z = r \oplus x$, and if $-1 < r < 0$, then $z = (1 + r) \oplus x$. In either case, there is some $q \in \mathbb{Q} \cap [0, 1)$ (either r or $1 + r$) such that $z = q \oplus x$, so $z \in \bigcup_{q \in \mathbb{Q} \cap [0,1)} q \oplus S$. □

Similarly to the last part of Definition 7.56, if $x \in \mathbb{R}$ and $A \subseteq \mathbb{R}$, then

$$x + A = \{x + a \colon a \in A\}.$$

Lemma 7.59. *Let $A \subseteq \mathbb{R}$ and $x \in \mathbb{R}$. If A is meager, so is $x + A$.*

Proof. If A is meager, then there is a countable collection $\mathcal{F}$ of nowhere dense closed sets with $A \subseteq \bigcup \mathcal{F}$. But, if F is closed and nowhere dense, so is $x + F$. Hence, $\mathcal{F}' = \{x + F \colon F \in \mathcal{F}\}$ is a countable collection of nowhere dense closed sets with $x + A \subseteq \bigcup \mathcal{F}'$. □

Lemma 7.60. *Let $A \subseteq [0, 1)$ and $x \in [0, 1)$. If A is meager, so is $x \oplus A$.*

Proof. Let $A \subseteq [0, 1)$ and $x \in [0, 1)$, and suppose that A is meager. Then,

$$\begin{aligned} x \oplus A &= \{x \oplus a \colon a \in A\} \\ &= \{x + a \colon a \in A \text{ and } x + a < 1\} \cup \{x + a - 1 \colon a \in A \text{ and } x + a \geq 1\} \\ &\subseteq (x + A) \cup \big((x - 1) + A\big). \end{aligned}$$

Because A is meager, both $x + A$ and $(x - 1) + A$ are meager by the previous lemma. Hence, $x \oplus A$ is meager. □

Theorem 7.61. *Vitali sets do not have the property of Baire.*

Proof. Aiming for a contradiction, suppose $S \subseteq [0, 1)$ is a Vitali set with the property of Baire. This means $S = U \Delta M$ for some open set U and meager set M. We consider two cases.

For the first case, suppose $U = \emptyset$. Then, $S = M$, so S is a meager set. By Lemma 7.60, $q \oplus S$ is meager for every $q \in \mathbb{Q} \cap [0, 1)$. By Lemma 7.58, this means that $[0, 1) = \bigcup_{q \in \mathbb{Q} \cap [0,1)} q \oplus S$ is a countable union of meager sets. But $[0, 1)$ is not meager (see Example 3.60), so this is a contradiction.

For the second case, suppose $U \neq \emptyset$. Then, there is an interval $(a, b) \subseteq U$ with $a < b$. Furthermore, because $S \subseteq [0, 1)$ and $(a, b) \nsubseteq M$ (again, see Example 3.60), the fact that $S = U \Delta M$ implies $0 \leq a < b \leq 1$. Let q be a rational number such that $0 < q < b - a$, and consider the open interval $I = q + (a, b) = (q + a, q + b)$. Because $0 < a < b - a$, we have $(a, b) \cap I = (q + a, b) \neq \emptyset$. Let $V = (q + a, b)$. Because $V \subseteq U$,

$$V \setminus M \subseteq U \setminus M \subseteq U \Delta M = S,$$

and because $V = (a + q, b) \subseteq q \oplus (a, b) \subseteq q \oplus U$,

$$V \setminus (q \oplus M) \subseteq (q \oplus U) \setminus (q \oplus M) \subseteq (q \oplus U) \Delta (q \oplus M) = q \oplus S.$$

By Lemma 7.57, $S \cap (q \oplus S) = (0 \oplus S) \cap (q \oplus S) = \emptyset$. Because $V \setminus M \subseteq S$ and $V \setminus (q \oplus M) \subseteq q \oplus S$ (as already shown), this implies

$$(V \setminus M) \cap \big(V \setminus (q \oplus M)\big) = \emptyset,$$

or equivalently, $V \subseteq M \cup (q \oplus M)$. But this is absurd: V is a non-empty open set, and hence non-meager, while M and $q \oplus M$ are both meager. □

Corollary 7.62. *Assuming the AC, there is a subset of $\mathbb{R}$ that does not have the property of Baire, hence is not Borel, and there is a set whose Banach–Mazur game is not determined.*

Proof. By Lemma 7.54, the AC implies there is a Vitali set $S \subseteq [0, 1)$. By Theorem 7.61, S does not have the property of Baire, and by Corollary 7.13, S is not Borel. By Theorem 7.19, there is some $A \subseteq S$ such that $\mathsf{BM}(\mathbb{R}, A)$ is not determined. □

Next, we show that Vitali sets are not Lebesgue measurable. To some extent, our proof of this parallels our proof of Theorem 7.61. We begin by proving analogues of Lemmas 7.59 and 7.60.

Lemma 7.63. *If $A \subseteq \mathbb{R}$ and $x \in \mathbb{R}$, then $\lambda^*(A) = \lambda^*(x + A)$. Furthermore, if A is Lebesgue measurable then so is $x + A$.*

Proof. Let $A \subseteq \mathbb{R}$, and let $x \in \mathbb{R}$. If $\mathcal{I}$ is a countable family of open intervals covering A with total length L, then $\{x + I\colon I \in \mathcal{I}\}$ is a countable family of open intervals covering $x + A$ with total length L. This implies $\lambda^*(x + A) \leq \lambda^*(A)$. Since $A = -x + (x + A)$, a similar argument shows $\lambda^*(A) \leq \lambda^*(x + A)$. Hence, $\lambda^*(x + A) = \lambda^*(A)$.

Now, suppose A is Lebesgue measurable: i.e., for every $\varepsilon > 0$ there is an open set U and a closed set F such that $F \subseteq A \subseteq U$ and $\lambda^*(U \setminus F) < \varepsilon$. But then, $x + U$ is open, $x + F$ is closed, $x + F \subseteq x + A \subseteq x + U$, and

$$\lambda^*\big((x + U) \setminus (x + F)\big) = \lambda^*\big(x + (U \setminus F)\big) = \lambda^*(U \setminus F) < \varepsilon.$$

It follows that $x + A$ is Lebesgue measurable. □

Lemma 7.64. *Let $A \subseteq [0, 1)$ and $x \in [0, 1)$. If A is Lebesgue measurable, then $x \oplus A$ is Lebesgue measurable and $\lambda^*(A) = \lambda^*(x \oplus A)$.*

Proof. Let $A \subseteq [0, 1)$ and $x \in [0, 1)$, and suppose $A \in \mathcal{L}$. Define

$$\begin{aligned} A_0 &= \{a \in A\colon a + x < 1\} = A \cap [0, 1 - x), \\ A_1 &= \{a \in A\colon a + x \geq 1\} = A \cap [1 - x, 1). \end{aligned}$$

Because $\mathcal{L}$ is a σ-algebra containing all the Borel sets, $A_0, A_1 \in \mathcal{L}$. Because $A_0 \cap A_1$, the countable additivity of Lebesgue measure (Theorem 7.43) implies $\lambda^*(A_0) + \lambda^*(A_1) = \lambda^*(A_0 \cup A_1) = \lambda^*(A)$. Now,

$$\begin{aligned} x \oplus A &= \{x \oplus a\colon a \in A\} \\ &= \{x + a\colon a \in A \text{ and } x + a < 1\} \cup \{x + a - 1\colon a \in A \text{ and } x + a \geq 1\} \\ &= (x + A_0) \cup \big((x - 1) + A_1\big). \end{aligned}$$

Using the previous lemma, this shows $x \oplus A$ is Lebesgue measurable. Furthermore, $x + A_0 \subseteq [x, 1)$ and $(x - 1) + A_1 \subseteq [0, x)$, and in particular $(x + A_0) \cap \big((1 - x) + A_1\big) = \emptyset$. Using countable additivity again, this implies $\lambda^*(x \oplus A) = \lambda^*(x + A_0) + \lambda^*\big((1 - x) + A_1\big) = \lambda^*(A_0) + \lambda^*(A_1) = \lambda^*(A)$. □

Theorem 7.65. *Vitali sets are not Lebesgue measurable.*

Proof. Aiming for a contradiction, suppose $S \subseteq [0, 1)$ is a Lebesgue measurable Vitali set. Let $m = \lambda^*(S)$. The set $\mathbb{Q} \cap [0, 1)$ is countable, so let us fix an enumeration of it: $\mathbb{Q} \cap [0, 1) = \{q_n\colon n \in \mathbb{N}\}$. The previous lemma shows that the sets $q_1 \oplus S, q_2 \oplus S, q_3 \oplus S, \ldots$ are all Lebesgue measurable and all have Lebesgue outer measure $\lambda^*(q_n \oplus S) = m$. By Lemmas 7.57 and 7.58, the sets $q_1 \oplus S, q_2 \oplus S, q_3 \oplus S, \ldots$ are pairwise disjoint, and $\bigcup_{n=1}^{\infty} q \oplus S = [0, 1)$. By the countable additivity of Lebesgue measure (Theorem 7.43),

$$\textstyle\sum_{n=1}^{\infty} \lambda^*(q_n \oplus S) = \lambda^*\big(\bigcup_{n=1}^{\infty} q_n \oplus S\big) = \lambda^*\big([0, 1)\big) = 1.$$

Because $\lambda^*(q_n \oplus S) = m$ for all n, if $m = 0$, then $\sum_{n=1}^{\infty} \lambda^*(q_n \oplus S) = 0$, and if $m > 0$, then $\sum_{n=1}^{\infty} \lambda^*(q_n \oplus S) = \infty$. In neither case does this series converge to 1, so we have a contradiction. □

Corollary 7.66. *Assuming the AC, there is a subset of* $\mathbb{R}$ *that is not Lebesgue measurable.*

Proof. By Lemma 7.54, the AC implies there is a Vitali set. □

Exercises

1. Let X be an infinite set. Prove that $\mathcal{C} = \{A \subseteq X\colon\ A \text{ is finite}\} \cup \{A \subseteq X\colon\ X \setminus A \text{ is finite}\}$ is not a σ-algebra on X.
2. Prove that for every $n \in \mathbb{N}$, a set is $\mathbf{\Sigma}_n^0$ if and only if its complement is $\mathbf{\Pi}_n^0$.
3. Prove that every closed subset of $\mathbb{R}$ is G_δ, and every open set is F_σ. More generally, prove that for every $n \in \mathbb{N}$, every $\mathbf{\Sigma}_n^0$ set is $\mathbf{\Pi}_{n+1}^0$, and every $\mathbf{\Pi}_n^0$ set is $\mathbf{\Sigma}_{n+1}^0$.
4. Give an example of a Borel set that is neither F_σ nor G_δ.
5. A function $f : \mathbb{R} \to \mathbb{R}$ is called *Borel measurable* if $f^{-1}(B)$ is a Borel set for every Borel $B \subseteq \mathbb{R}$. Prove that continuous functions $\mathbb{R} \to \mathbb{R}$ are Borel measurable.
6. Continuing the previous exercise, give an example of a noncontinuous Borel measurable function $\mathbb{R} \to \mathbb{R}$. In other words, give an example showing that Borel measurability does not imply continuity.
7. Give an example of a function $\mathbb{R} \to \mathbb{R}$ that is not Borel measurable.
8. Prove that all of the following sets are Borel:
   - **(a)** every closed subset of $\mathbb{R}$
   - **(b)** every interval in $\mathbb{R}$
   - **(c)** $\mathbb{Q}$ and $\mathbb{R} \setminus \mathbb{Q}$
   - **(d)** the set of all numbers in $\mathbb{R}$ whose usual base-10 expansion does not use the digit 7
   - **(e)** $\{x \in \mathbb{R} : f \text{ is continuous at } x\}$, where $f : \mathbb{R} \to \mathbb{R}$ is any function
   - **(f)** $\{x \in \mathbb{R} : f \text{ is differentiable at } x\}$, where $f : \mathbb{R} \to \mathbb{R}$ is any function
9. Prove the three facts about the symmetric difference operator that are listed just before Definition 7.10.
10. Find two subsets of $\mathbb{R}$ that are closed, but there is no positive "distance" between them in the sense of Lemma 7.29.
11. Suppose $A_1, A_2, A_3, \ldots$ is a sequence of Lebesgue measurable sets, with $A_1 \subseteq A_2 \subseteq A_3 \subseteq \ldots$. Prove that $\lambda^*\left(\bigcup_{n=1}^{\infty} A_n\right) = \lim_{n\to\infty} \lambda^*(A_n)$.
12. Suppose $A_1, A_2, A_3, \ldots$ is a sequence of Lebesgue measurable sets, with $A_1 \supseteq A_2 \supseteq A_3 \supseteq \ldots$. Prove that $\lambda^*\left(\bigcap_{n=1}^{\infty} A_n\right) = \lim_{n\to\infty} \lambda^*(A_n)$.
13. Let $\mathcal{B}$ denote the σ-algebra of all Borel subsets of $\mathbb{R}$. Prove that $\mathcal{R} = \{A \subseteq \mathbb{R}\colon\ A \text{ has the property of Baire}\}$ is the σ-algebra generated by $\mathcal{B} \cup \{M \subseteq \mathbb{R}\colon\ M \text{ is meager}\}$. Similarly, prove that $\mathcal{L}$ is the σ-algebra generated by $\mathcal{B} \cup \{N \subseteq \mathbb{R}\colon\ N \text{ is Lebesgue null}\}$.

14. Let $\mathcal{L}_{\text{fin}}$ denote the set of all Lebesgue measurable sets with finite outer measure. Prove that the function $d(A, B) = \lambda^*(A \Delta B)$ satisfies all but one of the requirements for being a metric on $\mathcal{L}_{\text{fin}}$.

15. Continuing the previous exercise, let $\sim$ be the equivalence relation on $\mathcal{L}_{\text{fin}}$ defined by taking $A \sim B$ if and only if $d(A, B) = 0$. Show that if $A \sim A'$ and $B \sim B'$, then $d(A, B) = d(A', B')$. Use this information to define a metric $\bar{d}$ on the set $\mathcal{P}_\sim$ of all $\sim$-equivalence classes.

16. Show that the metric $\bar{d}$ you defined in the previous exercise is complete.

17. Find an example of a subset of $\mathbb{R}$ that has the property of Baire but is not Lebesgue measurable. Deduce that there are non-Borel sets with the property of Baire.
(*Hint*: If M is a meager set and X does not have the property of Baire, then $X \setminus M$ does not have the property of Baire.)

18. Similarly, find an example of a subset of $\mathbb{R}$ that is Lebesgue measurable but does not have the property of Baire. Deduce that there are non-Borel Lebesgue measurable sets.
(*Hint*: If N is a Lebesgue null set and $X \notin \mathcal{L}$, then $X \setminus N \notin \mathcal{L}$.)

19. Prove that the interval $[0, 1)$ can be partitioned into finitely many pairwise disjoint sets $A_1, A_2, \ldots, A_n$ such that $\sum_{i=1}^{n} \lambda^*(A_i) > 1$. (Of course, not all of these sets can be Lebesgue measurable.)

20. Prove that there are two disjoint (nonmeasurable) sets, A and B, such that $\lambda^*(A) + \lambda^*(B) > \lambda^*(A \cup B)$.

CHAPTER 8

Function spaces

In this chapter, we explore the topic of function spaces, metric spaces (or more general topological spaces) whose points are functions. The idea of having a space of functions is suggested by the observation that a sequence of functions can seem to "converge" to a limit function, just like the points in a metric space can converge to a limit point.

Example 8.1. Fix some $f : \mathbb{R} \to \mathbb{R}$, and, for each $n \in \mathbb{N}$, let $f_n(x) = f(x) + \frac{1}{n}$. Then, the sequence of functions $\langle f_n \colon n \in \mathbb{N} \rangle$ converges (in some sense) to the limit function f.

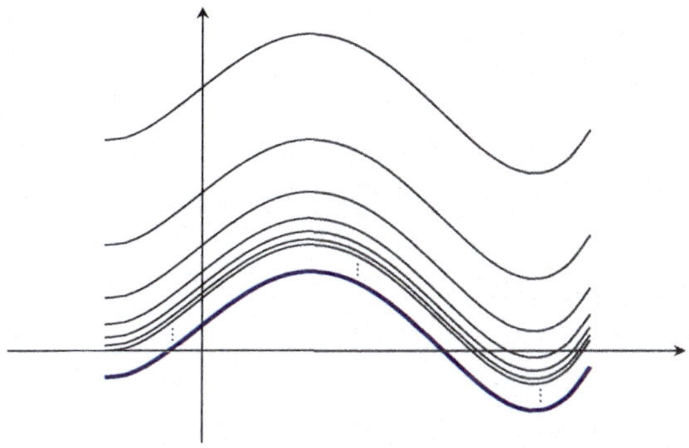

☕

This raises the question: How do we formalize this kind of thing? Given a sequence of functions, how can we define precisely when it has a limit and what that limit is? As it turns out, there are many different ways to do this, and the nuanced relationships between different modes of convergence are part of what the study of function spaces is about. In this chapter, we explore just two kinds of convergence for sequences of functions: pointwise and uniform convergence.

Pointwise and uniform convergence

Definition 8.2. Let X and Y be metric spaces, and let $\langle f_n \colon n \in \mathbb{N} \rangle$ be a sequence of functions $X \to Y$. This sequence *converges pointwise* to a function $f : X \to Y$ if $\lim_{n\to\infty} f_n(x) = f(x)$ for every $x \in X$. ☕

Definition 8.3. Let X and Y be metric spaces, and let $\langle f_n \colon n \in \mathbb{N} \rangle$ be a sequence of functions $X \to Y$. This sequence *converges uniformly* to a function $f : X \to Y$ if $\lim_{n\to\infty} \sup_{x\in X} d_Y(f_n(x), f(x)) = 0$. ☕

Foundations of Real Analysis
https://doi.org/10.1016/B978-0-44-348527-5.00012-3

These definitions can be rephrased as follows:

- $\langle f_n \colon n \in \mathbb{N}\rangle$ converges pointwise to f if, for every $\varepsilon > 0$ and $x \in X$, there is some $N \in \mathbb{N}$ such that if $n \geq N$, then $d_Y(f_n(x), f(x)) < \varepsilon$.
- $\langle f_n \colon n \in \mathbb{N}\rangle$ converges uniformly to f if, for every $\varepsilon > 0$, there is some $N \in \mathbb{N}$ such that if $n \geq N$, then $d_Y(f_n(x), f(x)) < \varepsilon$ for all $x \in X$.

The difference between pointwise and uniform convergence is that with pointwise convergence, the number N can depend on both ε and x, while with uniform convergence N depends only on ε: The same value of N must work for all $x \in X$ simultaneously. This is similar to the difference between continuity and uniform continuity.

Proposition 8.4. *Let X and Y be metric spaces, and let $\langle f_n \colon n \in \mathbb{N}\rangle$ be a sequence of functions $X \to Y$. If $\langle f_n \colon n \in \mathbb{N}\rangle$ converges to a function f uniformly, then it also converges to f pointwise.*

Given a sequence $\langle f_n \colon n \in \mathbb{N}\rangle$ of functions converging to some function f, one of our basic questions is: If all the f_n's have some given property, must f have that property, too? If P is a property of functions (like continuity, differentiability, boundedness, etc.), we say that property P is *preserved by pointwise limits* if whenever $\langle f_n \colon n \in \mathbb{N}\rangle$ is a sequence of functions with property P converging pointwise to a function f, then f has property P. Similarly, we say that P is *preserved by uniform limits* if whenever $\langle f_n \colon n \in \mathbb{N}\rangle$ is a sequence of functions with property P converging uniformly to a function f, then f has property P. For example, we show next that continuity is preserved by uniform limits (Theorem 8.5), but is not necessarily preserved by pointwise limits (Example 8.6).

Theorem 8.5. *Let X and Y be metric spaces, and suppose $\langle f_n \colon n \in \mathbb{N}\rangle$ is a sequence of functions $X \to Y$, converging uniformly to a function f. If all the f_n are continuous, then f is also continuous.*

Proof. Suppose $\langle f_n \colon n \in \mathbb{N}\rangle$ converges uniformly to f and f_n is continuous for every $n \in \mathbb{N}$. Let $\varepsilon > 0$, and let $x \in X$. By the definition of uniform convergence, there is some $N \in \mathbb{N}$ such that $\sup_{x \in X} d_Y(f_n(x), f(x)) < \varepsilon/3$ for all $n \geq N$. As f_N is continuous, there is some $\delta > 0$ such that if $t \in X$ and $d_X(x, t) < \delta$, then $d_Y(f_N(x), {}_N(y)) < \varepsilon/3$. If $t \in X$ and $d_X(x, t) < \delta$,

$$
\begin{aligned}
|f(x) - f(y)| &\leq |f(x) - f_N(x)| + |f_N(x) - f_N(y)| + |f_N(y) - f(y)| \\
&< \tfrac{\varepsilon}{3} + \tfrac{\varepsilon}{3} + \tfrac{\varepsilon}{3} = \varepsilon.
\end{aligned}
$$

As ε and x were arbitrary, this shows f is continuous. □

Example 8.6. Let $[a, b] = [0, 1]$, and for every $n \in \mathbb{N}$ define $f_n(x) = x^n$. Then, $\langle f_n \colon n \in \mathbb{N}\rangle$ converges pointwise to the function

$$
f(x) = \begin{cases} 0 & \text{if } x \in [0, 1) \\ 1 & \text{if } x = 1. \end{cases}
$$

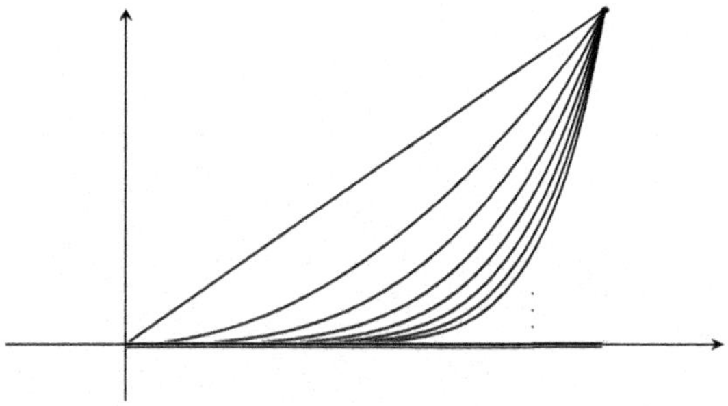

Even though f_n is continuous for every $n \in \mathbb{N}$, the limit f is not continuous. Note that while $\langle f_n \colon n \in \mathbb{N} \rangle$ converges pointwise to f, it does not converge uniformly: In fact, $\sup_{x \in [0,1]} |f_n(x) - f(x)| = 1$ for all n.

Our next example shows that differentiability is not necessarily preserved by either kind of limit.

Example 8.7. Define $f_n(x) = \sqrt{x^2 + 1/n}$ for every $n \in \mathbb{N}$. Each f_n is (one half of) a hyperbola with asymptotes $y = \pm x$, and $\langle f_n \colon n \in \mathbb{N} \rangle$ converges uniformly (hence also pointwise) to the function $f(x) = |x|$.

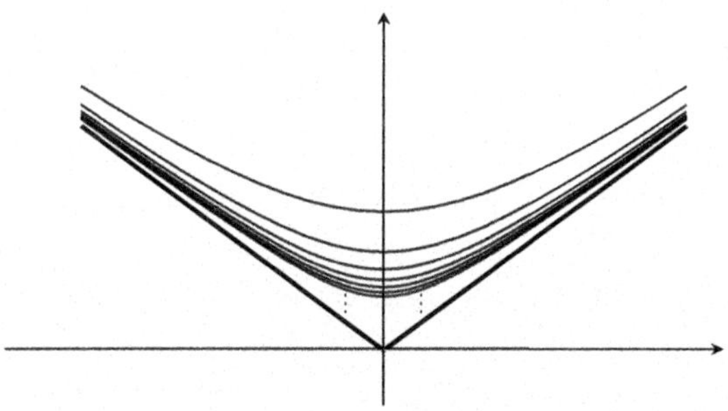

Clearly f_n is differentiable for every $n \in \mathbb{N}$, but f is not differentiable at 0. Thus, the uniform limit of differentiable functions need not be differentiable.

Later in this set of notes, we will demonstrate a more extreme example of this phenomenon: Any continuous function on $[a, b]$ (even a nowhere differentiable function like the Weierstrass function) is the uniform limit of a sequence of polynomials. Thus, differentiability may fail to be preserved not just at one point, but everywhere.

Example 8.8. For each $n \in \mathbb{N}$ define

$$f_n(x) = \begin{cases} -1/n & \text{if } x < -\pi/n \\ 1/n & \text{if } x > \pi/n \\ \frac{1}{n}\sin(nx) & \text{if } -\pi/n \le x \le \pi/n. \end{cases}$$

and let f be the constant function with value 0. Clearly, $\langle f_n : n \in \mathbb{N} \rangle$ converges uniformly (hence also pointwise) to f.

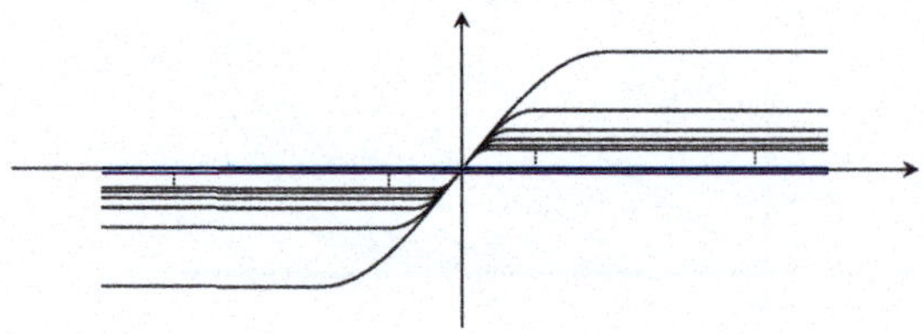

Each f_n is differentiable for every $n \in \mathbb{N}$, and $f_n'(0) = 1$. The pointwise limit f is still differentiable, but $f'(0) = 0$.

Thus, even when a sequence $\langle f_n : n \in \mathbb{N} \rangle$ of differentiable functions does converge uniformly to a differentiable function f, it is not necessarily the case that $f'(x) = \lim_{n\to\infty} f_n'(x)$.

Theorem 8.9. *Suppose $\langle f_n : n \in \mathbb{N} \rangle$ is a sequence of Riemann integrable functions $[a, b] \to \mathbb{R}$. If $\langle f_n : n \in \mathbb{N} \rangle$ converges uniformly to a function f, then f is Riemann integrable and $\int_a^b f(x)\,dx = \lim_{n\to\infty} \int_a^b f_n(x)\,dx$.*

Proof. Suppose $\langle f_n : n \in \mathbb{N} \rangle$ converges uniformly to f, and f_n is Riemann integrable for every $n \in \mathbb{N}$. Let $\varepsilon > 0$. By the definition of uniform convergence, there is some $N \in \mathbb{N}$ such that $\sup_{x\in[a,b]} |f_n(x) - f(x)| < \varepsilon/4(b-a)$ for all $n \geq N$. Let $k \geq N$. Because f_k is Riemann integrable, there is some partition $P = \langle x_0, x_1, \ldots, x_n \rangle$ of $[a, b]$ such that $U(P, f_k) - L(P, f_k) < \varepsilon/2$.

Because $\sup_{x\in[a,b]} |f_k(x) - f(x)| < \varepsilon/4(b-a)$, we have $m_i^f > m_i^{f_k} - \varepsilon/4(b-a)$ and $M_i^f < M_i^{f_k} + \varepsilon/4(b-a)$ for all $i \leq n$. It follows that

$$L(P,f) = \sum_{i=1}^{n} m_i^f \Delta x_i > \sum_{i=1}^{n} \left(m_i^{f_k} - \varepsilon/4(b-a) \right) \Delta x_i$$
$$= \sum_{i=1}^{n} m_i^{f_k} \Delta x_i - \sum_{i=1}^{n} \frac{\varepsilon}{4(b-a)} \Delta x_i = L(P, f_k) - \frac{\varepsilon}{4},$$

and a similar computation shows that $U(P,f) < U(P, f_k) + \varepsilon/4$.

It follows that $U(P,f) - L(P,f) < U(P,f_k) - L(P,f_k) + \varepsilon/2 = \varepsilon$, which shows that f satisfies the Cauchy criterion, and is therefore Riemann integrable. Because $L(P, f_k) \leq \int_a^b f_k(x)\,dx \leq U(P, f_k)$ and likewise $L(P,f) \leq \int_a^b f(x)\,dx \leq U(P,f)$, it also follows that $\left| \int_a^b f(x)\,dx - \int_a^b f_k(x)\,dx \right| < \varepsilon$. Thus, we have proved that for any $\varepsilon > 0$, there is some $N \in \mathbb{N}$ such that $\left| \int_a^b f(x)\,dx - \int_a^b f_k(x)\,dx \right| < \varepsilon$ for all $k \geq N$. In other words, we have proved that $\int_a^b f(x)\,dx = \lim_{n\to\infty} \int_a^b f_n(x)\,dx$. □

The next two examples show that the previous theorem fails completely for pointwise convergence: Not only is Riemann integrability not preserved in pointwise limits, but even when a sequence $\langle f_n : n \in \mathbb{N} \rangle$ of integrable functions does converge pointwise to a Riemann integrable limit f, it need not be true that $\int_a^b f(x)\, dx = \lim_{n\to\infty} f_n(x)\, dx$.

Example 8.10. Fix an enumeration of the rationals between 0 and 1, say $\mathbb{Q} \cap [0, 1] = \{q_n : n \in \mathbb{N}\}$. For every $n \in \mathbb{N}$, define $f_n : [0, 1] \to \mathbb{R}$ by setting

$$f_n(x) = \begin{cases} 1 & \text{if } x \in \{q_1, q_2, \ldots, q_n\} \\ 0 & \text{otherwise.} \end{cases}$$

Each f_n has only finitely many points of discontinuity, and is therefore Riemann integrable. However, the pointwise limit of the sequence $\langle f_n : n \in \mathbb{N} \rangle$ is the Dirichlet function

$$f(x) = \begin{cases} 1 & \text{if } x \in \mathbb{Q} \\ 0 & \text{if } x \notin \mathbb{Q}, \end{cases}$$

which is not Riemann integrable.

Example 8.11. For each $n \in \mathbb{N}$, define $f_n : [0, 1] \to \mathbb{R}$ by setting

$$f_n(x) = \begin{cases} n & \text{if } 0 < x < 1/n \\ 0 & \text{otherwise,} \end{cases}$$

and let f be the constant function with value 0.

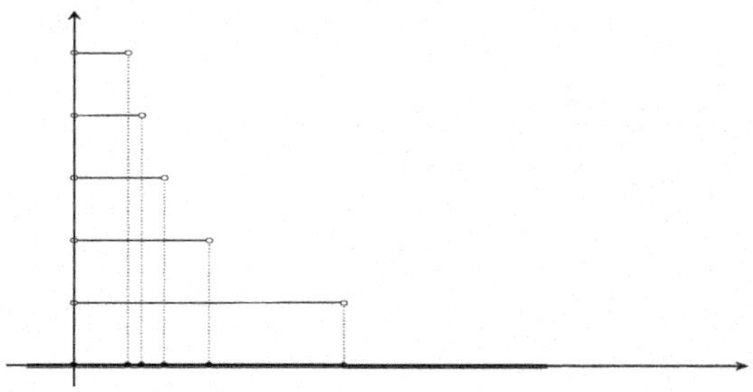

Then, $\langle f_n : n \in \mathbb{N} \rangle$ converges pointwise to f, and each f_n is Riemann integrable with $\int_0^1 f_n(x)\, dx = 1$, but of course $\int_0^1 f(x)\, dx = 0$. In particular, $\int_a^b f(x)\, dx \neq \lim_{n\to\infty} f_n(x)\, dx$.

The following table summarizes the results of this section, plus a few further results along these lines from the end-of-chapter exercises.

	pointwise limits	uniform limits
continuity	not preserved	preserved
uniform continuity	not preserved	preserved
boundedness	not preserved	preserved
Borel measurability	preserved	preserved
differentiability	not preserved	not preserved
integrability	not preserved	preserved
$\lim_{n\to\infty} f_n'(x) = f'(x)$	no	no
$\lim_{n\to\infty} \int_a^b f_n(x)\,dx = \int_a^b f(x)\,dx$	no	yes

A topological perspective on uniform convergence

Recall that a function $f : X \to Y$ between two metric spaces is *bounded* if its image $f[X]$ is a bounded subset of X.

Definition 8.12. If X and Y are metric spaces, then $B(X, Y)$ denotes the set of all bounded functions $X \to Y$. Given $f, g \in B(X, Y)$, define

$$\mathrm{dist}(f, g) = \sup_{x\in X} d_Y\big(f(x), g(x)\big).$$

This is a well-defined real number (not infinity): The fact that $f[X]$ and $g[X]$ are both bounded subsets of Y implies $\{d_Y(f(x), g(x)) : x \in X\}$ is a bounded subset of $\mathbb{R}$. (Proving this is an end-of-chapter exercise.)

In the special case $Y = \mathbb{R}$, we write $B(X)$ instead of $B(X, \mathbb{R})$, and we write $\|f - g\|$ instead of $\mathrm{dist}(f, g)$.

Theorem 8.13. *The function* $(f, g) \mapsto \mathrm{dist}(f, g)$ *is a metric on* $B(X, Y)$.

Proof. It is clear from the definition that $\mathrm{dist}(f, g) \geq 0$ for all $f, g \in B(X, Y)$, and that $\mathrm{dist}(f, g) = 0$ if and only if $f = g$, and that $\mathrm{dist}(f, g) = \mathrm{dist}(g, f)$ for all $f, g \in B(X, Y)$. It remains only to prove the triangle inequality. Let $f, g, h \in B(X, Y)$. Using the triangle inequality in Y,

$$d_Y(f(x), g(x)) \leq d_Y(f(x), h(x)) + d_Y(h(x), g(x))$$

for all $x \in X$. This implies

$$\sup_{x\in X} d_Y(f(x), g(x)) \leq \sup_{x\in X} d_Y(f(x), h(x)) + \sup_{x\in X} d_Y(h(x), g(x)),$$

which means $\text{dist}(f, g) \leq \text{dist}(f, h) + \text{dist}(h, g)$. □

This is the "standard" metric on $B(X, Y)$, and from now on, when we refer to $B(X, Y)$ as a metric space, this metric is meant. In the case $X = [a, b]$ and $Y = \mathbb{R}$, the ε-ball around some $f \in B([a, b])$ is the set of all $g : [a, b] \to \mathbb{R}$ such that $f(x) - \varepsilon < g(x) < f(x) + \varepsilon$ for all $x \in [a, b]$.

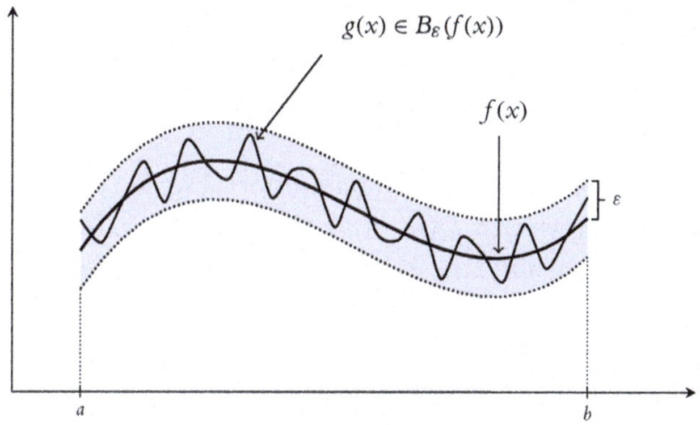

Proposition 8.14. *Let $\langle f_n : n \in \mathbb{N} \rangle$ be a sequence of functions in $B(X, Y)$, and let $f \in B(X, Y)$. Then $\langle f_n : n \in \mathbb{N} \rangle$ converges uniformly to f if and only if $\langle f_n : n \in \mathbb{N} \rangle$ converges to f in the metric space $B(X, Y)$.*

Proof. This is an immediate consequence of the definitions. □

Thinking of $B(X, Y)$ as a metric space gives us a new language in which to interpret the results of the previous section, as the next few results show.

Definition 8.15. If X and Y are metric spaces, let

$$C(X, Y) = \{f \in C(X, Y) : f \text{ is continuous}\}.$$

In the special case $Y = \mathbb{R}$, we write $C(X)$ instead of $C(X, \mathbb{R})$. ☕

Theorem 8.16. *For any spaces X and Y, $C(X, Y)$ is closed in $B(X, Y)$.*

Proof. Suppose f is a limit point of $C(X, Y)$. This means there is a sequence $\langle f_n : n \in \mathbb{N} \rangle$ of functions in $C(X, Y)$ converging to f in the metric space $B(X, Y)$. By the previous observation, this means $\langle f_n : n \in \mathbb{N} \rangle$ converges uniformly to f, and by Theorem 8.5, this implies f is continuous. Thus, every limit point of $C(X, Y)$ is a member of $C(X, Y)$, i.e., $C(X, Y)$ is closed. □

Example 8.17. In contrast with the previous theorem,

$$D([-1,1]) = \{f \in B([-1,1]) : f \text{ is differentiable}\}$$

is not a closed subset of $B([-1,1])$. This is follows from Example 8.7, which shows that the function $f(x) = |x|$ is a limit of a sequence of functions in $D([-1,1])$, which means $D([-1,1])$ does not contain all its limit points. Straightforward modifications to this example show that

$$D([a,b]) = \{f \in B([a,b]) : f \text{ is differentiable}\}$$

is not closed in $B([a,b])$ for any nondegenerate interval $[a,b]$.

In Theorem 8.21 we prove that $\overline{D([a,b])} = C([a,b])$. This also shows that $D([a,b])$ is not closed because we know $D([a,b]) \neq C([a,b])$ (because not every continuous function is differentiable).

Theorem 8.18. *Let $a, b \in \mathbb{R}$ with $a < b$. Then,*

$$R([a,b]) = \{f \in B([a,b]) : f \text{ is Riemann integrable}\}$$

is a closed subset of $B([a,b])$. Furthermore, the function mapping f to its integral $\int_a^b f(x)\,dx$ is a continuous function $R([a,b]) \to \mathbb{R}$.

Proof. Suppose f is a limit point of $R([a,b])$, i.e., there is a sequence $\langle f_n : n \in \mathbb{N}\rangle$ of functions in $R([a,b])$ converging to f. By Proposition 8.14 and Theorem 8.9, this implies f is Riemann integrable. Thus, every limit point of $R([a,b])$ is a member of $R([a,b])$, which means $R([a,b])$ is closed.

Let $\Omega : R([a,b]) \to \mathbb{R}$ denote the function defined by setting $\Omega(f) = \int_a^b f(x)\,dx$ for all $f \in R([a,b])$. Let $\langle f_n : n \in \mathbb{N}\rangle$ be a sequence in $R([a,b])$ that converges to some $f \in R([a,b])$. By Theorem 8.9,

$$\lim_{n\to\infty} \Omega(f_n) = \lim_{n\to\infty} \int_a^b f_n(x)\,dx = \int_a^b f(x)\,dx = \Omega(f).$$

By Theorem 4.5 in Chap. 4, this shows Ω is continuous on $R([a,b])$. □

Generally speaking, if P is a property of functions (continuity, differentiability, integrability, etc.), then $\{f \in B(X,Y) : f \text{ has property } P\}$ is a closed subset of $B(X,Y)$ if and only if P is preserved under taking uniform limits. Likewise, an operator $\Omega : A \to \mathbb{R}$ that maps functions in some $A \subseteq B(X,Y)$ to $\mathbb{R}$ is continuous on F if and only if $\Omega(f) = \lim_{n\to\infty} \Omega(f_n)$ whenever $\langle f_n : n \in \mathbb{N}\rangle$ converges uniformly to f.

$\{f \in B(X, Y) : f \text{ is continuous}\}$	closed
$\{f \in B(X, Y) : f \text{ is uniformly continuous}\}$	closed
$\{f \in B(X, Y) : f \text{ is Borel measurable}\}$	closed
$\{f \in B(X, Y) : f \text{ is differentiable}\}$	not closed
$\{f \in B(X, Y) : f \text{ is integrable}\}$	closed
$f \mapsto f'(x)$	not continuous
$f \mapsto \int_a^b f(x)\,dx$	continuous

We end this section by showing $B(X, Y)$ is complete if and only if Y is complete.

Lemma 8.19. *Let X and Y be metric spaces, and suppose $\langle f_n : n \in \mathbb{N} \rangle$ is a Cauchy sequence in $B(X, Y)$. If $\langle f_n : n \in \mathbb{N} \rangle$ converges pointwise to some function $f : X \to Y$, then it converges uniformly to f.*

Proof. Let $\langle f_n : n \in \mathbb{N} \rangle$ be a Cauchy sequence in $B(X, Y)$, and suppose that $\langle f_n : n \in \mathbb{N} \rangle$ converges pointwise to some $f : X \to Y$.

Let $\varepsilon > 0$. There is some $N \in \mathbb{N}$ such that $\text{dist}(f_m, f_n) < \varepsilon$ whenever $m, n \geq N$. Fix $k \geq N$. For any particular point $x \in X$, if $n \geq N$, then $\text{dist}(f_k, f_n) < \varepsilon$, which implies $d_Y(f_n(x), f_k(x)) < \varepsilon$, and because $f(x) = \lim_{n\to\infty} f_n(x)$, this means $d_Y(f(x), f_k(x)) \leq \varepsilon$. Because this is true for every $x \in X$, this means $\text{dist}(f, f_k) < \varepsilon$. But because this is true for every $k \geq N$, this shows $\langle f_n : n \in \mathbb{N} \rangle$ converges uniformly to f. □

Theorem 8.20. *Let X and Y be metric spaces with $X \neq \emptyset$. Then, $B(X, Y)$ is complete if and only if $C(X, Y)$ is complete if and only if Y is complete.*

Proof. We prove the theorem by showing that $B(X, Y)$ is complete implies $C(X, Y)$ is complete, $C(X, Y)$ is complete implies Y is complete, and Y is complete implies $C(X, Y)$ is complete.

For the first implication, suppose $B(X, Y)$ is complete. By Theorem 8.16, $C(X, Y)$ is a closed subset of $B(X, Y)$. By Exercise 6 from Chap. 3, closed subsets of complete spaces are complete. So, $C(X, Y)$ is complete.

For the next implication, suppose $C(X, Y)$ is complete. Let $\langle y_n : n \in \mathbb{N} \rangle$ be a Cauchy sequence in Y. For each $n \in \mathbb{N}$, let f_n denote the constant function $X \to Y$ defined by setting $f_n(x) = y_n$ for all $x \in X$. Because $C(X, Y)$ is complete, the sequence $\langle f_n : n \in \mathbb{N} \rangle$ converges uniformly to some $f \in B(X, Y)$. Uniform convergence implies pointwise convergence, so for any $x \in X$, we have $\lim_{n\to\infty} y_n = \lim_{n\to\infty} f_n(x) = y$. In particular, $\langle y_n : n \in \mathbb{N} \rangle$ converges in Y. Hence Y is complete.

For the third implication, suppose Y is complete. To show that $B(X, Y)$ is also complete, let $\langle f_n : n \in \mathbb{N}\rangle$ be a Cauchy sequence in $B(X, Y)$. For any particular $x \in X$, this implies $\langle f_n(x) : n \in \mathbb{N}\rangle$ is a Cauchy sequence in Y. Because Y is complete, each such sequence converges. Define a function $f : X \to Y$ by setting $f(x) = \lim_{n\to\infty} f_n(x)$ for each $x \in X$. By the definition of f, the sequence $\langle f_n : n \in \mathbb{N}\rangle$ converges pointwise to f. By the previous lemma, $\langle f_n : n \in \mathbb{N}\rangle$ converges uniformly to f. Using the fact that boundedness is preserved in uniform limits (which is proved in an end-of-chapter exercise), $f \in B(X, Y)$. Hence, every Cauchy sequence in $B(X, Y)$ converges to a point of $B(X, Y)$. □

The Weierstrass Approximation Theorem

The Weierstrass Approximation Theorem, first proved in 1885, shows that every continuous function $[a, b] \to \mathbb{R}$ can be approximated by a polynomial with an arbitrarily high degree of accuracy. This is somewhat reminiscent of Maclaurin/Taylor series, although it is important to note that Weierstrass's theorem applies to every continuous function, even the nowhere differentiable functions for which, of course, one cannot define a Taylor series. The theorem was later generalized by Marshall Stone, and its proof simplified, and even the original version is now sometimes referred to as the Stone–Weierstrass Theorem.

Theorem 8.21 (Weierstrass Approximation Theorem). *Let $f \in C([a, b])$. For any $\varepsilon > 0$, there is a polynomial function $p : [a, b] \to \mathbb{R}$ such that $\|f - p\| < \varepsilon$.*

The rough idea for the proof of this theorem is that a function f can be "smoothed out" by replacing the value $f(x)$ with a (possibly weighted) average of nearby values of f.

For example, consider the absolute value function $f(x) = |x|$, which is continuous but not differentiable. Define a new function $g : \mathbb{R} \to \mathbb{R}$ by setting $g(x)$ equal to the average value of f on $[x - 1, x + 1]$ as follows:

$$g(x) = \frac{1}{2}\int_{x-1}^{x+1} f(t)\, dt.$$

This function g is reasonably close to f (specifically $\|g - f\| = \frac{1}{2}$), but the sharp corner of f has been smoothed out, so that g is differentiable.

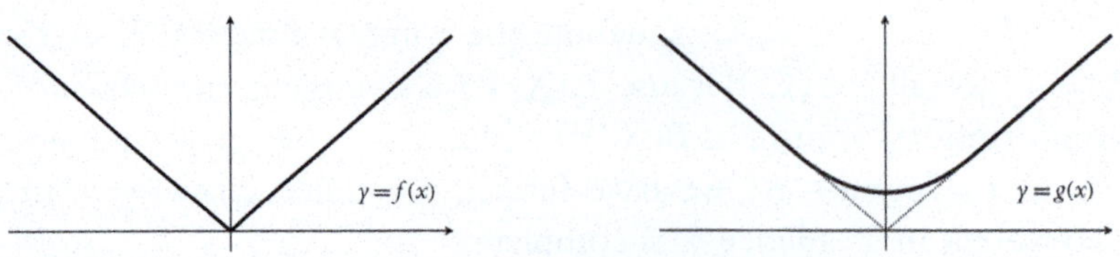

(This is meant merely as an example of the smoothing-out process. We are not claiming, and indeed it is not true, that the smoothed-out function g is a polynomial in this case.)

If we wish to obtain a better approximation of $f(x)$, then instead of setting $g(x)$ equal to the average of f on $[x-1, x+1]$, we can take the average of f over a smaller interval around x, say $[x-\frac{1}{10}, x+\frac{1}{10}]$. Define

$$h(x) = 5\int_{x-\frac{1}{10}}^{x+\frac{1}{10}} f(t)\,dt.$$

Note that h is differentiable and $\|f-h\| = \frac{1}{20}$.

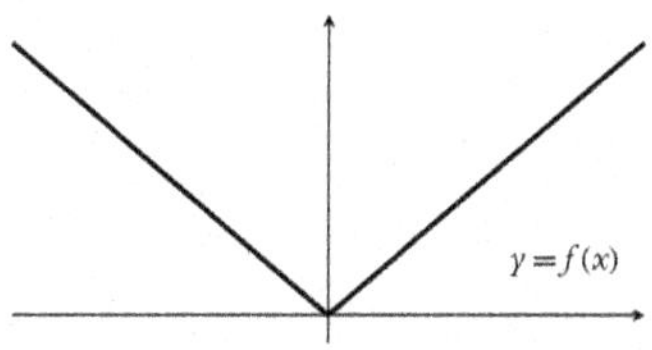

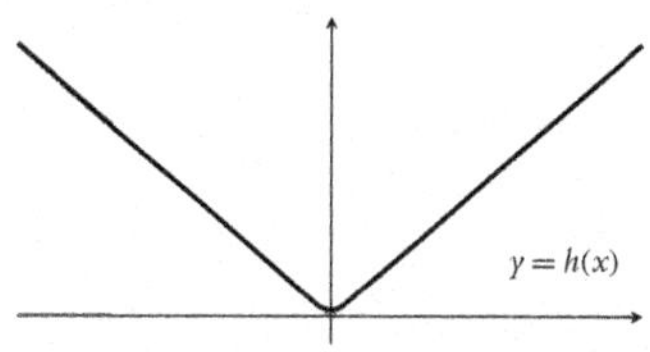

By taking averages over smaller and smaller intervals, one obtains differentiable functions closer and closer to f.

In the following proof, we obtain polynomial approximations to an arbitrary continuous function $f : [a, b] \to \mathbb{R}$ in much the same way. One crucial difference is that instead of "smoothing out" f by taking simple averages, we use weighted averages. Rather than averaging over smaller and smaller intervals around x to get better and better approximations of $f(x)$, we take weighted averages using weight functions assigning more and more weight to the points very close to x. The reason for this change is that, as we shall soon see, a judicious choice of weight functions results in our approximations being not merely differentiable, but polynomials.

Proof of Theorem 8.21. We begin by proving the special case of the theorem where $a = 1/3$ and $b = 2/3$. Afterward, we will show how the general theorem follows (relatively easily) from this special case.

For each $n \in \mathbb{N}$, define a weight function $w_n : [-1, 1] \to \mathbb{R}$ by setting

$$w_n(x) = c_n(1-x^2)^n,$$

where $c_n = 1/\int_{-1}^{1}(1-x^2)^n\,dx$. This choice of the constant c_n guarantees that $\int_{-1}^{1} w_n(x)\,dx = 1$ for all $n \in \mathbb{N}$.

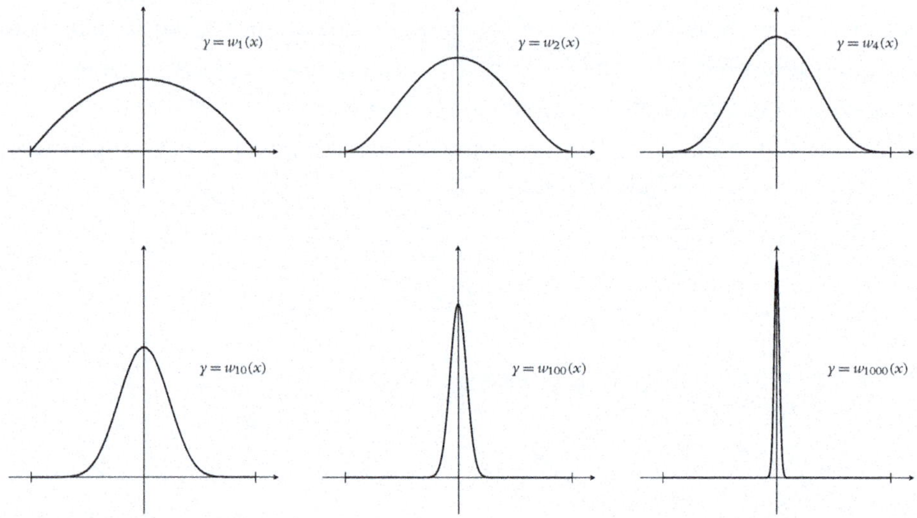

These graphs are not drawn to the same scale on the Y-axis. (This is obvious because our choice of c_n ensures area under the curve is 1 for all of these graphs.) For example, $w_1(0) = 3/4$, while $w_{1000}(0) \approx 17.85$.

Claim. *For each $n \in \mathbb{N}$, $c_n < \sqrt{n}$.*

Proof of Claim. Fix $n \in \mathbb{N}$. Define $g(x) = (1-x^2)^n - 1 + nx^2$. Note that $g(0) = 0$, and

$$g'(x) = -2nx(1-x^2)^{n-1} + 2nx = 2nx(1-(1-x^2)^{n-1}) \geq 0,$$

whenever $x \in [0,1]$. It follows that $g(x) \geq 0$ for all $x \in [0,1]$. In particular, this implies $(1-x^2)^n \geq 1 - nx^2$ for all $x \in [0,1]$. Consequently,

$$\begin{aligned}\int_{-1}^{1} (1-x^2)^n \, dx &= 2\int_0^1 (1-x^2)^n \, dx \geq 2\int_0^{1/\sqrt{n}} (1-x^2)^n \, dx \\ &\geq 2\int_0^{1/\sqrt{n}} 1 - nx^2 \, dx = \frac{4}{3\sqrt{n}},\end{aligned}$$

from which it follows that $c_n \leq 1/\frac{4}{3\sqrt{n}} = \frac{3\sqrt{n}}{4} < \sqrt{n}$. ⊣

Let $f \in C([1/3, 2/3])$, and let $\varepsilon > 0$. Let $\tilde{f} : \mathbb{R} \to \mathbb{R}$ be a continuous extension of f to $\mathbb{R}$ such that $\tilde{f}(x) = 0$ for all $x \notin (0,1)$. (Some such function exists by Theorem 4.37 in Chap. 4.)

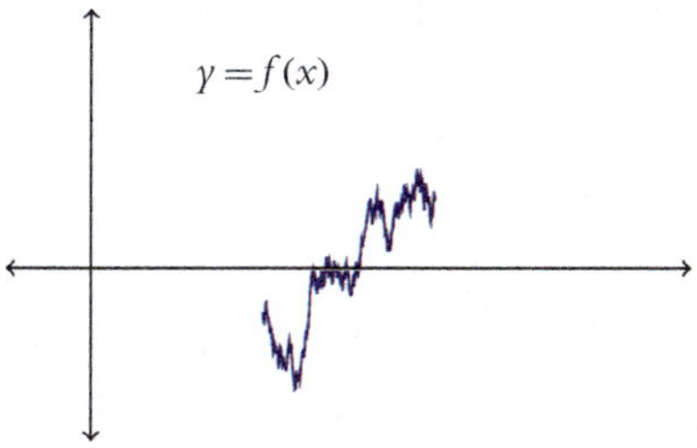

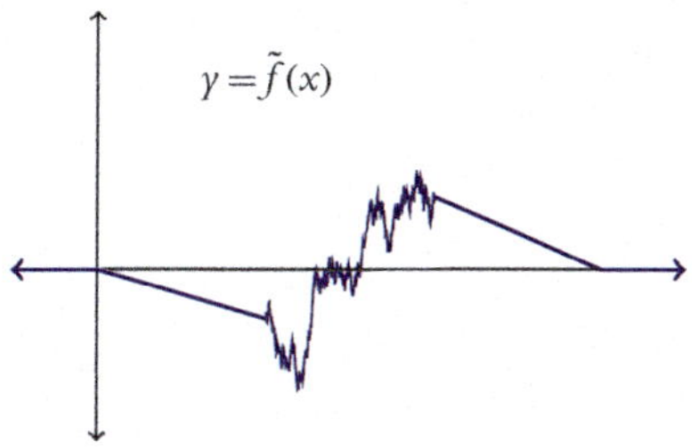

For each $n \in \mathbb{N}$, we now define a function $p_n : [1/3, 2/3] \to \mathbb{R}$ by defining $p_n(x)$ to be the weighted average of $\tilde{f}$ on the interval $[x-1, x+1]$, where the weight of each point is determined by w_n. More specifically, let

$$p_n(x) = \int_{x-1}^{x+1} \tilde{f}(t)\, w_n(t-x)\ dt$$

for all $x \in [1/3, 2/3]$. Notice that part of this integral takes place outside $[1/3, 2/3]$, which is why we use $\tilde{f}$ in this definition rather than f. To finish the proof of the theorem, we show that each p_n is a polynomial and that if N is large enough, then $\|f - p_N\| < \varepsilon$.

Claim. *For each $n \in \mathbb{N}$, the function $x \mapsto p_n(x)$ is a polynomial.*

Proof of Claim. By definition, $\tilde{f}(t) = 0$ whenever $t \notin (0, 1)$. In particular, if $x \in [1/3, 2/3]$, then $\tilde{f}(t) = 0$ for all $t \in [-1, 2] \setminus [x-1, x+1]$, and therefore

$$p_n(x) = \int_{x-1}^{x+1} \tilde{f}(t)\, w_n(t-x)\ dt = \int_{-1}^{2} \tilde{f}(t)\, w_n(t-x)\ dt.$$

Thus can we write the integral defining $p_n(x)$ in such a way that the bounds of integration do not depend on x. The function $w_n(x-t) = c_n(1-t+x)^n$ is a polynomial in two variables, x and t. Rearranging terms and factoring, we may (and do) think of $w_n(t-x)$ as a polynomial in x whose coefficients are polynomials in t. (Note that by rearranging terms and factoring, any polynomial in x and t can be written as a polynomial in x whose coefficients are polynomials in t.) Similarly, we may (and do) view the function $\tilde{f}(t)w_n(t-x)$ as a polynomial in x whose coefficients are continuous functions of t (a polynomial in t times $\tilde{f}(t)$). Integrating $\tilde{f}(t)w_n(t-x)$ with respect to t turns these coefficients (integrable functions of t) into constants, while leaving the powers of x as powers of x. That is, integrating $\tilde{f}(t)w_n(t-x)$ with respect to t yields a polynomial in x with real coefficients. $\dashv$

Because $\tilde{f}$ is continuous on the compact interval $[-1, 2]$, it is both uniformly continuous and bounded. Consequently, there is some $\delta > 0$ such that if $x, y \in [-1, 2]$ and $|x - y| \leq \delta$, then $|\tilde{f}(x) - \tilde{f}(y)| < \varepsilon/2$, and there is some $B > 0$ such that $|\tilde{f}(x)| < B$ for all $x \in [-1, 2]$.

Claim. *There is some* $N \in \mathbb{N}$ *such that* $\sup\{w_N(t) : t \in [\delta, 1]\} < \varepsilon/8B$.

Proof of Claim. By L'Hôpital's rule (specifically, the case of Exercise 15 from Chap. 5 with $g(x) = \left(\frac{1}{1-\delta^2}\right)^x$), $\lim_{n\to\infty} \sqrt{n}(1-\delta^2)^n = 0$. In particular, there is some $N \in \mathbb{N}$ such that $\sqrt{N}(1-\delta^2)^N < \varepsilon/8B$. If $t \in [\delta, 1]$, then $w_N(t) = c_n(1-t^2)^N < c_n(1-\delta^2)^N = w_N(\delta)$, so $\sup\{w_N(t) : t \in [\delta, 1]\} = w_N(\delta) = c_n(1-\delta^2)^N < \sqrt{N}(1-\delta^2)^N < \varepsilon/8B$. ⊣

Because $w_N(x) = w_N(-x)$ for all $x \in [-1, 1]$ (i.e., w_N is an even function), the previous claim also implies that $\sup\{w_N(t) : t \in [-1, -\delta]\} < \varepsilon/8B$. These inequalities are used to put upper bounds on some of the subsequent integrals.

Furthermore, to pass from the second to the third line in the following computation, recall that $\int_{-1}^{1} w_N(t)\,dt = 1$ by our choice of the constant c_N. To go from line 7 to line 10, note that by our choice of δ, if $|u| \le \delta$, then $|\tilde{f}(x) - \tilde{f}(x+u)| < \varepsilon/2$.

For any $x \in [1/3, 2/3]$,

$$
\begin{aligned}
|f(x) - p_N(x)| &= \left| f(x) - \int_{x-1}^{x+1} \tilde{f}(t)\, w_N(t-x)\, dt \right| \\
&= \left| \tilde{f}(x) - \int_{-1}^{1} \tilde{f}(x+u)\, w_N(u)\, du \right| \\
&= \left| \tilde{f}(x) \int_{-1}^{1} w_N(u)\, du - \int_{-1}^{1} \tilde{f}(x+u)\, w_N(u)\, du \right| \\
&= \left| \int_{-1}^{1} \left(\tilde{f}(x) - \tilde{f}(x+u) \right) w_N(u)\, du \right| \\
&\le \int_{-1}^{1} \left| \tilde{f}(x) - \tilde{f}(x+u) \right| w_N(u)\, du \\
&= \int_{-1}^{-\delta} \left| \tilde{f}(x) - \tilde{f}(x+u) \right| w_N(u)\, du \\
&\quad + \int_{-\delta}^{\delta} \left| \tilde{f}(x) - \tilde{f}(x+u) \right| w_N(u)\, du \\
&\quad + \int_{\delta}^{1} \left| \tilde{f}(x) - \tilde{f}(x+u) \right| w_N(u)\, du \\
&< \int_{-1}^{-\delta} \left| \tilde{f}(x) - \tilde{f}(x+u) \right| \frac{\varepsilon}{8B}\, du \\
&\quad + \int_{-\delta}^{\delta} \frac{\varepsilon}{2} w_N(u)\, du \\
&\quad + \int_{\delta}^{1} \left| \tilde{f}(x) - \tilde{f}(x+u) \right| \frac{\varepsilon}{8B}\, du \\
&< \int_{-1}^{-\delta} 2B\frac{\varepsilon}{8B}\, du + \frac{\varepsilon}{2} \int_{-\delta}^{\delta} w_N(u)\, du + \int_{\delta}^{1} 2B\frac{\varepsilon}{8B}\, du
\end{aligned}
$$

$$< \int_{-1}^{0} \frac{\varepsilon}{4}\, du + \frac{\varepsilon}{2} \int_{-1}^{1} w_N(u)\, du + \int_{0}^{1} \frac{\varepsilon}{4}\, du$$
$$= \varepsilon.$$

As x was an arbitrary member of $[a, b]$, this shows $\|f - p_N\| < \varepsilon$.

This finishes the proof of the theorem for the special case $[a, b] = [1/3, 2/3]$. For the general case, let $a, b \in \mathbb{R}$ with $a < b$. Fix some $f \in C([a, b])$, and let $\varepsilon > 0$. Define

$$h(t) = 3(b - a)(x - \tfrac{1}{3}) + a.$$

In other words, h is the linear function with $h(1/3) = a$ and $h(2/3) = b$. The function $f \circ h$ is a continuous function $[1/3, 2/3] \to \mathbb{R}$, so by the special case of the theorem we have already proved, there is a polynomial function $p \in C([1/3, 2/3])$ such that $\|p - f \circ h\| < \varepsilon$. But observe that h^{-1} is also a linear function, so that $p \circ h^{-1}$ is still a polynomial. Furthermore, it is not difficult to see that $\|p \circ h^{-1} - f\| = \|p - f \circ h\|$ (as the effect of h and h^{-1} is simply to rescale and translate the domains of the functions, without otherwise affecting the shapes of their graphs). Thus, $p \circ h^{-1}$ is a polynomial with $\|p \circ h^{-1} - f\| < \varepsilon$. □

Corollary 8.22. *For every continuous function $[a, b] \to \mathbb{R}$, there is a sequence of polynomials that converges uniformly to f on $[a, b]$.*

Corollary 8.23. *The set of polynomials is dense in $C([a, b])$.*

Example 8.24. Another corollary to the Weierstrass Approximation Theorem is that for any $f \in C(\mathbb{R})$, there is a sequence $\langle p_n : n \in \mathbb{N} \rangle$ of polynomials converging pointwise to f. To prove this, just let p_n be some polynomial such that $\|p_n - f\| < 1/n$ on $[-n, n]$.

However, the same is not true for uniform convergence. For example, let $f(x) = \sin(x)$. Every polynomial p is either constant, in which case $\|f - p\| \geq 1$, or unbounded on $\mathbb{R}$. In particular, there is no polynomial whose uniform distance from f is <1. ☕

The Arzelà–Ascoli Theorem

A function f is continuous if, for every $\varepsilon > 0$ and every point $x \in \mathrm{dom}(f)$, there is some $\delta > 0$ such that points within δ of x maps within ε of $f(x)$. The number δ may depend on both ε and x. A function is uniformly continuous, however, if it is possible to choose this δ so that it depends only on ε, but not on x: Uniform continuity means that (given some ε) the same δ can work for every x.

Now, instead of just one function f, suppose we have a set $\mathcal{F}$ of continuous functions, $\mathcal{F} \subseteq C(X, Y)$. The fact that each $f \in \mathcal{F}$ is continuous means that given some $\varepsilon > 0$, $x \in X$, and $f \in \mathcal{F}$, there is a $\delta > 0$ such that points within δ of x map to within ε of $f(x)$. Notice that now the number δ depends on three things rather than just two: ε, x, and f. Removing the dependence of δ on f, or both f and x, gives rise to the following definitions.

Definition 8.25. Suppose X and Y are metric spaces, and $\mathcal{F}$ is a set of functions $X \to Y$.

We say $\mathcal{F}$ is *equicontinuous* at some $x \in X$ if for any $\varepsilon > 0$, there is some $\delta > 0$ such that for every $f \in \mathcal{F}$ and every $z \in X$ with $d_X(x, z) < \delta$, $d_Y(f(x), f(z)) < \varepsilon$. We say $\mathcal{F}$ is *equicontinuous* if it is equicontinuous at every $x \in X$.

We say that $\mathcal{F}$ is *uniformly equicontinuous* if for every $\varepsilon > 0$, there is some $\delta > 0$ such that for every $f \in \mathcal{F}$ and every $x, z \in X$ with $d_X(x, z) < \delta$, $d_Y(f(x), f(z)) < \varepsilon$.

$$
\begin{aligned}
\text{each } f \in \mathcal{F} \text{ continuous}: &\quad \forall f, \varepsilon, x\ \exists \delta \quad d(x, z) < \delta \Rightarrow d(f(x), f(z)) < \varepsilon. \\
\text{each } f \in \mathcal{F} \text{ uniformly continuous}: &\quad \forall f, \varepsilon\ \exists \delta\ \forall x \quad d(x, z) < \delta \Rightarrow d(f(x), f(z)) < \varepsilon. \\
\mathcal{F} \text{ equicontinuous}: &\quad \forall \varepsilon, x\ \exists \delta\ \forall f \quad d(x, z) < \delta \Rightarrow d(f(x), f(z)) < \varepsilon \\
\mathcal{F} \text{ uniformly equicontinuous}: &\quad \forall \varepsilon\ \exists \delta\ \forall f, x \quad d(x, z) < \delta \Rightarrow d(f(x), f(z)) < \varepsilon.
\end{aligned}
$$

$\mathcal{F}$ uniformly equicontinuous

⇙ ⇘

$\mathcal{F}$ equicontinuous   every $f \in \mathcal{F}$ uniformly continuous

⇘ ⇙

every $f \in \mathcal{F}$ continuous

Example 8.26. Let X be a metric space, and let $\mathcal{F}$ denote the family of all isometries $X \to X$. Then, $\mathcal{F}$ is uniformly equicontinuous: For any $\varepsilon > 0$, setting $\delta = \varepsilon$ satisfies the definition.

Example 8.27. Let $\mathcal{F}$ denote the family of all linear functions $[0, 1] \to \mathbb{R}$, i.e., all functions of the form $x \mapsto mx$ for some $m \in \mathbb{R}$. Each individual $f \in \mathcal{F}$ is uniformly continuous. But, $\mathcal{F}$ is not equicontinuous (in short because the function $x \mapsto mx$ requires $\delta \leq \varepsilon/m$, which depends on m).

Example 8.28. For a (somewhat silly) example of a family of functions that is equicontinuous but not uniformly equicontinuous, let $f : \mathbb{R} \to \mathbb{R}$ be the function $f(x) = x^2$, or any other continuous but not uniformly continuous function, and take $\mathcal{F} = \{f\}$. A slightly less silly example is given by setting $f_c(x) = x^2 + c$ for all $c \in \mathbb{R}$, and taking $\mathcal{F} = \{f_c : c \in \mathbb{R}\}$.

The Heine–Borel theorem from Chap. 2 gives an exact characterization of the compact subsets of $\mathbb{R}$: A subset of $\mathbb{R}$ is compact if and only if it is closed and bounded.

The Arzelà–Ascoli Theorem is like the Heine–Borel theorem but for $C([a, b])$ instead of $\mathbb{R}$: It gives an exact characterization of the compact subsets of $C([a, b])$. Specifically, it states that $\mathcal{F} \subseteq C([a, b])$ is compact if and only if $\mathcal{F}$ is closed, bounded, and equicontinuous.

Example 8.29. Before proving the Arzelà–Ascoli Theorem, let us give an example showing that the "and equicontinuous" part of the theorem is really necessary. In other words, there are closed and bounded subsets of $C([a, b])$ that are not compact. To simplify things, we consider $C([0, 1])$ in this example (but the idea can be transferred to $C([a, b])$ by translation and scaling, much like in the last part of our proof of Theorem 8.21).

Just as in Example 8.6, let $f_n(x) = x^n$ for all n, and let $\mathcal{F} = \{f_n : n \in \mathbb{N}\}$. We claim $\mathcal{F}$ is closed and bounded in $C([0, 1])$, but not compact.

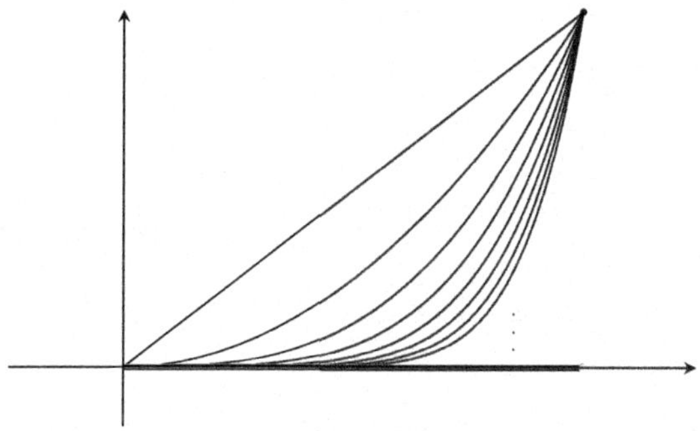

It is not hard to see that $\|f_1 - f_n\| < 1$ for all n, which means $\mathcal{F} \subseteq B_1(f_1)$. Hence, $\mathcal{F}$ is bounded.

To see that $\mathcal{F}$ is closed, suppose g is a limit point of $\mathcal{F}$. By Proposition 8.14, this means there is a sequence in $\mathcal{F}$ converging uniformly to g. But uniform convergence implies pointwise convergence, so there is a sequence in $\mathcal{F}$ converging pointwise to g. But this means $g(x) = \lim_{n\to\infty} f_n(x)$ for all $x \in [0, 1]$, so as we saw already in Example 8.6,

$$g(x) = \begin{cases} 0 & \text{if } x < 1, \\ 1 & \text{if } x = 1. \end{cases}$$

Thus, $g \notin C([0, 1])$. This shows that no $g \in C([0, 1])$ is a limit point of $\mathcal{F}$, and this implies $\mathcal{F}$ is closed in $C([0, 1])$.

To see that $\mathcal{F}$ is not a compact subset of $C([0, 1])$, let $\varepsilon_n = \left(\frac{1}{2}\right)^n - \left(\frac{1}{2}\right)^{n-1}$ for every $n \in \mathbb{N}$. If $m > n$, then

$$\|f_m - f_n\| \geq |f_m(1/2) - f_n(1/2)| = \left|\left(\tfrac{1}{2}\right)^n - \left(\tfrac{1}{2}\right)^m\right| \geq \varepsilon_n,$$

so $f_m \notin B_{\varepsilon_n}(f_n)$. Thus, $B_{\varepsilon_n}(f_n) \cap \mathcal{F}$ is finite for every n. This means that $\mathcal{U} = \{B_{\varepsilon_n}(f_n) : n \in \mathbb{N}\}$ is an open cover of $\mathcal{F}$ with no finite subcover. ☕

Definition 8.30. Let X be a metric space and let $A \subseteq X$. We say that A is *totally bounded* if for every $\varepsilon > 0$ there is a finite cover of A with open balls of radius ε.

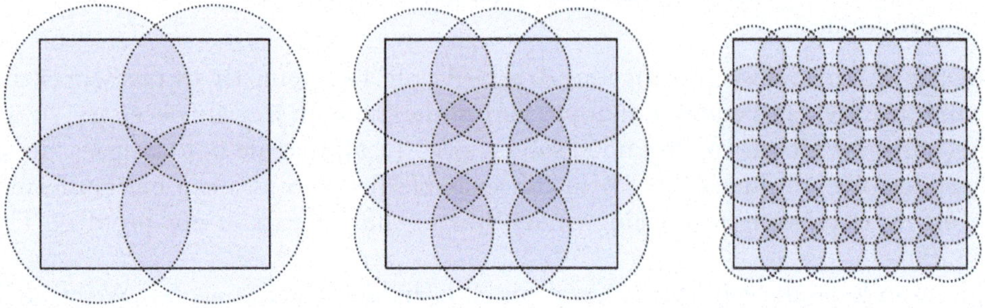

Proposition 8.31. *Compact subsets of a metric space are totally bounded.*

Proof. Suppose A is a compact subset of a metric space. Given $\varepsilon > 0$, let $\mathcal{U} = \{B_\varepsilon(x) : x \in A\}$. Because A is compact, $\mathcal{U}$ has a finite subcover, so there is a finite cover of A with open balls of radius ε. □

Proposition 8.32. *Every subset of a totally bounded set is totally bounded.*

Proof. If $B \subseteq A$, then any covering of A with finitely many ε-balls is also a covering of B with finitely many ε-balls. □

Theorem 8.33. *A subset of a metric space is compact if and only if it is both totally bounded and complete.*

Proof. We proved in Chap. 3 that compact sets are complete, and we proved in Proposition 8.31 that compact sets are totally bounded. To prove the theorem, we must now show that if a set is totally bounded and complete, then it is compact.

Aiming for a contradiction, suppose X is a metric space and $A \subseteq X$ is totally bounded and complete, but not compact.

Using the fact that A is not compact, let $\mathcal{U}$ be an open cover of A with no finite subcover. Let us say that $B \subseteq A$ is *$\mathcal{U}$-small* if some finite subset of $\mathcal{U}$ covers B, and let us say B is *$\mathcal{U}$-big* if no finite subset of $\mathcal{U}$ covers B. For example, A itself is $\mathcal{U}$-big because no finite subset of $\mathcal{U}$ covers A.

If B is a union of finitely many $\mathcal{U}$-small sets, say $B = B_1 \cup B_2 \cup \cdots \cup B_k$ where each B_i is $\mathcal{U}$-small, then B is $\mathcal{U}$-small. This is because for each $i \leq k$ there is some finite $\mathcal{U}_i \subseteq \mathcal{U}$ that covers B_i, but then $\bigcup_{i \leq k} \mathcal{U}_i$ is a finite subset of $\mathcal{U}$ that covers B. By taking the contrapositive of this assertion, we see that if a $\mathcal{U}$-big set B is a union of finitely many sets, then at least one of those sets must also be $\mathcal{U}$-big.

We now define, via recursion, a sequence $V_1, V_2, V_3, \ldots$ of $\mathcal{U}$-big subsets of A, as follows. For the base step, let $\mathcal{V}_1$ be a finite cover of A with open balls of radius

1. (Some such cover exists because A is totally bounded.) The set A is $\mathcal{U}$-big, and $A = \bigcup_{V \in \mathcal{V}_1} (A \cap V)$, so in particular, we have written A as a union of finitely many sets. By the previous paragraph, there is at least one $V \in \mathcal{V}_1$ such that $A \cap V$ is $\mathcal{U}$-big. Pick some such V, let $V_1 = A \cap V$.

For the general step of the recursion, suppose we are given $V_{n-1} \subseteq A$, and V_{n-1} is $\mathcal{U}$-big. Because A is totally bounded, V_{n-1} is too, by Proposition 8.32. Using this fact, let $\mathcal{V}_n$ be a finite cover of V_{n-1} with open balls of radius $1/n$. As in the base step, because V_{n-1} is $\mathcal{U}$-big and $V_{n-1} = \bigcup_{V \in \mathcal{V}_n} (V_{n-1} \cap V)$, there is at least one $V \in \mathcal{V}_n$ such that $V_{n-1} \cap V$ is $\mathcal{U}$-big. Pick some such V, let $V_n = V_{n-1} \cap V$.

This finishes the recursion. The result is a sequence $V_1 \supseteq V_2 \supseteq V_3 \supseteq \ldots$ of $\mathcal{U}$-big subsets of A such that for each n,

$$\operatorname{diam}(V_n) = \sup\left\{d(x, y) : x, y \in V_n\right\} \leq 2/n,$$

because (by construction) V_n is contained in a ball of radius $1/n$. For each n, choose a point $a_n \in V_n$.

We claim the sequence $\langle a_n : n \in \mathbb{N} \rangle$ is Cauchy. To see this, let $\varepsilon > 0$. Fix $N \in \mathbb{N}$ such that $2/N < \varepsilon$. If $m, n \geq N$, then $a_m \in V_m \subseteq V_N$ and $a_n \in V_n \subseteq V_N$, so $a_m, a_n \in V_N$ and $\operatorname{diam}(V_N) \leq 2/N < \varepsilon$.

Because A is complete, the Cauchy sequence $\langle a_n : n \in \mathbb{N} \rangle$ converges to some $a \in A$. Because $\mathcal{U}$ is an open cover of A, there is some $U \in \mathcal{U}$ with $a \in U$. Because U is open, there is some $\delta > 0$ such that $B_\delta(a) \subseteq U$. Fix $N \in \mathbb{N}$ large enough that $2/N < \delta$. As pointed out in the previous paragraph, we have $a_n \in V_N$ for all $n \geq N$. This implies $a = \lim_{n \to \infty} a_n \in \overline{V_N}$. In the end-of-chapter exercises, we see that the diameter of a set is equal to the diameter of its closure. In particular, $\operatorname{diam}(\overline{V_N}) = \operatorname{diam}(V_N) \leq 2/N$. Because $2/N < \delta$, it follows that $V_N \subseteq \overline{V_N} \subseteq B_\delta(a) \subseteq U$. This means V_N is $\mathcal{U}$-small because it is covered by $\{U\}$, a finite subset of $\mathcal{U}$. Contradiction! □

Recall that our goal in this section is to show a subset of $C([a, b])$ is compact if and only if it is closed, bounded, and uniformly equicontinuous. Now that we have proved compactness is equivalent to being complete and totally bounded, we can state our strategy for proving the "if" direction of this theorem: We shall show that closed subsets of $C([a, b])$ are complete and that bounded, uniformly equicontinuous subsets are totally bounded.

$$\begin{array}{ccc} \text{closed} & \Longrightarrow & \text{complete} \\ \left.\begin{array}{c}\text{bounded} \\ \text{uniformly} \\ \text{equicontinuous}\end{array}\right\} & \Longrightarrow & \text{totally bounded} \end{array} \Bigg\} \Longrightarrow \text{compact}$$

Theorem 8.34. *If a subset of $C([a, b])$ is closed, then it is complete.*

Proof. By Theorem 8.20, $C([a, b])$ is a complete metric space. By Exercise 6 in Chap. 3, closed subsets of complete spaces are complete. □

Theorem 8.35. *If a subset of $C([a, b])$ is bounded and uniformly equicontinuous, then it is totally bounded.*

Proof. Suppose $\mathcal{F}$ is a bounded, uniformly equicontinuous subset of $C([a, b])$. Let c_0 denote the constant function $x \mapsto 0$. Then, $c_0 \in C([a, b])$ and, because $\mathcal{F}$ is bounded, there is some $M > 0$ such that $\|f - c+_0\| < M$ for all $f \in \mathcal{F}$. But recalling the definition of $\|f - c_0\|$, this just means $|f(x)| < M$ for every $f \in \mathcal{F}$ and every $x \in [a, b]$.

Let $\varepsilon > 0$. As $\mathcal{F}$ is uniformly equicontinuous, there is some $\delta > 0$ such that if $x, y \in [a, b]$ and $|x - y| < \delta$, then $|f(x) - f(y)| < \varepsilon/2$ for all $f \in \mathcal{F}$.

Let $\langle d_0, d_1, \ldots, d_m \rangle$ be a partition of $[a, b]$ such that $d_{i+1} - d_i < \delta$ for all i, and let $D = \{d_0, d_1, \ldots, d_m\}$. Similarly, let $\langle r_0, r_1, \ldots, r_n \rangle$ be a partition of $[-M, M]$ such that $r_{i+1} - r_i < \varepsilon/2$ for all i, and let $R = \{r_0, r_1, \ldots, r_n\}$.

Let $\mathcal{G}$ be the set of all functions $D \to R$. Note that this is a finite set (in particular, $|\mathcal{G}| = n^m$). For each $g \in \mathcal{G}$, let $\bar{g}$ denote the piecewise linear extension of g to a function $[a, b] \to [-M, M]$.

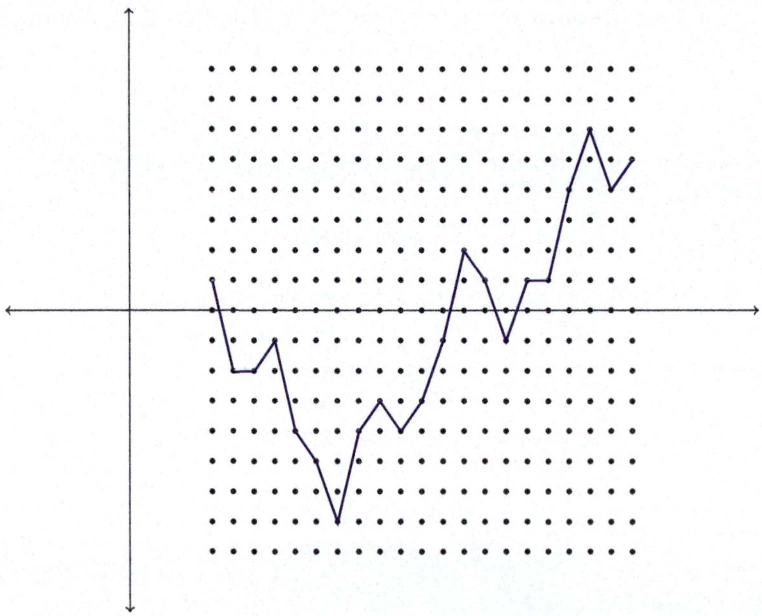

Note that each of these functions $\bar{g}$ is a member of $C([a, b])$. We claim that $\{B_\varepsilon(\bar{g}) : g \in \mathcal{G}\}$ is a finite set of ε-balls in $C([a, b])$ covering $\mathcal{F}$.

To see this, fix $f \in \mathcal{F}$. For each $d \in D$, there is some $r_d \in R$ such that $|f(d) - r_d| < \varepsilon/2$. Let $g \in \mathcal{G}$ be the function $r \mapsto r_d$. Now fix some $x \in [a, b]$. There is some $i < m$ such that $d_i \leq x \leq d_{i+1}$. By design, $d_{i+1} - d_i < \delta$, which implies both $|x - d_i| < \delta$ and $|x - d_{i+1}| < \delta$. By our choice of δ, this implies both $|f(x) - f(d_i)| < \varepsilon/2$ and $|f(x) - f(d_{i+1})| < \varepsilon/2$. Hence,

$$|f(x) - r_{d_i}| \leq |f(x) - f(d_i)| - |f(d_i) - r_{d_i}| < \varepsilon/2 + \varepsilon/2 = \varepsilon, \quad \text{and}$$
$$|f(x) - r_{d_{i+1}}| \leq |f(x) - f(d_{i+1})| - |f(d_{i+1}) - r_{d_{i+1}}| < \varepsilon/2 + \varepsilon/2 = \varepsilon.$$

Because $\bar{g}$ is the piecewise linear extension of g to $[a, b]$, $\bar{g}(x)$ is between $g(d_i) = r_{d_i}$ and $g(d_{i+1}) = r_{d_{i+1}}$. By these computations, it is thus implied that $f(x) - \varepsilon < \bar{g}(x) < f(x) + \varepsilon$, and in particular, $|\bar{g}(x) - f(x)| < \varepsilon$. Thus, $(\bar{g} - f)(x) < \varepsilon$ for all $x \in [a, b]$. By the Extreme Value Theorem, this means $\|\bar{g} - f\| = \sup_{x \in [a,b]} |(\bar{g} - f)(x)| < \varepsilon$. In other words, $f \in B_\varepsilon(\bar{g})$.

As f was arbitrary, this shows $\{B_\varepsilon(\bar{g}) : g \in \mathcal{G}\}$ covers $\mathcal{F}$, as claimed. As ε was arbitrary, this shows $\mathcal{F}$ is totally bounded. □

Theorem 8.36 (The Arzelà–Ascoli Theorem). *A subset of $C([a, b])$ is compact if and only if it is closed, bounded, and uniformly equicontinuous.*

Proof. The more difficult direction of the proof is nearly done already. If $\mathcal{F} \subseteq C([a, b])$ is closed, it is complete by Theorem 8.34. If $\mathcal{F}$ is bounded and uniformly equicontinuous, it is totally bounded by Theorem 8.35. Thus, if $\mathcal{F}$ is closed, bounded, and uniformly equicontinuous, it is complete and totally bounded, hence compact, by Theorem 8.33.

It remains to prove the converse direction of the theorem: If a subset of $C([a, b])$ is compact, then it is closed, bounded, and uniformly equicontinuous. We proved already in Chap. 2 that in any metric space, compact sets are closed and bounded. To finish the proof, we must show that compact subsets of $C([a, b])$ are uniformly equicontinuous.

Suppose $\mathcal{F} \subseteq C([a, b])$ is compact, and let $\varepsilon > 0$. Because $\mathcal{F}$ is compact, $\mathcal{U} = \{B_{\varepsilon/3}(f) : f \in \mathcal{F}\}$ has a finite subcover: In other words, there are some $f_1, f_2, \ldots, f_k \in \mathcal{F}$ such that $\mathcal{F} \subseteq \bigcup_{i=1}^{k} B_{\varepsilon/3}(f_i)$. For each $i \leq k$, f_i is continuous on the compact set $[a, b]$, hence uniformly continuous, so there is some $\delta_i > 0$ such that if $x, y \in [a, b]$ and $|x - y| < \delta_i$, then $|f_i(x) - f_i(y)| < \varepsilon/3$. Let $\delta = \min\{\delta_1, \delta_2, \ldots, \delta_k\}$. We claim δ witnesses the uniform equicontinuity of $\mathcal{F}$ for ε. To see this, fix $f \in \mathcal{F}$. There is some $i \leq k$ with $f \in B_{\varepsilon/3}(f_i)$, so

$$\begin{aligned} |f(x) - f(y)| &\leq |f(x) - f_i(x)| + |f_i(x) - f_i(y)| + |f_i(y) - f(y)| \\ &< \varepsilon/3 + \varepsilon/3 + \varepsilon/3 = \varepsilon \end{aligned}$$

whenever $|x - y| < \delta$. As ε was arbitrary, $\mathcal{F}$ is uniformly equicontinuous. □

The term "Arzelà–Ascoli Theorem" often refers not to the theorem we just proved, but rather to the following closely related corollary:

Corollary 8.37. *If a sequence of functions in* $C([a, b])$ *is bounded and uniformly equicontinuous, then it has a uniformly convergent subsequence.*

Of course, saying that a sequence $\langle f_n : n \in \mathbb{N} \rangle$ of functions is uniformly equicontinuous simply means that the corresponding set $\{f_n : n \in \mathbb{N}\}$ is uniformly equicontinuous, in the sense of Definition 8.25. Before proving this corollary, we need a lemma that states that uniform equicontinuity is preserved by taking closures.

Lemma 8.38. *If a subset of* $C([a, b])$ *is uniformly equicontinuous, then so is its closure.*

Proof. Suppose that $\mathcal{F} \subseteq C([a, b])$ is uniformly equicontinuous, and let $\varepsilon > 0$. Because $\mathcal{F}$ is uniformly equicontinuous, there is some $\delta > 0$ such that if $f \in \mathcal{F}$ and $|x - y| < \delta$, then $|f(x) - f(y)| < \varepsilon/3$.

Let $g \in \overline{\mathcal{F}}$. Because $g \in \overline{\mathcal{F}}$, there is some $f \in \mathcal{F}$ such that $\|g - f\| < \varepsilon/3$. If $|x - y| < \delta$, then

$$
\begin{aligned}
|g(x) - g(y)| &\leq |g(x) - f(x)| + |f(x) - f(y)| + |f(y) - g(y)| \\
&< \varepsilon/3 + \varepsilon/3 + \varepsilon/3 = \varepsilon
\end{aligned}
$$

As ε was arbitrary, this shows $\overline{\mathcal{F}}$ is uniformly equicontinuous. □

Proof of Corollary 8.37. Let $\langle f_n : n \in \mathbb{N} \rangle$ be a bounded, uniformly equicontinuous sequence of functions in $C([a, b])$. Let $\mathcal{F}$ denote the closure of $\{f_n : n \in \mathbb{N}\}$. In any metric space, the closure of a bounded set is bounded (see Exercise 19 in Chap. 2). Thus, $\mathcal{F}$ is closed and bounded, and $\mathcal{F}$ is uniformly equicontinuous by Lemma 8.38. By Theorem 8.36, $\mathcal{F}$ is compact. By Lemma 3.19 in Chap. 3, every sequence of points in a compact set has a subsequential limit in that set. Because $\mathcal{F}$ is compact, this means that $\langle f_n : n \in \mathbb{N} \rangle$ has a convergent subsequence. □

How discontinuous can a derivative be?

Most of this chapter so far has focused on nice results concerning uniform convergence, and a few counterexamples showing that the same nice results do not hold for pointwise convergence. Let us end with an interesting and important result showing something nice about pointwise convergence.

We know from Example 8.6 that a pointwise limit of continuous functions need not be continuous. But just how discontinuous can the limit be? Our example has only one point of discontinuity. Could we get more? What about infinitely many, or even uncountably many points of discontinuity? Taking it to the extreme, can a pointwise limit of continuous functions be everywhere discontinuous, something like the Dirichlet function?

We have also seen, in Chaps. 5 and 6, examples of differentiable functions with discontinuous derivatives. But again, this raises the question: Just how discontinuous can a derivative be? Volterra's function shows that a derivative can have a Lebesgue nonnull set of discontinuities. Can a derivative be everywhere discontinuous? These questions about derivatives are closely related to the questions in the previous paragraph, as the following observation shows.

Proposition 8.39. *If $f : \mathbb{R} \to \mathbb{R}$ is a differentiable function, then its derivative f' is the pointwise limit of a sequence of continuous functions.*

Proof. For each $n \in \mathbb{N}$, let $f_n(x) = n\big(f(x + \frac{1}{n}) - f(x)\big)$. Each f_n is continuous, because f is continuous. And because f is differentiable,

$$f'(x) = \lim_{t \to x} \tfrac{f(t)-f(x)}{t-x} = \lim_{n \to \infty} f_n(x)$$

for every $x \in \mathbb{R}$. □

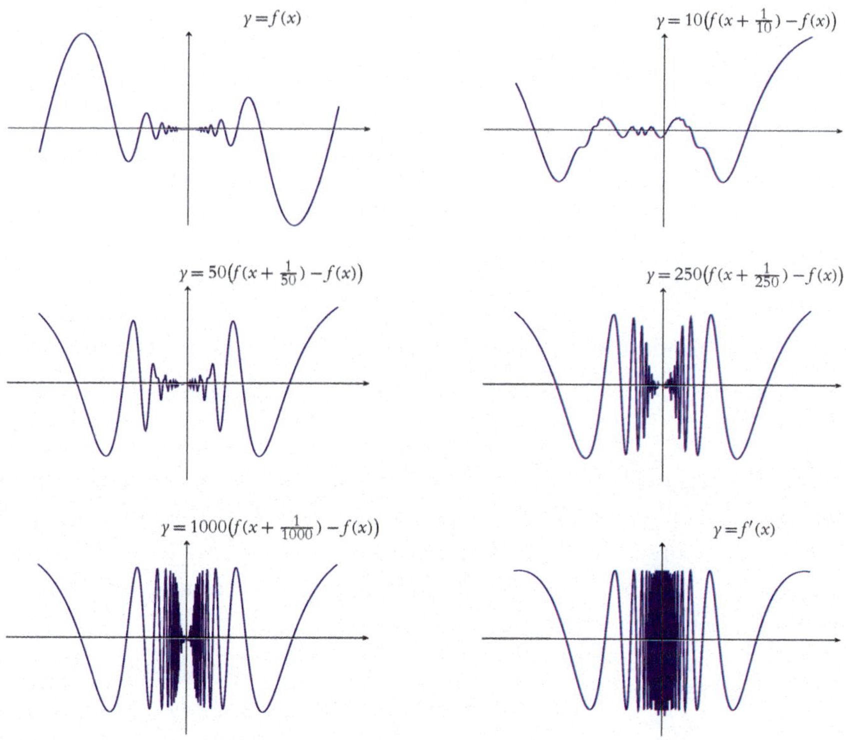

Lemma 8.40. *If F is a closed subset of a metric space X, then $F \setminus \operatorname{Int}(F)$ is nowhere dense.*

Proof. Let U be a non-empty open subset of X. If $U \subseteq F$, then $U \subseteq \operatorname{Int}(F)$, so $W = U$ is disjoint from $F \setminus \operatorname{Int}(F)$. If $U \not\subseteq F$, then $W = U \setminus F \neq \emptyset$, and W is a subset of U disjoint from $F \setminus \operatorname{Int}(F)$. Either way, U has a non-empty open subset disjoint from $F \setminus \operatorname{Int}(F)$. □

Lemma 8.41. *Let $\mathcal{B} = \{(p, q): p, q \in \mathbb{Q} \text{ and } p < q\}$, the set of all rational intervals, and let $f : \mathbb{R} \to \mathbb{R}$ be any function. Then, f is continuous at a point $x \in \mathbb{R}$ if and only if $x \in \operatorname{Int}(f^{-1}[I])$ for every $I \in \mathcal{B}$ with $x \in f^{-1}[I]$.*

Proof. Fix $f : \mathbb{R} \to \mathbb{R}$ and fix $x \in \mathbb{R}$.

Suppose f is continuous at x, and let $I \in \mathcal{B}$. If $x \in f^{-1}[I]$, then $f(x) \in I$. Because I is open, there is some $\varepsilon > 0$ such that $B_\varepsilon(f(x)) \subseteq I$. Because f is continuous, there is some $\delta > 0$ such that if $|y - x| < \delta$, then $|f(y) - f(x)| < \varepsilon$. Thus, $y \in B_\delta(x)$ implies $f(y) \in B_\varepsilon(f(x)) \subseteq I$, which means $B_\delta(x) \subseteq f^{-1}[I]$. Thus, $x \in \operatorname{Int}(f^{-1}[I])$.

Conversely, suppose $x \in \operatorname{Int}(f^{-1}[I])$ for every $I \in \mathcal{B}$ with $x \in f^{-1}[I]$. Let $\varepsilon > 0$. Using the fact that $\mathbb{Q}$ is dense in $\mathbb{R}$, there is some $I \in \mathcal{B}$ such that $f(x) \in I \subseteq B_\varepsilon(f(x))$. This means $x \in f^{-1}[I]$, so by assumption $x \in \operatorname{Int}(f^{-1}[I])$ as well. Thus, there is some $\delta > 0$ with $B_\delta(x) \subseteq f^{-1}[I]$. If $|y - x| < \delta$, then $y \in B_\delta(x) \subseteq f^{-1}[I]$, so $y \in I \subseteq B_\varepsilon(f(y))$, which means that $|f(y) - f(x)| < \varepsilon$. As ε was arbitrary, f is continuous at x. □

Lemma 8.42. *Suppose $f : \mathbb{R} \to \mathbb{R}$ is the pointwise limit of a sequence of continuous functions. For every $a, b \in \mathbb{R}$, $f^{-1}[(a, b)]$ is F_σ.*

Proof. Let $\langle f_n : n \in \mathbb{N} \rangle$ be a sequence of continuous functions that converges pointwise to f. For every point $x \in X$, we have $x \in f^{-1}[(a, b)]$ if and only if $\lim_{n\to\infty} f_n(x) = f(x) \in (a, b)$, if and only if $\lim_{n\to\infty} f_n(x) \in [a + \frac{1}{m}, b - \frac{1}{m}]$ for some m, if and only if for some m, there is some $N \in \mathbb{N}$ such that for all $n \geq N$, $f_n(x) \in [a + \frac{1}{m}, b - \frac{1}{m}]$. Consequently,

$$f^{-1}[(a, b)] = \bigcup_{m=1}^{\infty} \bigcup_{N=1}^{\infty} \bigcap_{n=N}^{\infty} f_n^{-1}\left[[a + \tfrac{1}{m}, b - \tfrac{1}{m}]\right].$$

Because each f_n is continuous, each of the sets $f_n^{-1}\left[[a + \frac{1}{m}, b - \frac{1}{m}]\right]$ is closed. Hence, $\bigcap_{n=N}^{\infty} f_n^{-1}\left[[a + \frac{1}{m}, b - \frac{1}{m}]\right]$ is closed for each $m, n \in \mathbb{N}$, and $\bigcup_{m=1}^{\infty} \bigcup_{N=1}^{\infty} \bigcap_{n=N}^{\infty} f_n^{-1}\left[[a + \frac{1}{m}, b - \frac{1}{m}]\right]$ is F_σ. □

The following theorem, which answers the questions at the beginning of this section, is known as *Baire's simple limit theorem*. It was proved by the same mathematician responsible for the Baire Category Theorem, and in fact, this theorem seems to have been the motivation for developing the Baire Category Theorem and the ideas of meager and comeager sets in the first place.

Theorem 8.43 (Baire's simple limit theorem). *Suppose $f : \mathbb{R} \to \mathbb{R}$ is the pointwise limit of a sequence of continuous functions. Then, the set of points at which f is discontinuous is a meager F_σ set.*

Proof. Suppose $f : \mathbb{R} \to \mathbb{R}$ is the pointwise limit of a sequence of continuous functions, and let

$$D = \{x \in \mathbb{R} : f \text{ is not continuous at } x\}.$$

Letting $\mathcal{B} = \{(p, q) : p, q \in \mathbb{Q} \text{ and } p < q\}$ as before, Lemma 8.41 states that $f \in D$ if and only if there is some $I \in \mathcal{B}$ such that $x \in f^{-1}[I] \setminus \mathrm{Int}(f^{-1}[I])$. More succinctly,

$$D = \bigcup_{I \in \mathcal{B}} \left(f^{-1}[I] \setminus \mathrm{Int}(f^{-1}[I])\right).$$

By Lemma 8.42, each of the sets $f^{-1}[I]$ is F_σ, meaning that there is a countable sequence $F_1^I, F_2^I, F_3^I, \ldots$ of closed sets $f^{-1}[I] = \bigcup_{n=1}^{\infty} F_n^I$. For each n, of course $F_n^I \subseteq f^{-1}[I]$, which implies

$$F_n^I \setminus \mathrm{Int}(f^{-1}[I]) \subseteq F_n^I \setminus \mathrm{Int}(F_n^I).$$

The latter of these two sets is nowhere dense by Lemma 8.40, therefore $F_n^I \setminus \mathrm{Int}(f^{-1}[I])$ is nowhere dense as well. Thus,

$$f^{-1}[I] \setminus \mathrm{Int}(f^{-1}[I]) = \bigcup_{n=1}^{\infty} \left(F_n^I \setminus \mathrm{Int}(f^{-1}[I])\right)$$

is a countable union of nowhere dense closed sets for every $I \in \mathcal{B}$. But $\mathcal{B}$ is also countable, so this means

$$D = \bigcup_{I \in \mathcal{B}} \left(f^{-1}[I] \setminus \mathrm{Int}(f^{-1}[I])\right)$$

is also a countable union of nowhere dense closed sets. □

Corollary 8.44. *If $f : \mathbb{R} \to \mathbb{R}$ is differentiable, then its derivative f' is continuous on a dense G_δ subset of $\mathbb{R}$.*

Proof. This follows from Proposition 8.39 and Theorem 8.43. □

Exercises

1. Prove that uniform continuity is preserved in uniform limits. Note that Example 8.6 shows that uniform continuity is not preserved in pointwise limits.
2. Prove that boundedness is preserved in uniform limits.
3. On the other hand, give an example showing that boundedness is not preserved in pointwise limits.
4. Recall that a function $f : \mathbb{R} \to \mathbb{R}$ is *Borel measurable* if $f^{-1}[B]$ is Borel for every Borel $B \subseteq \mathbb{R}$. Prove that f is Borel measurable if and only if $f^{-1}[(c, \infty)]$ is Borel for every $c \in \mathbb{R}$.
5. Prove that Borel measurability is preserved in pointwise limits and in uniform limits.
6. Prove that monotonicity is preserved in pointwise limits and in uniform limits.
7. Prove that every monotone function is a pointwise limit of continuous monotone functions. Then, provide an example showing the same is not true for uniform limits.

8. Let X and Y be metric spaces, and suppose $f, g \in B(X.Y)$. Prove that $\text{dist}(f, g)$ is well-defined, i.e., prove that $\{d_Y(f(x), g(x)) : x \in X\}$ is bounded in $\mathbb{R}$, and therefore has a supremum.

9. Suppose $\emptyset \neq D \subseteq \mathbb{R}$. Find an example of a nonconstant continuous function $B(D) \to \mathbb{R}$.

10. Given any two non-empty metric spaces X and Y, prove there is a subset $\mathcal{Y} \subseteq B(X, Y)$ that is isometric to Y.

11. The space $B(\mathbb{N})$ of bounded functions $\mathbb{N} \to \mathbb{R}$ is often denoted ℓ^∞. Identifying a function $\mathbb{N} \to \mathbb{R}$ with a sequence of real numbers in the natural way, ℓ^∞ is the space of all bounded sequences of real numbers. Let c denote the subspace of ℓ^∞ consisting of convergent sequences. Show that c is closed in ℓ^∞.

12. Continuing the previous question, show that c is separable, i.e., there is a countable $Q \subseteq c$ such that $\overline{Q} = c$.

13. On the other hand, prove that ℓ^∞ is not separable.

14. Suppose f is a Riemann integrable function $[a, b] \to \mathbb{R}$. Prove that there is a sequence of continuous functions $\langle f_n : n \in \mathbb{N} \rangle$ that converges pointwise to a function g such that $g(x) = f(x)$ whenever f is continuous at x.
(*Hint:* Define $f_n(x)$ to be the average of f from $x - \frac{1}{n}$ to $x + \frac{1}{n}$.)

15. Prove, as a corollary to the Weierstrass Approximation Theorem, that the space $C([a, b])$ is separable.

16. Prove that every bounded subset of $\mathbb{R}$ is totally bounded.

17. Let X and Y be metric spaces. Prove that every finite $\mathcal{F} \subseteq C(X, Y)$ is equicontinuous.

18. Let X and Y be metric spaces, and let $\mathcal{F}$ be a family of functions $X \to Y$. Prove that if X is compact and $\mathcal{F}$ is equicontinuous, then $\mathcal{F}$ is uniformly equicontinuous.

19. Suppose $\mathcal{F}$ is a set of differentiable functions $[a, b] \to \mathbb{R}$. Prove that if there is some bound $B \in \mathbb{R}$ such that $|f'(x)| < B$ for all $x \in [a, b]$ and $f \in \mathcal{F}$, then $\mathcal{F}$ is equicontinuous.

20. Show that the converse of the previous exercise is not true. That is, find a uniformly equicontinuous set $\mathcal{F}$ of differentiable functions $[a, b] \to \mathbb{R}$ such that there is no bound $B \in \mathbb{R}$ with $|f'(x)| < B$ for all $x \in [a, b]$ and $f \in \mathcal{F}$.

21. Suppose X is a metric space and A is a bounded subset of X. Recall that $\text{diam}(A) = \sup\{d(x, y) : x, y \in A\}$. Prove $\text{diam}(A) = \text{diam}(\overline{A})$.

22. Let X be a metric space. Prove that $A \subseteq X$ is totally bounded if and only if for every $\varepsilon > 0$, there is a cover of A with (not necessarily open) sets of diameter $\leq \varepsilon$.

23. Prove the converse of Theorem 8.35: If a subset of $C([a, b])$ is totally bounded, then it is bounded and uniformly equicontinuous.

24. Find an example of a continuous function between two metric spaces mapping a totally bounded set to a set that is not totally bounded.

25. On the other hand, prove that uniformly continuous functions map totally bounded sets to totally bounded sets.

26. Suppose $f : \mathbb{R} \to \mathbb{R}$ has finitely many points of discontinuity. Prove there is a sequence of continuous functions converging pointwise to f.

27. Give an example of a function $f : \mathbb{R} \to \mathbb{R}$ such that the set of points at which f is discontinuous is closed and nowhere dense, but there is no sequence of continuous functions converging pointwise to f.
(*Hint*: First prove there is a subset that is not F_σ.)

28. Suppose $\langle f_n : n \in \mathbb{N} \rangle$ is a sequence of differentiable functions $\mathbb{R} \to \mathbb{R}$, converging pointwise to a function f. Furthermore, suppose the sequence $\langle f_n' : n \in \mathbb{N} \rangle$ of derivatives converges uniformly. Prove that $f = \lim_{n\to\infty} f_n$ is differentiable, and its derivative is $f' = \lim_{n\to\infty} f_n'$.

29. Using the previous exercise, prove the power rule: the derivative of $f(x) = x^r$ is $f'(x) = rx^{r-1}$ for any $r \in \mathbb{R}$.

Index

Symbols

σ-algebra, 177

A

Absolutely convergent series, 68
Algebraic Limit Theorem, 59, 61, 91
Algebraic number, 23
Almost everywhere, 163
Anti-derivative, 155, 159, 160
Archimedean property, 8, 11
Arzelà–Ascoli Theorem, 219, 225
Automatically true, 25
Axiom of Choice (AC), 195
Axiom of Completeness, 1, 4
Axioms, 1

B

Baire Category Theorem (BCT), 72, 74, 77, 228
Baire property, 177
Baire's Simple Limit Theorem, 228
Ball (metric space), 26
Banach–Mazur game, 77, 78, 182, 183
Betweenness, 36
Borel
  σ-algebra, 179
  hierarchy, 180
  measurable function, 202, 229
  set, 177, 179, 180, 194
  subsets, 179, 202
Bounded, 43
  function, 114, 137, 210
  intervals, 36
  (sequence), 59
  set, 41
$B(X, Y)$, 211

C

Cantor
  set, 45–48, 108, 110, 163
  space, 45, 108, 111
Cantor, Georg, 20, 21
Cantor–Schröder–Bernstein Theorem, 23
Cantor's function, 108
Cantor's staircase function, 110
Cardinality, 15
Cauchy
  completion, 88
  sequence, 57, 58, 86
Cauchy criterion, 66, 140, 142
  (for infinite series), 64
  (integration), 142
Chain rule, 119
Characteristic function, 50
Choice function, 195
Closed, 28, 43
  set, 28
  subsets, 39
  under complementation, 177
  under countable unions, 177
Closed Graph Theorem, 114
Closure, 28, 54
Cofinite subsets, 179
Comeager, 76, 77
  set, 75, 77, 228
  subsets, 76
Common refinement, 142
Compact, 38, 58, 93, 222
Compactness, 38, 57
Comparison test, 66
Complete metric space, 53, 57–59, 74, 213, 222, 224
Complete separable metric space, 82, 85, 185
Completely metrizable, 87
Completeness, 53
Composition of integrable functions, 152
Conditionally convergent series, 68, 69, 72
Congruence classes, 196
Connected, 34, 35
Connectedness, 34
Continuity, 89
Continuous, 30, 31, 36, 89, 91–93, 100, 109, 113, 116, 117, 121, 122, 124, 129, 131, 155, 158, 206, 217, 228
  almost everywhere, 164

bijection, 114, 122
extension, 105, 107, 114, 216
functions, 30–32, 35, 89, 92, 121, 126, 144, 155, 207, 226
sequence, 227, 228, 231
image, 36, 40, 45
injections, 121, 122
monotone functions, 229
surjection, 103, 108, 110, 111
Contraction mapping, 88
Convergence, 53
Convergent sequences, 57, 59–61, 230
Convergent subsequences, 58, 226
Converging sequences, 91
Convex
function, 134
set, 36
subsets, 36
Countable, 16, 18
Countable additivity, 188, 193
Countable subadditivity, 186
Cover (of a set), 38
$C(X, Y)$, 211

D

D([a, b]), 212
Darboux, Gaston, 137
Darboux function, 128
Darboux–Riemann integration, 170
Darboux's Theorem, 128
De Morgan's laws, 48
Degenerate intervals, 36
Dense, 9, 72
Derivative, 115, 116, 126, 128, 155, 162, 226, 227, 229
discontinuous, 127, 227
noncontinuous, 126
Descriptive set theory, 180
Determined (Banach-Mazur games), 183, 200
Devil's staircase, 110
Diameter, 82
Differentiable, 116–119, 124, 129, 155, 161, 207, 212, 229
Differentiable function, 123, 126, 128, 129, 158, 161, 207, 208, 227
Differentiation, 116
Differentiation rules, 117
Dirichlet function, 50, 95, 140
Disconnected, 34
Discontinuities of the first and second kind, 95
Discontinuity, 30, 95, 126, 127, 146, 158, 226, 231
Discontinuity points, 103
Discontinuous, 30, 94, 95, 126, 127, 174, 226, 228
derivative, 126, 127, 227
function, 30, 94, 98
Discrete metric, 26
Disjoint, 106, 107, 146, 169, 227
closed intervals, 165
closures, 168
compact subsets, 188
intervals, 47
open sets, 185
open subsets, 80, 184
subsets, 105, 186
union, 46

E

Embedding, 7
Equal almost everywhere, 164
Equicontinuous, 220, 221, 230
Equivalence class, 195
Equivalence relation, 23, 32, 34, 88, 189, 195
Essentially unique, 88
Extended Algebraic Limit Theorem, 61, 62, 65
Extended real line, 59, 60
Extending continuous functions, 104
Extension (of a function), 104
Extreme Value Theorem (EVT), 43, 44, 135, 144

F

F_σ, 228
F_σ set, 101, 194
Fat Cantor set, 154, 165, 173, 175
Finite
partial sum, 64
sequences, 108, 137
sets, 43, 166
subcover, 38–40, 93, 101, 164, 168, 221, 222
unions, 178, 187, 190
First Fundamental Theorem of Calculus, 155
Fixed point property, 51
Froda's Theorem, 97
Function
discontinuous, 98
noncontinuous, 126
Fundamental Theorems of Calculus, 155

G

G_δ set, 101, 179, 194

Generate (σ-algebra), 179
Graph of a function, 114

H

Harmonic series, 65
Height control group, 146
Heine–Borel Theorem, 43, 44, 220
Homeomorphic, 33, 34, 36, 48, 87, 108, 111
Homeomorphic metric spaces, 36, 41, 87
Homeomorphism, 33
Huntington's Theorem, 12

I

Improper integration, 139, 175
Infimum, 3, 4
Infinite
  sequence, 55, 86, 111
  series, 63–65, 67, 68, 109, 190
Integrable function, 137
Integration by parts, 175
Integration rules, 148
Interior, 26
Intermediate Value Theorem (IVT), 34, 37, 114, 128
Interval, 36
  bounded, 36
  degenerate, 36
  unbounded, 36
Inverse function rule, 120
Inverse operation
  differentiation, 162, 169
  integration, 162, 169
Irrational numbers, 5, 22, 102, 165
Irrationality of *e*, 87
Isolated point, 29
Isometric, 32, 33, 88, 230
Isometric metric space, 87
Isometry, 32
Isomorphism, 7

J

Jump, 30
Jump discontinuity, 96, 158

L

Lebesgue
  measurable, 186, 187, 191, 192, 201, 202
  measurable set, 177, 186, 192, 194, 200, 201
  measurable sets sequence, 193, 202
  measurable Vitali set, 201
  null, 76, 163, 164, 187, 203
  null set, 163, 187, 194
  null subsets, 163
  outer measure, 165, 166, 186
Lebesgue, Henri, 167
Lebesgue-Vitali Theorem, 167
L'Hôpital's rule, 135, 218
Limit, 89
  (one-sided), 94
  point, 29, 89, 90, 205, 212
Linearly, 106
Local maximum, 123
Lower bound, 3

M

Meager, 75–77, 126, 181, 183, 199, 228
  set, 75, 77, 176, 181, 182, 185, 194, 199, 200
  subset, 75, 79, 181
Meagerness, 75
Mean Value Theorem (MVT), 123, 124
Metric property, 32
Metric space, 25, 26, 28, 31, 32, 34, 41, 43, 50, 53, 54, 56–58, 72, 74–76, 88–91, 93, 99–101, 181, 183, 205, 206, 213, 220, 222
Microscopically small number, 115
Monotone function, 97, 145

N

Negative part of an infinite series, 68
Nested Interval Property, 48, 49, 109
Nicely definable subsets, 180
Non-Archimedean ordered field, 11
Non-empty metric spaces, 230
Noncontinuous
  derivative, 126
  function, 126
Nondecreasing, 63, 97, 125, 126, 145
  function, 97, 125
  sequence, 63
Nondegenerate intervals, 36, 164
Nonincreasing, 63, 68, 71, 87, 125, 126, 175
  function, 97, 125, 146
  sequences, 63
Nonmeager subset, 75
Nonnull subsets, 165
Normal topological space, 107
Nowhere
  dense, 74–76, 181, 199, 227, 229, 231
  dense subset, 74, 75

differentiable function, 129, 207, 214
Null set, 76, 163, 168, 176, 194

O
Open, 28
cover, 38, 39
set, 26
Ordered field, 1
Oscillation, 98
(of a function), 99, 143
(series divergence), 64
Outer measure, 165, 186

P
Pairwise disjoint, 80, 185, 189, 190, 192, 201
Lebesgue measurable sets, 193
sets, 203
subsets, 189
Partition, 138, 140, 195–197, 224
(integration), 137
(of a set), 195
Peano, 103
Pointwise convergence, 205
Pointwise limits, 206
Polynomial function, 92
Polynomials, 92, 118, 214
Positive part of an infinite series, 67
Possible future moves, 80
Power rule, 118, 119, 122
Power rule (derivatives), 118, 119
Power set, 21
Preservation of a property in pointwise/uniform limits, 206
Product rule (derivatives), 117
Projection mapping, 111
Property of Baire, 180, 181, 199, 200
Pythagoreans, 5

Q
Quotient rule (derivatives), 118

R
R($[a, b]$), 212
Rational functions, 11, 92, 119
Rational intervals, 228
Rational numbers, 18
Real line, 57
Rearrangement (infinite series), 68
Recursion, 46, 48, 70, 71, 81, 222, 223
Refinement (partition), 140
Removable discontinuity, 96, 99, 157
Restriction, 26
Riemann
integrable, 137, 138, 143–145, 147, 152, 153, 155, 161, 175, 208, 209, 212
integrable function, 138, 152, 155, 157, 159, 161, 162, 167, 176, 208, 230
Riemann rearrangement theorem, 69
Riemann–Darboux integrable, 137
Riemann–Darboux integrable functions, 138
Riemann–Darboux integration, 147

S
Secant lines, 116
Second Fundamental Theorem of Calculus, 160
Selector (for a partition), 196
Separable, 80, 230
Separation, 35
Sequence, 53, 55, 61, 81, 82, 208, 219, 222, 226
continuous functions, 227, 228, 231
infinite, 55, 86
Sets, 35, 43, 177, 179, 180, 182, 183, 185, 188, 194
finite, 43
infinite sequence, 46
meager, 75, 77, 181, 182, 185, 194, 200
pairwise disjoint, 203
Sine function, 134
Smallest σ-algebra, 178
Smith-Volterra-Cantor set, 165
Smoothed out, 214
Smoothing out, 215
Solovay, Robert, 195
Space-filling curve, 103, 110, 111
Standard metric, 25, 26, 211
Step functions, 30
Stone-Weierstrass Theorem, 214
Strategy-stealing argument, 84
Strictly
decreasing function, 121
increasing function, 110, 121
Subcover, 38
Subsequence, 53, 55
Subsequential limit, 55, 56, 58, 86
Subsets, 38, 43, 49, 75, 77, 177, 178, 186, 188
disjoint, 105, 186
pairwise disjoint, 189
Subspace, 26

Sums rule (derivatives), 117
Sup metric, 51
Supersets, 166
Supremum, 3
Symmetric difference, 180

T

Tactic (for a Banach-Mazur game), 78
Taxicab metric, 51
Thomae's function, 51, 152, 174
Tietze Extension Theorem, 107
Topological property, 36
Total length, 163
Totally bounded, 222
Totally bounded sets, 230
Transcendental number, 23
Triangle inequality, 25
True area, 139
True continuous functions, 160

U

U-substitution, 175
Unbounded intervals, 36
Uncountable, 16, 20, 21
Uncountable meager subset, 75
Undetermined Banach-Mazur game, 183, 185
Uniform
  continuity, 92, 113
  convergence, 205
  limits, 206
Uniformly
  continuous functions, 92
  equicontinuous, 220
Union, 26, 47, 75, 81, 178, 188, 222, 223
Union disjoint, 46
Upper bound, 3
Usual metric, 45

V

Vertical asymptote, 96
Vitali, Giuseppe, 167, 197
Vitali set, 197–199, 201
Volterra, Vito, 170
Volterra's function, 162, 170, 173

W

Weierstrass Approximation Theorem, 214, 219, 230
Weierstrass function, 129
Weight function, 215
Well-behaved collection
  of sets, 177
  of subsets, 177
Width control group, 146
Winning tactic, 78, 81, 83–85, 183
Winning tactic (for a Banach-Mazur game), 78

Printed and bound by CPI Group (UK) Ltd, Croydon, CR0 4YY
08/09/2026
02212889-0002